Posttranslational Covalent Modifications of Proteins

Academic Press Rapid Manuscript Reproduction

International Conference on Posttranslational Covalent Modifications of Proteins for Function Held in Oklahoma City, Oklahoma, November 7–10, 1982, sponsored by Oklahoma Medical Research Foundation

Posttranslational Covalent Modifications of Proteins

EDITED BY

B. Connor Johnson
Ben C. Wileman Learning Center
Oklahoma Medical Research Foundation
Oklahoma City, Oklahoma

1983

ACADEMIC PRESS
A Subsidiary of Harcourt Brace Jovanovich, Publishers
NEW YORK LONDON
PARIS SAN DIEGO SAN FRANSISCO SÃO PAULO SYDNEY TOKYO TORONTO

ACADEMIC PRESS, INC.
111 Fifth Avenue, New York, New York 10003

United Kingdom Edition published by
ACADEMIC PRESS, INC. (LONDON) LTD.
24/28 Oval Road, London NW1 7DX

Library of Congress Cataloging in Publication Data
Main entry under title:

Posttranslational covalent modifications of proteins.

Includes index.
1. Proteins--Metabolism--Congresses. 2. Genetic translation--Congresses. I. Johnson, B. Connor.
QP551.P645 599'.019245 83-17181
ISBN 0-12-387560-9 (alk. paper)

PRINTED IN THE UNITED STATES OF AMERICA

83 84 85 86 9 8 7 6 5 4 3 2 1

To the Advisory Group for the Conference

Dr. Vincent G. Allfrey (Rockefeller)
Dr. Mary P. Carpenter (OMRF)
Dr. Robert Delaney (OUHSC)
Dr. Martin J. Griffin (OMRF)
Dr. John Suttie (Wisconsin)
Dr. Martha Vaughn (NHLBI)

Contents

Contributors

Numbers in parentheses indicate the pages on which the authors' contributions begin.

A. Aitken (19), *Department of Biochemistry, Medical Sciences Institute, University of Dundee, Dundee, Scotland, United Kingdom*

Vincent G. Allfrey (181), *The Rockefeller University, New York, New York*

David L. Anton (281), *Department of Pharmacology, Harvard Medical School; and Charles A. Dana Research Institute, Harvard–Thorndike Laboratory of Beth Israel Hospital, Department of Medicine, Beth Israel Hospital and Harvard Medical School, Boston, Massachusetts*

Nathan A. Berger (359), *Washington University School of Medicine, Division of Hematology/Oncology, The Jewish Hospital of St. Louis, St. Louis, Missouri*

William J. Bettger (295), *Department of Biochemistry, St. Louis University School of Medicine, St. Louis, Missouri; and University of Pittsburgh School of Medicine, Pittsburgh, Pennsylvania*

Gary M. Bokoch (373), *Department of Pharmacology, University of Texas Health Science Center at Dallas, Dallas, Texas*

Barbara A. Booth (359), *Washington University School of Medicine, Division of Hematology/Oncology, The Jewish Hospital of St. Louis, St. Louis, Missouri*

P. Cohen (19), *Department of Biochemistry, Medical Sciences Institute, University of Dundee, Dundee, Scotland, United Kingdom*

Z. Damuni (19), *Department of Biochemistry, Medical Sciences Institute, University of Dundee, Dundee, Scotland, United Kingdom*

M. de Metz (231), *Department of Biochemistry, State University of Limburg, Maastricht, The Netherlands*

Eleanor Erikson (39), *Department of Pathology, University of Colorado School of Medicine, Denver, Colorado*

Raymond L. Erikson (39), *Department of Pathology, University of Colorado School of Medicine, Denver, Colorado*

Jack Erlichman (81), *Departments of Molecular Pharmacology, Medicine, and Biochemistry, Albert Einstein College of Medicine, Bronx, New York*

Paul A. Friedman (281), *Department of Pharmacology, Harvard Medical School; and Charles A. Dana Research Institute, Harvard-Thorndike Laboratory of Beth Israel Hospital, Department of Medicine, Beth Israel Hospital and Harvard Medical School, Boston, Massachusetts*

Alfred G. Gilman (373), *Department of Pharmacology, University of Texas Health Science Center at Dallas, Dallas, Texas*

Jean-Marie Girardot (205), *Oklahoma Medical Research Foundation, Oklahoma City, Oklahoma*

Yosef Graziani (39), *Department of Pathology, University of Colorado School of Medicine, Denver, Colorado*

Anne L. Hall (295), *Department of Biochemistry, St. Louis University School of Medicine, St. Louis, Missouri; and University of Pittsburgh School of Medicine, Pittsburgh, Pennsylvania*

B. A. Hemmings (19), *Department of Biochemistry, Medical Sciences Institute, University of Dundee, Dundee, Scotland, United Kingdom*

Philip Hohmann (143), *Department of Experimental Biology, Roswell Park Memorial Institute, Buffalo, New York*

Masahiko Hoshijima (105), *Department of Biochemistry, Kobe University School of Medicine, Kobe, Japan; and Department of Cell Biology, National Institute for Basic Biology, Okazaki, Japan*

T. S. Ingebritsen (19), *Department of Biochemistry, Medical Sciences Institute, University of Dundee, Dundee, Scotland, United Kingdom*

Elaine L. Jacobson (343), *Department of Biology, Texas Woman's University, Denton, Texas*

Myron K. Jacobson (343), *Departments of Chemistry and Biochemistry, North Texas State University; and Texas College of Osteophathic Medicine, Denton, Texas*

Rosemary Jagus[1] (159), *Laboratory of Molecular Hematology, Section on Protein Biosynthesis, National Heart, Lung, and Blood Institute, National Institute of Health, Bethesda, Maryland*

B. Connor Johnson (205), *Oklahoma Medical Research Foundation, Oklahoma City, Oklahoma*

Richard A. Kahn (373), *Department of Pharmacology, University of Texas Health Science Center at Dallas, Dallas, Texas*

Kozo Kaibuchi (105), *Department of Biochemistry, Kobe University School of Medicine, Kobe, Japan; and Department of Cell Biology, National Institute for Basic Biology, Okazaki, Japan*

Jolanta M. Kanabus-Kaminska (205), *Oklahoma Medical Research Foundation, Oklahoma City, Oklahoma*

William K. Kappel (295), *Department of Biochemistry, St. Louis University School of Medicine, St. Louis, Missouri; and University of Pittsburgh School of Medicine, Pittsburgh, Pennsylvania*

Toshiaki Katada (373), *Department of Pharmacology, University of Texas Health Science Center at Dallas, Dallas, Texas*

[1] Present address: Laboratory of Cellular and Developmental Biology, Bldg. 6, Room B1-07, NIADDK, NIH, Bethesda, Maryland.

Ushio Kikkawa (105), *Department of Biochemistry, Kobe University School of Medicine, Kobe, Japan; and Department of Cell Biology, National Institute for Basic Biology, Okazaki, Japan*

Claude B. Klee (61), *Laboratory of Biochemistry, National Cancer Institute, National Institutes of Health, Bethesda, Maryland*

Ernest Kun (397), *Department of Pharmacology, The University of California; and The Cardiovascular Research Institute, San Francisco, California*

Stephen K. Kwan (281), *Department of Pharmacology, Harvard Medical School; and Charles A. Dana Research Institute, Harvard-Thorndike Labratory of Beth Israel Hospital, Department of Medicine, Beth Israel Hospital and Harvard Medical School, Boston, Massachusetts*

Florence C. Lee (295), *Department of Biochemistry, St. Louis University School of Medicine, St. Louis, Missouri; and University of Pittsburgh School of Medicine, Pittsburgh, Pennsylvania*

J. J. McTigue (253), *Department of Biochemistry, College of Agricultural and Life Sciences, University of Wisconsin-Madison, Madison, Wisconsin*

Robert G. Meyer (295), *Department of Biochemistry, St. Louis University School of Medicine, St. Louis, Missouri; and University of Pittsburgh School of Medicine, Pittsburgh, Pennsylvania*

Joel Moss (321, 329), *Laboratory of Cellular Metabolism, National Heart, Lung, and Blood Institute, National Institutes of Health, Bethesda, Maryland*

Dianne L. Newton (61), *Laboratory of Biochemistry, National Cancer Institute, National Institutes of Health, Bethesda, Maryland*

Yasutomi Nishizuka (105), *Department of Biochemistry, Kobe University School of Medicine, Kobe, Japan; and Department of Cell Biology, National Institute for Basic Biology, Okazaki, Japan*

John K. Northup (373), *Department of Pharmacology, University of Texas Health Science Center at Dallas, Dallas, Texas*

Robert E. Olson (295), *Department of Biochemistry, St. Louis University School of Medicine, St. Louis, Missouri; and University of Pittsburgh School of Medicine, Pittsburgh, Pennsylvania*

P. J. Parker (19), *Department of Biochemistry, Medical Sciences Institute, University of Dundee, Dundee, Scotland, United Kingdom*

D. Michael Payne (343), *Departments of Chemistry and Biochemistry, North Texas State University, Texas College of Osteophathic Medicine, Denton, Texas*

Shirley J. Petzold (359), *Washington University School of Medicine, Division of Hematology/Oncology, The Jewish Hospital of St. Louis, St. Louis, Missouri*

C. Picton (19), *Department of Biochemistry, Medical Sciences Institute, University of Dundee, Dundee, Scotland, United Kingdom*

P. C. Preusch (253), *Department of Biochemistry, College of Agriculture and Life Sciences, University of Wisconsin-Madison, Madison, Wisconsin*

Christiane Quirin-Stricker (125), *Laboratoire de Génétique Moléculaire des Eucaryotes du CNRS, Faculté de Medecine, Institut de Chimie, Strasbourg, France*

T. Resink (19), *Department of Biochemistry, Medical Sciences Institute, University of Dundee, Dundee, Scotland, United Kingdom*

Charles S. Rubin (81), *Departments of Molecular Pharmacology, Medicine, and Biochemistry, Albert Einstein College of Medicine, Bronx, New York*

Kimihiko Sano(105),*Department of Biochemistry, Kobe University School of Medicine, Kobe, Japan; and Department of Cell Biology, National Institute for Basic Biology, Okazaki, Japan*

Dwijen Sarkar (81), *Departments of Molecular Pharmacology, Medicine, and Biochemistry, Albert Einstein College of Medicine, Bronx, New York*

James L. Sims (359), *Washington University School of Medicine, Division of Hematology/Oncology, The Jewish Hospital of St. Louis, St. Louis, Missouri*

B. A. M. Soute (231), *Department of Biochemistry, State University of Limburg, Maastricht, The Netherlands*

Richard Sterner (181), *The Rockefeller University, New York, New York*

A. A. Stewart (19), *Department of Biochemistry, Medical Sciences Institute, University of Dundee, Dundee, Scotland, United Kingdom*

Jill C. Stein (81), *Departments of Molecular Pharmacology, Medicine, and Biochemistry, Albert Einstein College of Medicine, Bronx, New York*

Irene Y. C. Sun (181), *The Rockefeller University, New York, New York*

J. W. Suttie (253), *Department of Biochemistry, College of Agricultural and Life Sciences, University of Wisconsin-Madison, Madison, Wisconsin*

Yoshimi Takai (105),*Department of Biochemistry, Kobe University School of Medicine, Kobe, Japan; and Department of Cell Biology, National Institute for Basic Biology, Okazaki, Japan*

N. K. Tonks (19), *Department of Biochemistry, Medical Sciences Institute, University of Dundee, Dundee, Scotland, United Kingdom*

Martha Vaughan (321, 329), *Laboratory of Cellular Metabolism, National Heart, Lung, and Blood Institute, National Institutes of Health, Bethesda, Maryland*

C. Vermeer (231), *Department of Biochemistry, State University of Limburg, Maastricht, The Netherlands*

Finn Wold (1), *Department of Biochemistry and Molecular Biology, The University of Texas Medical School, Houston, Texas*

J. Woodgett (19), *Department of Biochemistry, Medical Sciences Institute, University of Dundee, Dundee, Scotland, United Kingdom*

Junji Yamanishi (105),*Department of Biochemistry, Kobe University School of Medicine, Kobe, Japan; and Department of Cell Biology, National Institute for Basic Biology, Okazaki, Japan*

Foreword

As is true in most fields of medical science, the last several years have seen significant strides in the area of phosphorylation and carboxylation. The effect of these processes on proteins and their function is of such interest to the scientific community at large that the Oklahoma Medical Research Foundation organized an international conference in order to bring together distinguished scientists from the United States and abroad to exchange information, projections, and ideas for both the present and the future. Dr. B. Connor Johnson ably organized and directed the conference.

The Oklahoma Medical Research Foundation sponsors one such conference annually. Such conferences require not only the active leadership of scientists at the Foundation but also the participation of well-respected individuals from all parts of the world. The Foundation is grateful to all of those who participated and particularly to Dr. Johnson for not only his organization of the conference but also the editing of these proceedings. The role of the staff of Academic Press in making the information quickly available to scientists throughout the world is also appreciated.

William G. Thurman, M.D.
President, Oklahoma Medical Research Foundation

Preface

In November 1982, a conference was held at the Oklahoma Medical Research Foundation on the regulation of protein function by covalent change after peptide bond formation. Since this field covers such a large number of reactions, it was decided to limit the conference to four areas of regulation, namely, phosphorylation–dephosphorylation, carboxylation of glutamyl residues, ADP-ribosylation, and acetylation. These are relatively recent discoveries and important mechanisms for the regulation of protein function, and all are being actively studied at the present time. In the case of carboxylation, the emphasis is on the molecular mechanism of the vitamin K hydroquinone–requiring reaction. In the other three cases, which have been more widely studied, the emphasis is on functional changes brought about by the modification, but again the specific nature of the modification is also considered.

I would like to express my thanks to each of the participants, many of whom came a long distance, for their contributions, and to Diane Morris for all her help during the conference and during the editing of this book. It is my hope that all those working at the level of post peptide-bond-formation protein modification in the regulation of protein function will find this volume of value. The rapidity of publication will make the information available while it is still current for those working in the field and will serve to bring the rest of us up-to-date on the advances being made in these areas.

This conference was made possible by the support of the Board of Directors of the Oklahoma Medical Research Foundation.

B. Connor Johnson

List of Speakers

Vincent G. Allfrey, *The Rockefeller University, 1230 York Avenue, New York, New York 10021*

Nathan A. Berger, *Washington University School of Medicine, The Jewish Hospital of St. Louis, 216 South Kingshighway, St. Louis, Missouri 63110*

Phillip Cohen, *Department of Biochemistry, Medical Sciences Institute, University of Dundee, Dundee, Scotland, United Kingdom DD1 4HN*

Paul Friedman, *Department of Pharmacology, Harvard Medical School, 25 Shattuck Street, Boston, Massachusetts 02115*

Jean-Marie Girardot, *Biomembrane Research Laboratory, Oklahoma Medical Research Foundation, 825 N.E. 13th Street, Oklahoma City, Oklahoma 73104*

Yosef Graziani, *University of Colorado Medical School, Box B-216, 4200 E. Ninth Avenue, Denver, Colorado 80262*

Philip G. Hohmann, *Department of Experimental Biology, Roswell Park Memorial Institute, 666 Elm Street, Buffalo, New York 14263*

Myron Jacobson, *Department of Chemistry and Biochemistry, Texas College of Osteopathic Medicine, North Texas State University, Denton, Texas 76203*

Rosemary Jagus, *Laboratory of Cellular and Developmental Biology, NIADDK, National Institutes of Health, 9000 Rockville Pike, Building 6, Room B1-07, Bethesda, Maryland 20205*

Richard A. Kahn, *Department of Pharmacology, University of Texas Health Science Center at Dallas, 5323 Harry Hines Boulevard, Dallas, Texas 75235*

Claude B. Klee, *Macromolecular Interactions Section, National Cancer Institute, National Institutes of Health, 9000 Rockville Pike, Building 37, Room 4C-06, Bethesda, Maryland 20205*

Ernest Kun, *Department of Pharmacology, University of California, San Francisco, California 94143*

Joel Moss, *Section on Molecular Mechanisms, Laboratory of Cellular Metabolism, National Heart, Lung, and Blood Institute, National Institutes of Health, 9000 Rockville Pike, Building 10, Room 5N-307, Bethesda, Maryland 20205*

Yasutomi Nishizuka, *Department of Biochemistry, Kobe University School of Medicine, Ikuta-ku, Kobe 650, Japan*

Robert E. Olson, *Department of Biochemistry, University of Pittsburgh School of Medicine, Pittsburgh, Pennsylvania 15261*

Christiane Quirin-Stricker, *Laboratoire de Génétique Moléculaire des Eucaryotes du CNRS, Institut de Chimie Biologique, 11 Rue Humann, Strasbourg, France 67085*

Charles S. Rubin, *Department of Molecular Pharmacology, Albert Einstein College of Medicine, 1300 Morris Park Avenue, Bronx, New York 10461*

J. W. Suttie, *Department of Biochemistry, University of Wisconsin, Madison, Wisconsin 53706*

Martha Vaughn, *Laboratory of Cellular Metabolism, National Heart, Lung, and Blood Institute, National Institutes of Health, 9000 Rockville Pike, Building 10, Room 5N-307, Bethesda, Maryland 20205*

C. Vermeer, *Department of Biochemistry, State University of Limburg, P.O. Box 616, Maastricht, The Netherlands 6200 MD*

Finn Wold, *Department of Biochemistry and Molecular Biology, University of Texas Medical School, P.O. Box 20708, Houston, Texas 77096*

Posttranslational Covalent Modifications of Proteins

POSTTRANSLATIONAL PROTEIN MODIFICATIONS: PERSPECTIVES AND PROSPECTIVES[1]

Finn Wold

Department of Biochemistry and Molecular Biology
The University of Texas Medical School
Houston, Texas

This brief overview of posttranslational (and cotranslational) modifications of proteins will be organized in terms of the common interrogatives of all scientific inquiries: *What* are the chemical reactions involved? *How, where* and *when* do they occur? *Why* do they occur? In the first part the general types of reactions that have been established to date are presented, and the long list of known amino acid derivatives is brought up to date. In the second part the complex set of interdependent questions related to chemical mechanisms, substrate specificity determinants, compartment location and regulation are introduced without many answers to go with the questions. In the third section under the question of why the derivatives are made, a key question of this conference, namely the function of specific covalent protein modifications is considered, again with a rather slim yield of answers. A fourth section, devoted to the "prospectives" of the title is included to dwell briefly on some of the practical problems that have been and will need to be tackled in this area in the immediate future.

Perhaps the only precise definition of *in vivo* covalent modifications of proteins requires thorough knowledge of both the entire base sequence of the mRNA transcript of a structural gene and the complete covalent structure of the protein product of that gene. With this knowledge, any difference

[1]*Aspects of the preparation of this chapter were in part supported by research grants from the USPHS (GM 31305) and from the Robert A. Welch Foundation (AU 916).*

ISBN 0-12-387560-9

between the predicted structure and that actually observed can be concluded to be the result of the in vivo chemical modification of either the constituent amino acids or of the polypeptide chain itself. The chemical modification of individual amino acids may occur at either of three distinct levels:

1. Amino Acyl-tRNA (AA-tRNA $\longrightarrow$ AA^{*}-tRNA)
2. During polymerization ($\sim AA \sim$ ribosome $\longrightarrow \sim AA^{*} \sim$ ribosome)
3. After polymerization ($\sim AA \sim \longrightarrow \sim AA^{*} \sim$)

Since only levels 2 and 3 contain any peptide bonds, modifications involving peptide bond cleavage need only be considered at these two levels. Similarly, while modifications of the N-terminal alpha-amino group can occur at either of the three levels, modifications of the C-terminal α-carboxylate obviously can only take place after the last amino acid has been incorporated at level 3.

If the entire process of synthesizing the polypeptide chain according to the genetic specifications transcribed in the mRNA is designated by the useful descriptive term "translation", then all the above reactions could be classified as posttranslational reactions. However, according to the collective wisdom of our scientific community, "translation" has been defined as being synonymous to "polymerization" and numbers 2 and 3 of the above three levels must now be designated, respectively, as the cotranslational and posttranslational steps of protein synthesis, leaving number 1 totally devoid of any adjectival designation. Regardless of one's reactions to this kind of nomenclature, these terms are now firmly established as part of the protein processing language, and consequently must be defined here.

The covalent modifications involving peptide bond cleavage will not be considered here. They have been treated in many recent reviews (e.g. Reid and Porter, 1981; Steiner and Docherty, 1982; Ondetti and Cushman, 1982). The main concern of this conference is the type of chemical modifications that convert the 20 "primary" amino acids into the many derivatized "secondary" amino acids actually found in proteins. "Primary" here signifies the 20 amino acids that have been established to be specified by the genetic code and to be the substrates in the commitment step in the translation process, namely the formation of amino acyl-tRNAs, catalyzed by the appropriate transferases. It is probably fair to say that the evidence that there are precisely 20 primary amino acids is less than perfect in that it is unlikely that all possible amino acid derivatives have been tested as substrates for amino acyl-tRNA synthases. A recent suggestion (Tappel,

1981) that selenocysteine may be an acceptable substrate for a synthase to form a Se-Cys-tRNA derivative with one specific $tRNA_{cyst}$, as the activation step for its incorporation into proteins in rat tissues has not yet been substantiated by direct, unequivocal evidence, but it serves as a reminder that we probably do not have sufficient data to exclude the possibility that the genetic code may specify more than 20 amino acids, and that we should keep an open mind on this point a while longer.

The actual known derivatives of amino acids in proteins have been presented in a recent review (Wold, 1981) and rather than repeating the long list of individual derivatives and their protein sources, a summary illustration of the derivatives has been presented in Figure 1. The main purpose of the figure is to convey in a glance the impressive number of known derivatives. For a more detailed consideration of these derivatives, where they occur, and how they are recognized, the reader is referred to the above reference.

Regarding the questions of how, where and when the modification reactions occur, it may be useful to consider them both in terms of molecular and cell biological aspects. The molecular aspects include the catalysis and chemical mechanisms, the specificity determinants that select a given residue for modification and the reversibility of the modification reaction, while the cell biology is concerned with the compartment in which the reactions occur and the specificity determinants that guide the proteins along specific pathways through these compartments. Since presence in a given compartment could be the single most important prerequisite for a modification reaction to occur, the means by which that compartment was selected for a given protein is obviously as important as the processes by which a certain domain or sequence in that protein is recognized as an appropriate substrate for the molecular modification. Thus, in considering the overall specificity of a given modification reaction, the recognition signals by which a protein gains residence in a given compartment must be considered in parallel with those that are recognized by the enzymes involved in the reaction.

In the most general terms the types of chemical transformations involved in posttranslational modifications can be categorized as illustrated in Scheme 1. Reaction a, a simple reversible reaction catalyzed by a single enzyme, is probably very rare. The reaction by which sulfhydryls and disulfides may be interconverted reversibly in response to variations in the environmental redox potential could be an example of such a reaction type. Reaction b, in which two separate enzymes are involved in a reversible modification reaction, is more

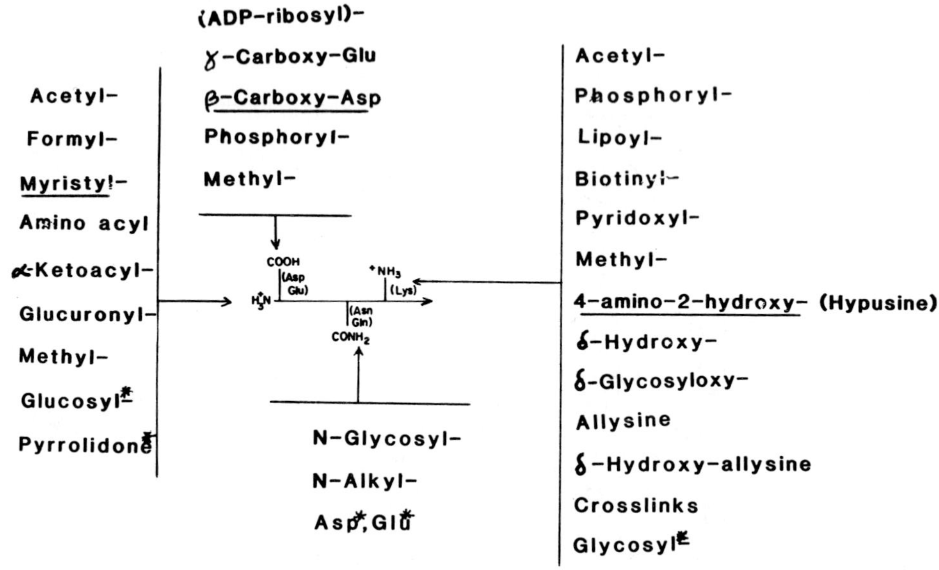

FIGURE 1a

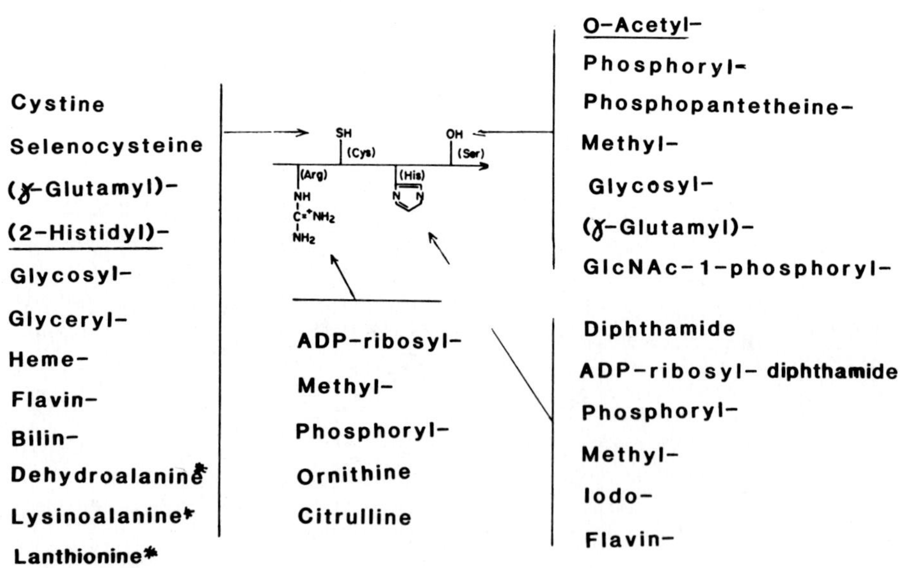

FIGURE 1b

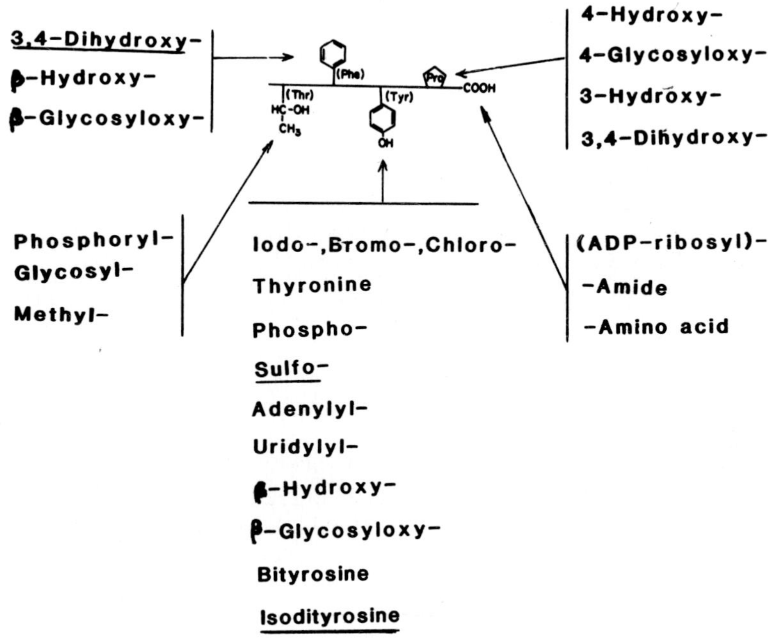

FIGURE 1c

FIGURE 1. An overview of most of the known types of amino acid derivatives in proteins. The derivatives in the figure are taken from a recent review (Wold, 1981); those underlined in the figure were not included in the review: S-(2-histidyl)-Cys (Lerch, 1982); Nα-myristyl proteins (Carr et al.*, 1982; Henderson* et al.*, 1983; Cohen, 1983); β-carboxy-Asp (Richey* et al.*, 1982); Nε-(4-amino-2-hydroxybutyl)-Lys (hypusine) (Park* et al.*, 1981); 3,4-dihydroxy-Phe (Waite and Tanzer, 1980); O-Acetyl-Ser (Rudman* et al.*, 1979); isodiTyr (Fry, 1982); and one derivative, O-sulfo-Tyr, now appears to be much more common than originally believed (Huttner, 1982). Derivatives marked with an asterisk are known products of nonenzymatic reactions; this does not eliminate the possibility that they could also be formed enzymatically.*

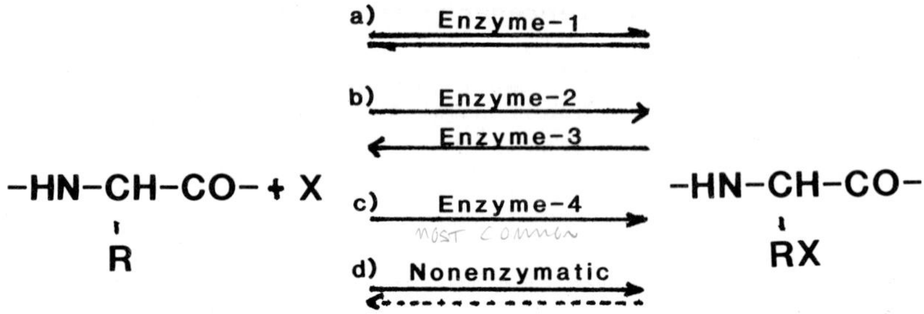

Scheme 1

common; certain phosphorylation/dephosphorylation, methylation/demethylation and acetylation/deacetylation reactions are typical examples in this category. The reversible nature of the reactions in a and b can at this stage best be associated with a regulatory effect on the proteins modified, turning their activity "on" in one direction and "off" in the other direction. Reaction c in which the modification reaction is irreversible appears to be by far the most common reaction type, and most of the reactions in Figure 1, including examples of disulfide bond formation, phosphorylations, acetylations and methylations fit in this category. The product of this reaction is a new amino acid which through its unique structure presumably has bestowed the protein in which it resides with new unique properties. The irreversible nature of the reaction may be determined by the reaction itself or by the fact that the product is rapidly removed from the environment of the catalyst. (Cleavage of the peptide bond is also an obvious example of reaction type c). Reaction type d specifying nonenzymatic modifications is somewhat controversial in that many feel that these reactions are not true posttranslational modifications in that no enzyme is involved to make the reactions fast and precise. It would seem, however, that any reaction whether fast or slow by which a primary amino acid incorporated into a protein is converted to a derivatized amino acid under physiological conditions should be included when *in vivo* chemical modification is considered. Although the slow spontaneous conversions may not be the most exciting components of cellular dynamics and biological regulation, they nevertheless proceed in all living cells as steadily and inevitably as the living and aging process itself.

The reason the consideration of specificity has become such a major issue in the field of posttranslational modification is partly that it represents the major common meeting place for all the separate groups that have traditionally split into specialty areas of individual modification reactions (the consideration of several reactions such as phosphorylation, acetylation, carboxylation, ADP-ribosylation, all at one meeting as we have here is virtually a unique event), and partly that it involves a whole new set of concepts that we do not understand well in terms of the traditional "lock-key" models of enzyme specificity. The new concepts of specificity needed to understand these reactions may be illustrated by superimposing a complicated "shooting gallery" mechanism onto the lock-key fit of substrate and enzyme (gun and target). In this complicated shooting gallery, the targets passing across the line of fire are constantly changing in color and shape, and a hit is only recorded if the proper shape and color is displayed when the shot is made, and each compartment of the cell appears to have its own set of game rules. The targets for all this firing, the proteins, thus have to encounter the proper enzyme, in the proper compartment, at exactly the time when they display the proper structural features for a "hit" to be recorded, and the understanding of the specificity process becomes a complex coordination of time, location and enzyme-(continuously changing) substrate fit. Figure 2 represents an attempt to illustrate the various choices that have to be made at different stages of protein processing. It seems reasonable to propose that amino acid sequence is the primary specificity determinant for all the various branch points of

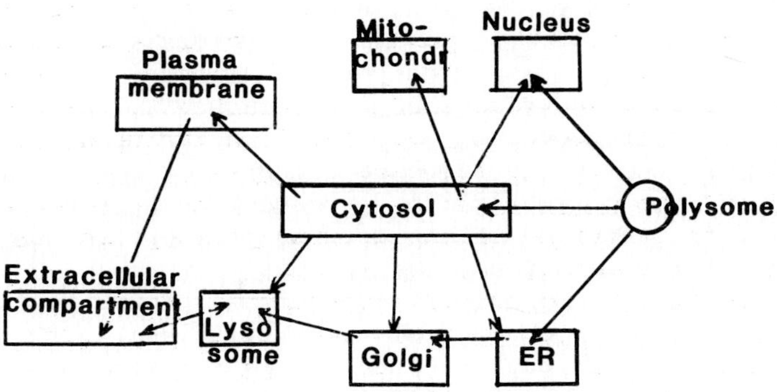

FIGURE 2. A schematic illustration of the various "choices" that have to be made in the synthesis of proteins. Starting at the site of synthesis, the protein may be destined for any of the major compartments indicated in the figure, either by direct insertion or by transfer through other compartments.

the protein processing paths outlined in Figure 2, but all aspects of sequence information should be included, from simple, exposed, short sequence [i.e., -Asn-X-Ser(Thr)- for Asn-glycosylation sites (Eylar, 1966; Struck and Lennarz, 1980) and -Arg(Lys)-Arg(Lys)-X-Y-Ser(Thr)- for Ser(Thr) phosphorylation sites (Cohen, 1983)] to complex helical segments or domain-determined juxtaposition of residues or sequences. For all of these, the concept of a continuously changing structure in which a given recognition signal is alternately exposed and masked as the processing goes on is probably an essential component.

Recent developments lend support to the proposal that some expression of amino acid sequence is universally involved, as even the specificity of the very first segregation step in protein synthesis is based on sequence recognition. According to these findings the decision of whether a protein is to be synthesized by membrane-associated ribosomes (i.e., for insertion in the lumen of the RER) or by free ribosomes is at least in part determined by the initial sequence information. As illustrated in Figure 3, it appears that the synthesis of all proteins may be initiated on free ribosomes; after the polymerization of the first 70-80 amino acids, certain ones of these sequences are recognized by a soluble sequence-recognizing protein (Warren and Dobberstein, 1978; Walter and Blobel, 1981), a protein complex which contains RNA (!) (Walter and Blobel, 1982), and which attaches itself to these amino-terminal sequences and arrests any further polymerization, while the rest, which are not recognized by the binding protein, are permitted to proceed to completion on the free polysomes. Protein synthesis on the arrested ones can be initiated by exposure to a second protein which is membrane-bound. Thus, whenever the arrested ribosome-protein synthesizing system becomes attached to the so-called "docking protein" (Meyer *et al.*, 1982) on the membrane, the inhibitory action of the sequence-recognizing protein is released and the polymerization proceeds on the membrane surface, with insertion of the growing polypeptide chain into the lumen of the endoplasmic reticulum.

In case of the nonenzymatic reactions (Figure 1), the reaction specificity is probably less stringent, but considering all the common factors determining the rate of chemical reactions, reagent concentration, solvent system (environment) and inherent chemical reactivity of the reactants, it is natural to conclude that amino acid sequence in determining the immediate environment of a reactive residue should be a key factor. The elegant studies on the rates of spontaneous hydrolysis of glutamine and asparagine under physiological conditions, convincingly illustrate the effect of neigh-

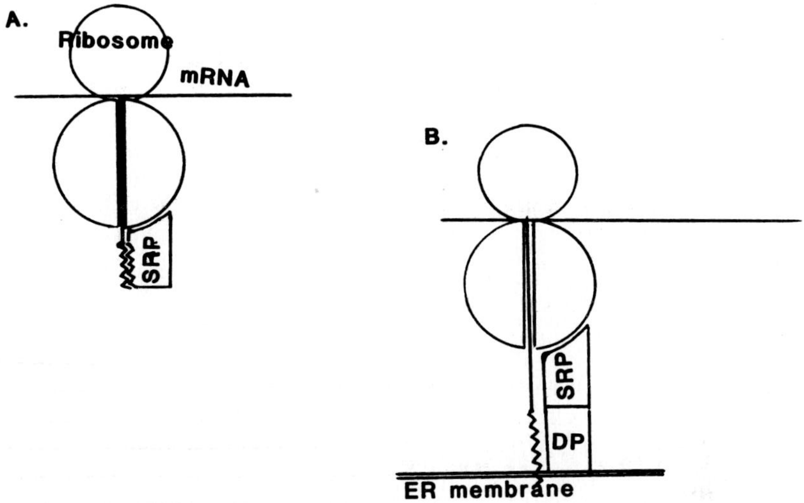

FIGURE 3. A proposed mechanism for one of the earliest decisions in protein synthesis: Is the product to be released to the cytosol as a soluble protein synthesized on free polysomes or is it to be inserted into the lumen of the ER as a protein destined, for example, for secretion or for residence in lysosomes? According to the mechanism, the synthesis of a protein in the latter category is arrested during the early stages of the polymerization process by combining with the (N-terminal) sequence-recognizing protein (SRP) as indicated in A. The synthesis can be resumed as indicated in B, if the complex encounters a protein ("docking protein" - DP) which is associated with the ER membranes and is known to bind to SRP. [Redrawn from the illustrations and data in Meyer et al.*, 1982].*

boring amino acids on the relative rate of hydrolysis of these amides (Robinson *et al.*, 1973).

Clearly, the issues of how, where and when the *in vivo* covalent modifications take place are still only in the early stages of development. As stated above, this is the one part of this whole field where the same or similar questions are asked about all the very different kinds of reactions, and consequently, where the broadest possible exchange of ideas and findings could have the most impressive and exciting impact on our understanding of the entire process by which proteins are sorted, transported and chemically modified on the path from the site of synthesis to the final compartment of action. Interactions such as the ones fostered at this symposium thus appear to represent unique and important opportunities to develop new concepts and ideas in this area.

As the science of biochemistry ultimately attempts to understand biological function, expressions of life processes, in terms of the chemical structures involved, the question of why the individual derivatives are made probably constitutes the real biochemical focus of all the exciting activities in this area over the last couple of decades. Whereas the questions about the site and specificity of reactions have major common components regardless of the nature of the reaction, the questions of function may well turn out to be a less obvious meeting place for people working on different reactions. First, the various functions that have been assigned to different modifications cover an extremely broad range of expressions, and second, any one modification may reasonably be assigned widely different biological "functions" in different systems, even if these different functions eventually may be established to express a single or only a few molecular properties of that modification. To illustrate the dilemmas of this latter point, protein glycosylation provides a useful example. Glycosylated proteins, along with glycolipids and other glycoconjugates appear to fulfill a major general molecular role as the ligands that interact with lectins, proteins that recognize and bind oligosaccharides. Lectin-sugar interactions in turn have been implicated in a large number of rather diversified biological processes - sperm-egg recognition, legume-symbiont interaction, circulatory glycoprotein clearance, intracellular targeting of lysosomal enzymes, microbial toxins-host cell susceptibility, and cell differentiation. The general biological "language" expressed by these interactions furthermore appears to extend over the entire biosphere as illustrated by the broad practical use of lectins from, for example, plants or invertebrates to sort and distinguish between animal cell types, blood groups and bacterial cell surfaces. The way each new biological function is presented, one sometimes gets the impression that it represents an either-or alternative to other proposed or established functions, which should not be the case. Similar apparent conflicts could easily arise for other modification reactions, including some of the ones to be considered at this symposium, and it seems prudent to adopt the most flexible and open-minded attitude in considering the functions of posttranslational modifications. The basis for this attitude could for example follow the argument that any one modification originally gave a distinct biological advantage for some totally unknown reason, and that that advantage was enforced through evolution into an essential requirement for life in certain forms. Once established as essential, the necessary machinery for the production of this modification must always be present, and it then seems reasonable to propose that this machinery in turn may be used for additional specialized functions at no great cost, simply by adopting the reaction and its components to new functions. Thus, in the case of the many putative major functions of lectin-sugar interaction, it seems important to recognize that they all could be secondary adaptations of an existing functional capability which by virtue of its universal nature probably was established quite early in evolution and has survived in all living forms not because of any of these adaptations but because it is required in a much more universal and fundamental function. The main points here are that the adaptation to multiple functions for a single modification type need not in any way be mutually exclusive or contradictory, that all forms need not use any particular adaptation, and finally, having established a specific function for a given modification does not necessarily permit one to stop looking for others.

With reference to the reaction types in Scheme 1, it should be possible to establish that in general the reversible reactions in a) and b) are involved in regulatory mechanisms, while the irreversible ones in c) represent the introduction of permanent structural or functional features into specific proteins. Clearly, this is no more than a generalization, containing a large gray area of overlapping functions of both types. Thus, glycosylation which to the first approximation is an irreversible modification reaction, can undergo a variety of reversible changes in the outer oligosaccharide chain, which must definitely be considered part of cellular dynamics, and protein phosphorylations and acetylation can be either permanent or reversible depending on the proteins and the systems involved. Another approach to classification of function, again rather superficial and general, is to presume understanding of certain molecular designs and on that shaky basis assign the biological significance of specific reactions within that design. Thus, we may claim that as a design strategy it makes biochemically economic sense to attach covalently coenzymes and prosthetic groups to proteins (Table I) or to covalently crosslink certain proteins for structural stability (Table II), but that claim represents an incomplete understanding of the total design strategy of cells if it does not also explain the many cases where prosthetic groups and coenzymes are not covalently bound or where structural stability is achieved without crosslinking. It does not seem unrealistic to conclude at this stage that our general understanding of the functions of the covalent modifications at best is fragmentary and incom-

TABLE I. Covalent Prosthetic Group Attachments in Proteins

Derivative	*Where R (R-CO-) is*
R-CO-NH-Lys	*Biotinyl, Lipoyl*
R-OPO$_2$-O-Ser	*Pantetheine*
R=N-Lys	*Pyridoxal, Retinal*
R-S-Cys	*Heme, Flavin, Bilin*
R-N(1,3)-His	*Flavin*
R-O(4)-Tyr	*Flavin*

TABLE II. Natural Crosslinks in Proteins

Cys-S——S-Cys	*Glu-γCO——NH-Lys*
Cys-S——C(2)-His	*Glu-γCO——O-Ser*
Cys-S——C(3)-Ala(Lanthionine)	*Glu-γCO——S-Cys*
Lys-NH——C(3)-Ala	*(Ser-O-PO2——O-Ser)*[a]
Lys-NH——Norleu[b]	*Allysine aldol*[b]
Tyr-C(3)——C(3^1)Tyr	*Tyr-C(3)——O(4)Tyr(isodiTyr)*

[a]*This proposed structure is based on the NMR spectral evidence for phosphate diester in certain flavoproteins (Edmondson and James, 1979). (The nature of the residues involved in the diester linkage is not known).*

[b]*Only two of the "parent" compounds from the large family of crosslinks in structural proteins such as collagen and elastin are listed here. For a recent overview see Guay and Lamy (1979).*

plete; the subsequent chapters from this symposium should enforce this view in considering phosphorylation, acetylation, and carboxylation reactions in which many but not all derivatives can be associated with very specific biological effects, and ADP-ribosylation reactions for which an understanding of the functional significance is still in the early exploratory stages.

Regarding the prospects of posttranslational modification, there can be no doubt that this whole area of understanding the functional significance of *in vivo* chemical modifications will continue to be the focus of active investigation in the years to come. There are lots of obvious problems to solve beyond those that are pursued today, and the explosion of new developments in both cell biology, molecular biology, and biochemistry suggests new approaches and new significance to these problems. While these investigations continue and expand, new derivatives will be uncovered and characterized to present new puzzles, and hopefully, new insight. There are many reasons for this last prediction, among them are the impressive methodological advances over the last years and their proven success in the structural analysis of proteins. The availability of high resolution nuclear magnetic resonance and mass spectrometry along with the improved methodology in protein sequencing, peptide and amino acid separation and characterization and the impressive resolving power of one- and two-dimensional gel-electrophoresis have all contributed in making seemingly impossible tasks quite feasible. Thus, the recent discovery of the myristate N-terminal blocking group (Carr *et al.*, 1982; Henderson *et al.*, 1983; Cohen, 1983) is probably a direct consequence of the resolving power of reverse phase HPLC. Another example is the isolation and characterization of the modified histidine derivative (Scheme 2) 2-[3-carboxamido-3-(trimethylammonio)propyl]histidine (Van Ness *et al.*, 1980). The evidence is excellent that this derivative is present as a single modification in a single protein, elongation factor 2, in eukaryotes (Dunlap and Bodley, 1983). It was isolated from yeast, and its structure tentatively identified by nuclear magnetic resonance spectroscopy, and subsequently confirmed by fast atom bombardment mass spectroscopy. The biological

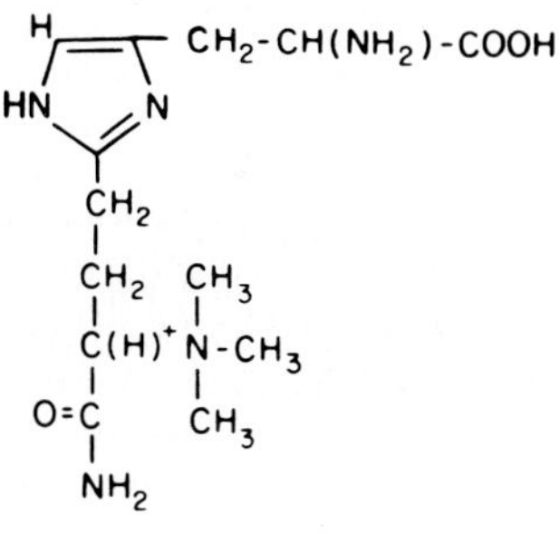

Diphthamide

2-(3-carboxyamido-3-(trimethylammonio)propyl)histidine

Scheme 2

function of this derivative is totally unknown at this stage; it is of interest, however, that the basis for its identification as a unique amino acid derivative was the fact that it is the site at which diphtheria toxin specifically incorporates an ADP-ribosyl group (at the N_1 of the modified histidine ring (Oppenheimer and Bodley, 1981) and as a consequence kills the infected cell. Because of this fact, the derivative has been given the trivial name, diphthamide.

Other similar examples of the identification of unique amino acid derivatives occurring at very low levels and even transiently in the cell cycle could be cited [e.g. hypusine (Park *et al.*, 1981)], but these examples almost always involve derivatives that are stable enough to withstand acid hydrolysis as the first step toward their isolation and characterization. (In the case of diphthamide, the hydrolysis product is obviously not the amide but the corresponding carboxylic acid derivative (diphthine), which is still a suitable derivative to identify as a unique amino acid after hydrolysis). Any derivative which is destroyed by routine hydrolysis methods, for example, γ-carboxyglutamate destruction by the usual acid hydrolysis, will remain undetected even if it occurs at reasonably high levels, and is probably totally unapproachable with our current methods if it represents a rare event in protein processing. What is needed is clearly a simple and reproducible method of hydrolysis which will cleave the peptide bonds, but leave the amino acid derivatives intact. This is by no means a new idea; in fact, proteases are used quite extensively in most laboratories engaged in the isolation and characterization of modified amino acids. The current protease applications are in general *ad hoc*, however, and a good deal of time and effort is spent on developing the proper hydrolysis procedure for each particular situation. It would therefore seem to be worthwhle to continue the search for a standardized set of proteases and hydrolysis conditions which would have a high probability of giving complete hydrolysis of any unknown protein. To illustrate both past efforts in this area and some of the difficulties involved, I thought I would close my remarks by briefly summarizing some work in our laboratory several years ago (Chin and Wold, 1974). With the goal of achieving complete hydrolysis of proteins using only proteases and excluding any harsh pretreatment (oxidation, reduction) of the substrate, we established that in order to obtain reasonable hydrolysis rates, the initial proteolysis had to be done under denaturing conditions, which for most proteins could be achieved most readily at low pH. The first step consequently was digestion with an acid protease (Aspergillopeptidase) with a pH optimum of 2-3. After fairly extensive cleavage at the low pH, the substrate was subjected to continued digestion with crude pronase, aminopeptidase and prolidase either in sequence or simultaneously. To achieve complete hydrolysis of the substrate, it was also found to be necessary to use high levels of proteases, and we consequently had to prepare insoluble derivatives of the enzymes (by attaching them to Sepharose or glass beads) in order to avoid contaminating the substrates with amino acids from the proteases themselves. With this procedure, using protease:substrate ratios of 0.2-1, extensive hydrolysis was indeed observed, and the main differences between the acid hydrolysis and the protease-catalyzed hydrolysis products could be explained by the resistance of the peptide bonds adjacent to disulfide bonds. When glycoproteins were checked as substrates, a good yield of free glycosylasparagine was, as expected, obtained from ovalbumin, a lower yield from γ-globulin, while an extensively glycosylated polypeptide, submaxillary mucin, was found to be essentially resistant to hydrolysis. We feel that in spite of the obvious deficiencies of our early methodology, the results strongly suggest that it should be possible to develop a "universally" applicable protease mixture, especially for the specific purpose of isolating unstable amino acid derivatives resulting from *in vivo* modification. When a set of proteases has been well characterized and demonstrated to be free of derivatized amino acids, the fear of contaminating the substrate with proteases has been eliminated, making it possible to use large quantities of soluble enzymes in such hydrolysis experiments. On the other hand our experience with refractive disulfide-containing peptides and glycosylated peptides suggest that the modified residues themselves may well decrease the efficiency of protease action on the peptides containing them. Nevertheless, the prospects of being able to procure the proposed set of well characterized proteases and apply them to protein hydrolysis as reproducibly as our current chemical methods are now applied, represent an attractive possibility which could have a major impact on future developments in the area of posttranslational modification.

ACKNOWLEDGMENT

Anne DeForest's expert contributions in typing the manuscript are gratefully acknowledged.

REFERENCES

Carr, S.A., Biemann, K., Shoji, S., Parmelee, D.C., and Titani, K. (1982). *Proc. Natl. Acad. Sci. USA 79,* 6128.
Chin, C.C.Q., and Wold, F. (1974). *Anal. Biochem. 61,* 379.
Cohen, P. (1983). This volume.
Dunlop, P.C., and Bodley, J.W. (1983). *J. Biol. Chem. 258,* in press.
Edmondson, D.E., and James, T.L. (1979). *Proc. Natl. Acad. Sci. USA 76,* 3786.
Eylar, E.H. (1966). *J. Theor. Biol. 10,* 89.
Fry, S.C. (1982). *Biochem. J. 204,* 449.
Guay, M., and Lamy, F. (1979). *Trends Biochem. Sci. 4,* 160.
Henderson, L.E., Krutzch, H.C., and Oroszlan, S. (1983). *Proc. Natl. Acad. Sci. USA 80,* in press.
Huttner, W. (1982). *Nature 299,* 273.
Lerch, K. (1982). *J. Biol. Chem. 257,* 6414.
Meyer, D.I., Krause, E., and Dobberstein, B. (1982). *Nature 297,* 647.
Ondetti, M.A., and Cushman, D.W. (1982). *Annu. Rev. Biochem. 51,* 283.
Oppenheimer, N.J., and Bodley, J.W. (1981). *J. Biol. Chem. 256,* 8579.
Park, M.H., Cooper, H.L., and Folk, J.E. (1981). *Proc. Natl. Acad. Sci. USA 78,* 2869.
Reid, K.B.M., and Porter, R.R. (1981). *Annu. Rev. Biochem. 50,* 433.
Richey, B., Christy, M.R., Haltiwanger, R.C., Koch, T.H., and Gill, S.J. (1982). *Biochemistry 21,* 4819.
Robinson, A.B., Scotchler, J.W., and McKerron, J.H. (1973). *J. Am. Chem. Soc. 95,* 8156.
Rudman, D., Chawla, R.K., and Hollis, B.M. (1979). *J. Biol. Chem. 254,* 10102.
Steiner,D.F., and Docherty, K., (1982). *Annu. Rev. Physiol. 44,* 625.
Struck, D.K., and Lennarz, W.J. (1980). *In* "The Biochemistry of Glycoproteins and Proteoglycans" (W.J. Lennarz, ed.), p. 35. Plenum Press, New York.
Tappel, A.L. (1981). *In* "Selenium in Biology and Medicine" (J.E. Spallholz, J.L. Martin, and E.E. Ganther, eds.), p. 44. Avi Press, Connecticut.
Van Ness, B.G., Howard, J.B., and Bodley, J.W. (1980). *J. Biol. Chem. 255,* 10710.
Waite, J.H., and Tanzer, M.L. (1980). *Biochem. Biophys. Res. Commun. 96,* 1554.
Walter, P., and Blobel, G. (1981). *J. Cell. Biol. 91,* 557.

Walter, P., and Blobel, G. (1982). *Nature 299*, 691.
Warren, G., and Dobberstein, B. (1978). *Nature 273*, 569.
Wold, F. (1981). *Annu. Rev. Biochem. 50*, 783.

PROTEIN PHOSPHORYLATION AND THE NEURAL AND HORMONAL CONTROL OF ENZYME ACTIVITY[1]

P. Cohen
A. Aitken
Z. Damuni
B.A. Hemmings
T.S. Ingebritsen
P.J. Parker
C. Picton
T. Resink
A.A. Stewart
N.K. Tonks
J. Woodgett

Department of Biochemistry
Medical Sciences Institute
University of Dundee
Dundee, Scotland

CYCLIC AMP AND Ca^{2+}-DEPENDENT PHOSPHORYLATION

It has become obvious over the past 10 years that the phosphorylation and dephosphorylation of enzymes represents

[1]The work carried out in this laboratory has been supported by a Programme Grant from the Medical Research Council, London, U.K., and by grants from the British Diabetic Association, British Heart Foundation and Cancer Research Campaign. We also thank the Medical Research Council for providing post-graduate studentships for Zahi Damuni, Nicholas Tonks and James Woodgett. Peter Parker was the recipient of a Medical Research Council Postdoctoral Fellowship and Thomas Ingebritsen acknowledges postdoctoral fellowships from the National Science Foundation and National Institutes of Health, U.S.A.

ISBN 0-12-387560-9

TABLE I. List of Enzymes in Mammalian Tissues Reported to be Regulated by Cyclic AMP Dependent Protein Kinase[a]

Enzyme[b]	*Metabolic process affected*	*Activated (A) or inhibited (I)*	*Major tissue*
Phosphorylase kinase	*Glycogenolysis*	*A*	*Muscle, liver*
Glycogen synthase	*Glycogen synthesis*	*I*	*Muscle, liver*
Triglyceride lipase	*Lipolysis*	*A*	*Adipose tissue*
Glycerol phosphate acyl transferase	*Triglyceride synthesis*	*I*	*Adipose tissue*
Cholesterol esterase	*Steroidogenesis*	*A*	*Adrenal cortex*
Acetyl-CoA carboxylase	*Fatty acid synthesis*	*I*	*Liver, adipose tissue mammary gland*
Pyruvate kinase	*Glycolysis*	*I*	*Liver*
6-phosphofructo-2-kinase	*Glycolysis/gluconeogenesis*	*I*	*Liver*
Fructose-2,6-bisphosphatase	*Glucolysis/gluconeogenesis*	*A*	*Liver*
Fructose-1,6-bisphosphatase	*Gluconeogenesis*	*A*	*Liver*
Phenylalanine hydroxylase	*Aromatic amino acid degradation*	*A*	*Liver*
Tyrosine hydroxylase	*Catecholamine and neuro-transmitter synthesis*	*A*	*Adrenal medulla, brain*
Myosin light chain kinase	*Muscle contraction*	*I*	*Smooth muscle*

[a]*Cohen, 1982; Van Shaftinger* et al., *1982; El-Maghrabi* et al., *1982.*

[b]*It should be emphasized that some of these enzymes do not yet meet all the criteria (Nimmo and Cohen, 1977) necessary to establish that they are physiological substrates for cAMP-Prk* in vivo.

the major general mechanism by which neural and hormonal signals control cellular metabolism (Cohen, 1982). Many hormones exert their effects by triggering the activation of adenylate cyclase. This results in an elevation of the intracellular concentration of cyclic AMP, which transmits the hormonal signal by activating cyclic AMP dependent protein kinase (cAMP-PrK). The specificity of the response is therefore determined by whether receptors for the hormone are present on the plasma membranes of target cells, and which substrate proteins for cAMP-PrK are located within those cells. A list of the enzymes that are known to be regulated by cAMP-PrK are given in Table I. It is, however, important to stress that cAMP-PrK is a very specific enzyme that phosphorylates relatively few proteins at significant rates. Even physiological substrates are phosphorylated at only one or two sites out of a large number of serine and threonine residues that are potentially available. The basis for this specificity has become clearer following the discovery that two adjacent basic amino acids are usually present just N-terminal to the serine (or occasionally threonine) that is phosphorylated (Table II). This structural feature is now known to be critical for specific substrate recognition.

Neural or hormonal excitation of a variety of cells produces transient rises in the cytoplasmic concentration of Ca^{2+}. This can arise from an accelerated influx of Ca^{2+} across the plasma membrane, or from its mobilization from intracellular stores, such as the mitochondrion or sarcoplasmic reticulum. Over the past five years it has become clear that many of the biological actions of Ca^{2+} are mediated by calmodulin (or structurally related Ca^{2+}-binding proteins), in a manner that closely resembles the action of cAMP-PrK (Figure 1). The ability of calmodulin to fuction as a Ca^{2+}-dependent regulator of enzyme activity is a consequence of the conformational changes that follow its binding of Ca^{2+}, leading to the formation of specific interaction domains. A list of enzymes that are activated by Ca^{2+}-calmodulin are given in Table III. This list includes a number of protein kinases, and one protein phosphatase that is discussed in more detail below.

Although the cyclic AMP and Ca^{2+}-calmodulin systems are formally analogous in the way that they produce phosphorylation of a number of proteins (Figure 1), a major difference between the two systems is that mammalian cells contain just one catalytic subunit of cAMP-PrK that phosphorylates a variety of proteins, whereas there appear to be many different calmodulin dependent protein kinases. Nevertheless, it remains possible that a calmodulin dependent protein kinase(s) of much broader specificity also exists.

TABLE II. Amino Acid Sequences at the Phosphorylation Sites of Substrates for cAMP-Dependent Protein Kinase[a]

Substrate[b]	*Sequence*
Glycogen synthase (site-1A)	*Pro-Gln-Trp-Pro-Arg-Arg-Ala-Ser(P)-Cys-Thr-Ser*
Pyruvate kinase, rat liver	*Gly-Tyr-Leu-Arg-Arg-Ala-Ser(P)-Val-Ala-Gln*
Ribosomal protein S6 (rat liver)	*Arg-Arg-Leu-Ser(P)-Ser-Leu-Arg*
Phosphorylase kinase (α-subunit)	*Phe-Arg-Arg-Leu-Ser(P)-Ile-Ser-Thr*
cAMP-dependent protein kinase (bovine heart)	*Asp-Arg-Arg-Val-Ser(P)-Val*
Phosphorylase kinase (β-subunit)	*Ala Arg-Thr-Lys-Arg-Ser-Gly-Ser(P)-Val-Tyr-Glu*
Glycogen synthase (site-1B)	*Gly-Gly-Ser-Lys-Arg-Ser-Asn-Ser(P)-Val-Asp-Thr*
Histone H1, rat liver	*Ala-Lys-Arg-Lys-Ala-Ser(P)-Gly-Pro-Pro*
Phenylalanine hydroxylase, rat liver	*Ser-Arg-Lys-Leu-Ser(P)-Asx-Phe-Gly*
Inhibitor-1	*Ile Arg-Arg-Arg-Arg-Pro-Thr(P)-Pro-Ala-Thr*
Troponin I, rabbit heart	*Val-Arg-Arg-Ser(P)-Asp-Arg-Ala*
Glycogen synthase (site-2)	*Pro-Leu-Ser Arg-Thr-Leu-Ser(P)-Val-Ser-Ser*
Fructose Bisphosphatase, rat liver	*Ser-Arg-Tyr-Ser(P)-Leu-Pro-Leu*

[a]*Cohen, 1980; Picton* et al., *1982; Embi* et al., *1982; Wettenhall and Cohen, 1981; Humble* et al., *1979; Humble, 1980; Wretborn* et al., *1980.*

[b]*Each protein was from rabbit skeletal muscle unless otherwise stated.*

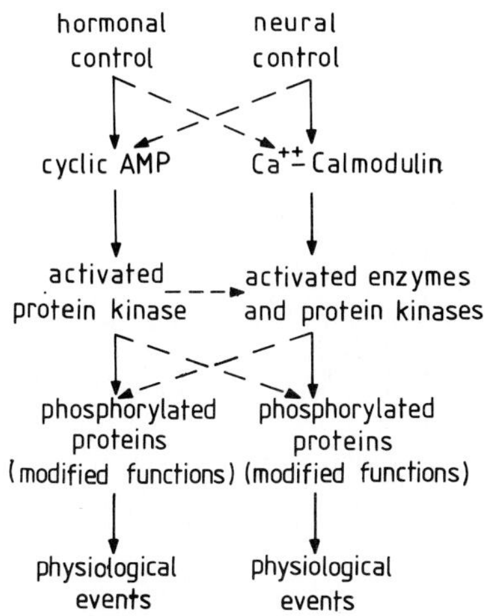

FIGURE 1. Some interrelationships between the cyclic AMP and Ca^{2+}-calmodulin systems in the neural and hormonal control of cellular activity (Cohen, 1982).

Another point illustrated by the broken lines in Figure 1 is that the cAMP and Ca^{2+}-calmodulin pathways are closely interlinked, and may often be interchangeable. For example, the stimulation of glycogenolysis by epinephrine in skeletal muscle is purely a β-adrenergic effect mediated by cyclic AMP, whereas in adult rat liver it is an α-adrenergic effect mediated, at least in part, by Ca^{2+} (Exton *et al.*, 1981). Electrical excitation of skeletal muscle elevates the cytoplasmic concentration of Ca^{2+}, whereas in the brain it increases cyclic AMP as well as Ca^{2+}. This may be a consequence of the activation of the Ca^{2+}-calmodulin dependent adenylate cyclase (Table III) that is present in brain, but not, apparently, in most other tissues. cAMP-PrK is known to regulate two calmodulin dependent enzymes, namely phosphorylase kinase and myosin light chain kinase (Tables I and III). cAMP-PrK and Ca^{2+}-calmodulin dependent protein kinases also frequently phosphorylate the same proteins, although usually at distinct sites due to the high specificity of protein kinases. Examples of this phenomenon include the enzymes glycogen synthase and tyrosine hydroxylase (Tables I and III) and the synaptic membrane protein, termed protein-I (Table III).

A number of physiological stimuli, such as muscarinic-cholinergic stimulators, vasopresin, angiotensin and α-adrenergic stimulation induce a rapid hydrolysis of phosphatidyl-

TABLE III. List of Enzymes in Mammalian Tissues that are Activated by Calmodulin[a]

Enzyme	Metabolic process	Major tissue
High Km cyclic nucleotide phosphodiesterase	Cyclic AMP and/or cyclic GMP hydrolysis	Brain (neuronal cells), heart
Adenylate cyclase	Cyclic AMP formation	Brain (glial cells)
Ca^{2+}-Mg^{2+} ATPase	Calcium transport	Erythrocyte and other plasma membranes
Phosphorylase kinase	Glycogen metabolism	Skeletal muscle
Myosin light chain kinase	Muscle contraction	Muscle and non muscle tissues
Glycogen synthase kinase	Glycogen synthesis	Muscle, liver
Tyrosine/tryptophan hydroxylase kinase	Neurotransmitter synthesis	Brain
Phospholamban kinase	Ca^{2+}-uptake into the sarcoplasmic reticulum	Heart
Protein-1 kinases	Control of transmitter release?	Brain
Protein phosphatase 2B	Protein dephosphorylation	Skeletal muscle and brain

[a]Cohen, 1982; Woodgett et al., 1982.

inositol in the plasma membranes of their target cells to yield inositol phosphates and diacylglycerol. It is therefore of particular interest that a new type of protein kinase has been discovered that has an absolute requirement for diacylglycerol, phosphatidylserine and physiological concentrations of Ca^{2+} (Takai *et al.*, 1981). This enzyme, termed protein kinase-C, exists in an inactive form in the soluble fraction of all mammalian cells, but in the presence of Ca^{2+} associates with membranes to exhibit full catalytic activity. The Ca^{2+} dependency of this enzyme is not conferred by calmodulin or related Ca^{2+}-binding proteins. It is attractive to speculate that protein kinase-C mediates the actions of several hormones with diacylglycerol acting as a second messenger. There is now quite good evidence that protein kinase-C is involved in the thrombin-induced release of serotonin from platelets, and a protein of M_r = 40,000 appears to be the major intracellular target for protein kinase-C in this cell (Nishizuka *et al.*, 1983) The role of protein kinase-C in hormone action is however difficult to evaluate at present, since almost no physiological substrates have been identified. There is currently no evidence that protein kinase-C is involved in regulating enzyme activities or in the control of cellular metabolism.

THE MECHANISM OF ACTION OF INSULIN

Attempts to elucidate the mechanism by which insulin influences cellular metabolism stem from the work of Larner over 20 years ago, when he showed that insulin activated glycogen synthase in muscle (Villar-Palasi and Larner, 1960). This enzyme could be isolated as a phosphorylated form(s) dependent on glucose-6P for activity, or as a dephosphorylated form that was almost fully active in the absence of glucose-6P (Friedman and Larner, 1963). Insulin increased the proportion of the glucose-6P independent form within minutes in the perfused rat diaphragm indicating that dephosphorylation of the enzyme had taken place (Villar-Palasi and Larner, 1963). This, in turn, implied that insulin had either decreased the activity of a glycogen synthase kinase or increased the activity of a glycogen synthase phosphatase. The effect of insulin altered when glucose was omitted from the perfusion medium (Villar-Palasi and Larner, 1961), demonstrating that it was not a consequence of the increased rate of transport of glucose into muscle that is also stimulated by insulin.

Following the discovery of cAMP-PrK (Walsh *et al.*, 1968), it became clear that this protein kinase catalyzed the phosphorylation of skeletal muscle glycogen synthase (Schlender *et al.*, 1969; Soderling *et al.*, 1970). Since insulin was known to decrease cAMP levels in adipose tissue and liver that had previously been elevated by adrenalin and glucagon respectively, an obvious possibility was that insulin activated muscle glycogen synthase by the same mechanism, resulting in decreased cAMP-PrK activity. However, insulin activated glycogen synthase in the absence of cAMP elevating agents, and no change in the concentration of cAMP in muscle could be detected under conditions where glycogen synthase activity was stimulated (Goldberg *et al.*, 1967; Craig *et al.*, 1969). This observation led Larner to postulate that the interaction of insulin with its receptors triggered the formation of a "second messenger" distinct from cAMP that bound to cAMP-PrK, increasing its Ka for cAMP (Larner, 1972). Such a compound was subsequently reported to exist (Larner *et al.*, 1979).

An alternative idea for insulin's action was suggested in 1974, with the discovery of a second glycogen synthase kinase (Nimmo and Cohen, 1974), and it is now realized that the regulation of glycogen synthase is exceedingly complex involving at least six protein kinases and seven phosphorylation sites (Figure 2). Recently, the activation of glycogen synthase by insulin has been shown to result from the specific dephosphorylation of sites (3a+3b+3c), the serine residues phosphorylated by glycogen synthase kinase-3 (GSK-3) (Figures 2, 3 and Table IV). Since the residues phosphorylated by cAMP-PrK (sites 1a, 1b and 2 - Figure 2) are unaffected by insulin, and the activity of GSK-3 is unaffected by cAMP, the results exclude the possibility that insulin acts by decreasing cAMP or inhibiting cAMP-PrK by any other mechanism. Insulin must therefore either decrease the activity of GSK-3 or increase the activity of the protein phosphatase(s) that dephosphorylates (sites 3a+3b+3c). Two enzymes termed protein phosphatase-1 (PrP-1) and phosphatase 2A (PrP-2A) are capable of dephosphorylating sites (3a+3b+3c) *in vitro*, PrP-1 being the dominant activity (section 3). However, neither enzyme is specific for sites (3a+3b+ 3c), and site-2 is dephosphorylated with almost equal rapidity *in vitro* (Parker *et al.*, 1982). Thus current evidence suggests that the activation of glycogen synthase by insulin should involve a specific inhibition of GSK-3.

GSK-3 has been purified to a state approaching homogeneity (Hemmings *et al.*, 1981) and its intracellular concentration is ~2×10^{-8} M, or ~50-fold higher than the normal circulating levels of insulin. The effect of insulin would

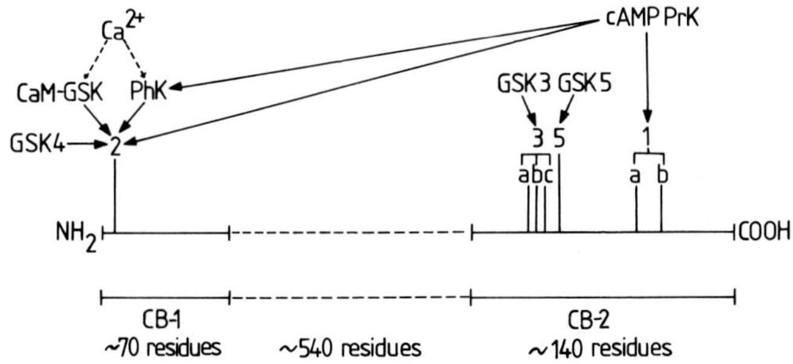

FIGURE 2. Organization of phosphorylation sites in rabbit skeletal muscle glycogen synthase. Site-2 is seven residues from the N-terminus and sites 3a, 3b, 3c, 5, 1a and 1b are 30, 34, 38, 46, 87 and 100 residues from the N-terminus of a large CNBr-peptide (CB-2) at the C-terminal end of the molecule (Picton et al.*, 1982; Embi* et al.*, 1981). Sites 3a, 3b and 3c are located in the same tryptic peptide. Abbreviations: cAMP-PrK, cyclic AMP dependent protein kinase; PhK, phosphorylase kinase; CaM, calmodulin; GSK, glycogen synthase kinase.*

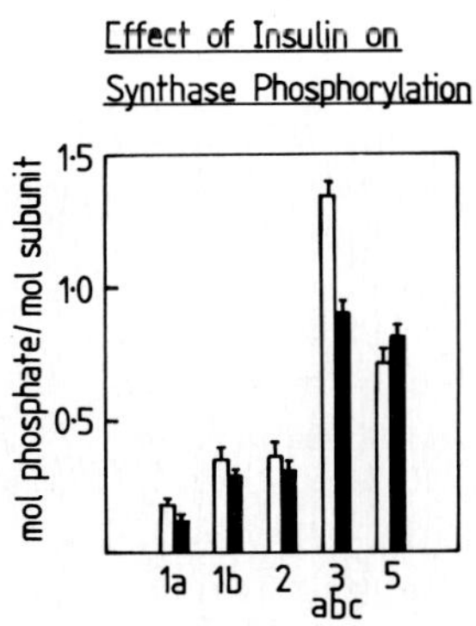

FIGURE 3. Effect of insulin on the phosphorylation of the seven phosphoserine residues of rabbit skeletal muscle glycogen synthase (Figure 2). The open bars show the phosphate content of the various sites in animals injected with propranolol, and the closed bars the phosphate contents after injection of propranolol and insulin. Further experimental details are given in the legend to Table IV. The data are taken from Parker et al. *[1982b].*

TABLE IV. Effect of Insulin on the Kinetic Properties and Phosphate Content of Skeletal Muscle Glycogen Synthase[a,b]

Treatment	*Activity ratio ± glucose-6P*	*Ka glucose-6P (mM)*	*Phosphate content (mol/mol subunit)*
L-Propranolol	*0.18 ± 0.02 (11)*	*1.2 ± 0.1 (11)*	*2.74 ± 0.09 (14)*
L-Propranolol + insulin	*0.35 ± 0.02 (12)*	*0.6 ± 0.05 (12)*	*2.33 ± 0.09 (14)*

[a]*Taken from* Parker et al., *1982b.*

[b]*Female New Zealand white rabbits were starved for 24 h. Propranolol (2 mg/kg) and insulin (16.5 μg/kg) were administered intravenously 15 min prior to sacrifice, and a further injection of insulin (16.5 μg/kg) was given 7.5 min prior to sacrifice. The animals were killed with a lethal dose of 6% sodium pentobarbitone (4.0 ml) and the muscle from the hind limbs and back homogenized in the presence of EDTA and NaF to freeze the state of phosphorylation. After centrifugation for 45 min at 6000 x g the supernatant was decanted, and aliquots passed through Sephadex G-25 and assayed for glycogen synthase activity. The activity ratio ± glucose-6P and Ka for glucose-6P were measured. Glycogen synthase was then purified to homogeneity, and the alkali-labile phosphate bound covalently to the enzyme determined. The values are given ± s.e.m. The values in parentheses indicate the number of determinations.*

therefore have to be indirect. One possibility is that the insulin-receptor interaction triggers the formation of a "second messenger" that inhibits GSK-3. However, an intriguing alternative possibility is that the insulin receptor is a protein kinase that only becomes activated when it interacts with insulin. The precedent for this idea is epidermal growth factor (EGF), a hormone that promotes the growth of a variety of cells. The binding of EGF to its receptor causes the latter protein to be phosphorylated by a protein kinase that is closely associated with the receptor (Cohen *et al.*, 1981), and there is compelling evidence that the protein kinase is indeed the receptor itself (Buhrow *et al.*, 1982). Similar results have now been reported for the platelet derived growth factor (Ek and Heldin, 1982; Nishimura *et al.*, 1982) and insulin (Kasuga *et al.*, 1982). In all these cases, phosphorylation of the receptors occurs on tyrosine residues. This is most unusual, since phosphotyrosine represents only 0.03% of the phosphorylated amino acids in normal cells, phosphoserine and phosphothreonine accounting for 99.97% (Sefton *et al.*, 1980). These findings raise the possibility that hormones like EGF and insulin do not act through second messengers. The idea would be that the receptor is a transmembrane protein, the binding site(s) for insulin facing the outer surface, and the protein kinase activity the inner surface of the plasma membrane. The binding of insulin would then trigger a conformational change in the receptor, leading to activation of the protein kinase activity, and the phosphorylation of intracellular proteins. An important possibility that therefore needs to be examined is whether GSK-3 can be phosphorylated (on a tyrosine residue) and inactivated by the insulin receptor.

Insulin influences a variety of cellular processes. It not only stimulates glycogen synthesis, but also fatty acid, cholesterol and protein synthesis. The enzymes that are likely to control these pathways, acetyl-CoA carboxylase (fatty acid synthesis), hydroxymethylglutaryl (HMG) CoA reductase (cholesterol synthesis), and the aminoacyl-tRNA synthetase complex (protein synthesis) can each be inactivated *in vitro* by protein kinases distinct from cAMP-PrK (Hardie and Guy, 1980; Ingebritsen and Gibson, 1980; Damuni *et al.*, (1982). It has been established that insulin can activate HMG-CoA reductase (Ingebritsen and Gibson, 1980) and the aminoacyl-tRNA synthetase complex (Damuni *et al.*, 1982) by promoting the dephosphorylation of these enzymes. These findings therefore raise the possibility that several of the actions of insulin may involve inhibition of protein kinases distinct from cAMP-PrK.

THE PROTEIN PHOSPHATASES INVOLVED IN CELLULAR REGULATION

An important generality to have emerged is that rate determining enzymes in biodegradative pathways are activated, whereas those in biosynthetic pathways are inactivated by phosphorylation (Tables I and III and Cohen, 1980). This finding has been important conceptually, since it suggests that different metabolic pathways are either regulated by the same protein kinases and phosphatases, or by different interconverting enzymes that respond to identical effector molecules. These ideas are already established for protein kinases, since cAMP-PrK phosphorylates both biodegradative and biosynthetic enzymes (Table I), while a number of different protein kinases are activated by the Ca^{2+}-calmodulin complex (Table III).

However, the phosphorylation state of any protein represents a balance between the activities of the protein kinase(s) and phosphatase(s) that acts on that protein, raising the question of the number of protein phosphatases involved in cellular regulation. We have recently examined this question and reached the conclusion that only four protein phosphatase catalytic subunits participate in the control of glycogenolysis, glycogen synthesis, glycolysis/gluconeogenesis, fatty acid synthesis, cholesterol synthesis and protein synthesis, in mammalian skeletal muscle and liver (Ingebritson and Cohen, 1983a,b,c; Ingebritson *et al.*, 1983a, 1983b; Pato *et al.*, 1983; Stewart *et al.*, 1983). The four enzymes can be conveniently grouped into two classes (Table V) depending on whether they dephosphorylate the β-subunit of phosphorylase kinase and are inhibited by two thermostable proteins termed inhibitor-1 and inhibitor-2 [protein phosphatase-1], or whether they dephosphorylate the α-subunit of phosphorylase kinase and are insensitive to the inhibitor proteins (type-2 protein phosphatases). Each phosphatase is present in all mammalian tissues examined, although the relative concentrations differ considerable (Ingebritsen *et al.*, 1983c). In mammalian skeletal muscle about half of the protein phosphatase-1 (PrP-1) is associated with the protein-glycogen complex. In the livers of fed animals only 10% of the PrP-1 is bound to glycogen, but 20% is attached to microsomes. In contrast, protein phosphatase-2A (PrP-2A), 2B (PrP-2B) and 2C (PrP-2C) are exclusively cytosolic enzymes (Ingebritsen *et al.*, 1983c). The substrate specificities of each enzyme have been examined in detail, and these studies have demonstrated that PrP-1, PrP-2A and PrP-2C each have very broad, but distinct, substrate specificities (Ingebritsen and Cohen, 1983b). Consequently, it is difficult, at present, to

TABLE V. Classification of Protein Phosphatases in Cellular Regulation[a]

Type	*Protein phosphatase*	*Inhibited by inhibitor-1 and inhibitor-2*	*Specificity for phosphorylase kinase*
1	*Protein phosphatase-1*	*Yes*	*β-subunit*
2	*Protein phosphatase-2A*	*No*	*α-subunit*
2	*Protein phosphatase-2B*	*No*	*α-subunit*
2	*Protein phosphatase-2C*	*No*	*α-subunit*

[a]*Ingebritsen and Cohen, 1983a,b.*

be certain of their relative importance in the regulation of different metabolic pathways. The most clearcut situation is that of glycogen metabolism in skeletal muscle, where PrP-1 is the dominant phosphatase (Ingebritsen *et al.*, (1983a,c). However, PrP-2A, and perhaps PrP-2C, may also regulate this process in hepatic tissue. PrP-2A appears more active than PrP-2C and PrP-1 in regulating the enzymes of glycolysis/gluconeogenesis and fatty acid synthesis, while PrP-2C is most active in the regulation of cholesterol synthesis (Ingebritsen *et al.*, 1983c). Nevertheless, a role for PrP-1 in the control of cholesterol synthesis is not excluded in view of the specific association of this enzyme with microsomes. PrP-1 and PrP-2A account for virtually all the phosphatase activity towards protein synthesis initiation factor eIF-2 (Ingebritsen and Cohen, 1983b), and both these enzymes are very effective in reactivating the aminoacyl-tRNA synthetase complex (Damuni *et al.*, 1982). They may therefore be the most important protein phosphatases in the regulation of protein synthesis. The ability of inhibitor-2 to promote eIF-2 phosphorylation when added to reticulocyte lysates (Ernest *et al.*, 1982), and the specific association of PrP-1 with ribosomes (Foulkes *et al.*, 1982a), also suggest a role for this enzyme in the control of protein synthesis.

The substrate specificity of PrP-2B is very restricted, at least towards the substrates that have so far been tested (Ingebritsen and Cohen, 1983b; Stewart *et al.*, 1983). Its biological role is discussed below.

REGULATION OF PROTEIN PHOSPHATASES

Protein phosphatase-1 is inhibited by nanomolar concentrations of the two thermostable proteins inhibitor-1 and inhibitor-2 (Huang and Glinsman, 1976) that have been purified to homogeneity from skeletal muscle and characterized extensively (Nimmo and Cohen, 1978; Foulkes and Cohen, 1979, 1980; Aitken *et al.*, 1982; Aitken and Cohen, 1982). Inhibitor-1 is only an inhibitor after phosphorylation on a specific threonine residue by cAMP-PrK. In skeletal muscle, the extent of phosphorylation of inhibitor-1 is increased by epinephrine (Foulkes and Cohen, 1979; Khatra *et al.*, 1980; Foulkes *et al.*, 1982) and decreased by insulin which antagonizes the effects of low concentrations of β-adrenergic agonist isoproterenol (Foulkes *et al.*, 1982). Since PrP-1 is active on many proteins that are phosphorylated by cAMP-PrK (Ingebritsen and Cohen, 1983b), the activation of inhibitor-1 may prevent the dephosphorylation of these proteins and thus amplify the effects of cAMP on their phosphorylation. PrP-1 is also active on a number of proteins phosphorylated by protein kinases distinct from cAMP-PrK (Ingebritson and Gibson, 1980; Damuni *et al.*, 1982; Ingebritsen and Cohen, (1983b). The inhibition of PrP-1 by phosphoinhibitor-1 may therefore allow cAMP to modulate the phosphorylation of these substrates. This idea is supported by recent studies of the effects of epinephrine on the phosphorylation (inactivation) of glycogen synthase in skeletal muscle (Parker *et al.*, (1982). In normally fed animals, glycogen synthase contains approximately 3 moles of phosphate bound covalently to each subunit. Following administration of epinephrine this value increases to 5 moles/mole subunit, with most of the increase occurring on the serine residues termed site-2 and sites (3a+3b+3c) (Figure 2). The increased phosphorylation of site-2 was not unexpected, since it is phosphorylated by both cAMP-PrK and phosphorylase kinase. However, the increased phosphorylation of sites (3a+3b+3c) was surprising, as these sites are phosphorylated by GSK-3. Since PrP-1 is the major protein phosphatase acting on glycogen synthase (Ingebritsen *et al.*, 1983c), inhibition of this enzyme by phosphoinhibitor-1 may explain the enhanced phosphorylation of sites (3a+3b+3c) in response to epinephrine. The same mechanism could also explain the enhanced phosphorylation of HMG-CoA reductase (Ingebritsen and Gibson, 1980) and the aminoacyl-tRNA synthetase complex (Damuni *et al.*, 1982) in rat liver in response to glucagon, provided that PrP-1 is a major phosphatase acting on these enzymes *in vivo*.

PrP-1 can be isolated as an inactive form, termed the Mg-ATP dependent protein phosphatase (Stewart *et al.*, 1981). Activation of this species requires Mg-ATP and another protein factor (Goris *et al.*, 1979, 1980) that has been identified as GSK-3 (Hemmings *et al.*, 1981). The Mg-ATP dependent protein phosphatase consists of a 1:1 complex of the catalytic subunit of protein phosphatase-1 and inhibitor-2. Activation by GSK-3 and Mg-ATP involves the phosphorylation of inhibitor-2 on a threonine residue and is accompanied by the dissociation of the complex (Hemmings *et al.*, 1982). Since the activity of GSK-3 appears to be under the control of insulin (section 2), the phosphorylation state of inhibitor-2, and activation of the Mg-ATP dependent protein phosphatase, may also be under hormonal control. However, the physiological role of this control mechanism is still unclear, because the two actions of GSK-3, namely to phosphorylate glycogen synthase and to activate the Mg-ATP dependent protein phosphatase, are antagonistic. GSK-3 has also been shown to phosphorylate the type-II regulatory subunit of cAMP-PrK at a significant rate, but the functional significance of this reaction also remains to be established (Hemmings *et al.*, 1982).

PrP-2B is completely dependent on micromolar concentrations of Ca^{2+}, and its activity is stimulated 10-fold by calmodulin. The enzyme has been purified to homogeneity, and is composed of two subunits, A (M_r ∼60,000) and B (M_r ∼15,000) in a molar ratio 1:1 (Stewart *et al.*, 1982, 1983. It is identical to a calmodulin binding protein, termed calcineurin, first identified in neural tissue (Wang and Desai, 1977; Klee and Krinks, 1978). It is known that the A-subunit of calcineurin binds calmodulin, while the B-subunit is a Ca^{2+}-binding protein which, like calmodulin, binds four moles of Ca^{2+} with affinities in the micromolar range (Klee *et al.*, 1979). These structural features suggest that the A-subunit is catalytically active, and that the B-subunit is responsible for conferring Ca^{2+}-sensitivity to the enzyme in the absence of calmodulin. The interaction of calmodulin with the A-subunit increases the catalytic activity ∼10-fold.

The determination of the primary structure of the B-subunit (A. Aitken, P. Cohen and C.B. Klee, manuscript in preparation) has demonstrated that it belongs to the same family of Ca^{2+}-binding proteins as calmodulin and troponin-C. The sequence identity is 35% with calmodulin and 29% with troponin-C, with most of the identity occurring in the regions of the four Ca^{2+}-binding sites.

The NH_2-terminus of the B-subunit is blocked, and the blocking group has recently been identified as the C_{14} saturated fatty acid, myristic acid, which is linked by an amide bond to the NH_2-terminal glycine residue (Aitken *et al.*, 1982). This is of considerable interest, since the only other protein known to possess such a blocking group is the catalytic subunit of cAMP-PrK (Carr *et al.*, 1982). Whether this unusual NH_2-terminal blocking group is confined to protein kinases and protein phosphatases remains to be determined. However, if this were the case, it would imply that the myristyl group plays a role in the recognition of protein substrates. PrP-2B and the catalytic subunit of cAMP-PrK both start with the NH_2-terminal sequence myristyl-gly-asn-x-ala, suggesting that the enzyme that links myristyl-CoA (presumably) to the protein may recognize this sequence.

Only three phosphoproteins have so far been found that are dephosphorylated rapidly by PrP-2B *in vitro*, namely the α-subunit of phosphorylase kinase, inhibitor-1 and myosin light chains (Ingebritsen and Cohen, 1983b; Stewart *et al.*, 1983), but which, if any, of these represent physiological, substrates is unclear (discussed in [Ingebritsen and Cohen, 1983a; Stewart *et al.*, 1983]). The high levels of PrP-2B in brain, particularly in the neostriatum, and its immunohistochemical localization at postsynaptic densities and microtubules of dendrites (Wallace *et al.*, 1980) suggest a role in neurotransmitter action and microtubular function. If PrP-2B can dephosphorylate a number of proteins phosphorylated by cAMP-PrK, it might function with the calmodulin-dependent cyclic AMP phosphodiesterase (Table 3) as part of a general mechanism for enhancing the Ca^{2+} signal at the expense of the cyclic AMP signal (but see [Klee *et al.*] this volume).

The mechanisms for regulating PrP-2A and PrP-2C are not presently understood. However, PrP-2A can be resolved into three forms termed PrP-$2A_0$, $2A_1$ and $2A_2$ by chromatography on DEAE-cellulose (Ingebritsen *et al.*, 1983a), which all contain the same M_r = 35,000 catalytic subunit. PrP-$2A_1$ contains two further subunits M_r = 60,000 and 55,000, while Prp-$2A_2$ only contains the M_r = 60,000 subunit complexed with the catalytic subunit (Pato *et al.*, 1983a). PrP-$2A_0$ is catalytically inactive and its activity is only expressed following dissociation to the Mr = 35,000 catalytic subunit. It has not yet been obtained in pure form, and hence its subunit composition is unknown. Dissociation of the catalytic subunits results in a several-fold activation of PrP-$2A_1$, and a much smaller increase in the activity of PrP-$2A_2$ (Ingebritsen *et al.*, 1983a). PrP-2C is a Mg^{2+}-dependent enzyme ($A_{0.5}$ = 1.0 mM), and is composed of a single polypeptide chain M_r ~43,000 (Ingebritsen and Cohen, 1983a; Pato *et al.*, 1983). The mechanism(s) by which its activity is regulated is unknown.

In summary, considerable progress has recently been made towards understanding the nature of the protein phosphatases involved in cellular regulation, and it is now apparent that the mechanisms for regulating their activities are likely to be as sophisticated as those of the protein kinases. Although PrP-1, PrP-2A, PrP-2B and PrP-2C account for virtually all the protein phosphatase activity towards phosphoenzymes involved in the control of glycogen metabolism, glycolysis/glyconeogenesis, fatty acid synthesis, cholesterol synthesis and protein synthesis, it should be mentioned that distinct protein phosphatases participate in at least two other phosphorylation systems. The dephosphorylation of the mitochondrial pyruvate dehydrogenase is catalyzed by a Ca^{2+}-dependent protein phosphatase that is clearly distinct from PrP-2B (Reed *et al.*, 1981; Pratt *et al.*, 1982), while the protein phosphatases that dephosphorylate phosphotyrosine residues also appear to be unique (Foulkes *et al.*, 1982b).

CONCLUDING REMARKS

A striking feature of recent studies on phosphorylation-dephosphorylation systems is the finding that many of the protein kinases and phosphatases have multiple functions. cAMP-PrK, calmodulin, PrP-1, PrP-2A, PrP-2B and PrP-2C clearly have multiple roles in cellular regulation. Thus evidence is emerging, that the major pathways of intermediary metabolism are controlled by common protein kinases, protein phosphatases and the small modulator proteins that regulate their activities, creating an integrated network of regulatory pathways by which diverse cellular events are co-ordinated by neural and hormonal stimuli. A major goal of future research must be to establish the structure of this regulatory network.

REFERENCES

Aitken, A., Bilham, T., and Cohen, P. (1982). *Eur. J. Biochem. 126*, 235.

Aitken, A., and Cohen, P. (1982). *FEBS Lett. 147*, 54.

Aitken, A., Cohen, P., Santikarn, S., Williams, D.H., Calder, H.G., Smith, A., and Klee, C.B. (1982). *FEBS Lett.*, *150*, 314.

Buhrow, S.A., Cohen, S., and Staros, J.V. (1982). *J. Biol. Chem. 257*, 4019.

Carr, S.A., Biemann, K., Shoji, S., Parmelee, D.C., and

Titani, K. (1982). *Proc. Natl. Acad. Sci. USA 79*, 6128.

Cohen, P. (1980). *In* "Molecular Aspects of Cell Regulation", Vol. 1 (P. Cohen, ed.), p. 255. Elsevier-Biomedical, Amsterdam.

Cohen, P. (1982). *Nature 296*, 613.

Cohen, S., Chinkers, M., and Ushiro, H. (1981). *In* "Cold Spring Harbor Conferences on Cell Proliferation, Volume 8, Protein Phosphorylation" (O.M. Rosen, and E.G. Krebs, eds.), p. 801.

Craig, J.W., Rall, T.W., and Larner, J. (1969). *Biochim. Biophys. Acta 177*, 213.

Damuni, Z., Caudwell, F.B., and Cohen, P. (1982). *Eur. J. Biochem. 129*, 57.

Ek, B., and Heldin, C.H. (1982). *J. Biol. Chem. 257*, 10486.

El-Maghrabi, M.R., Fox E., Pilkis, J., and Pilkis, S.J. (1982). *Biochem. Biophys. Res. Commun. 106*, 794.

Embi, N., Parker, P.J., and Cohen, P. (1981). *Eur. J. Biochem. 115*, 405.

Ernst, V., Levin, D.H., Foulkes, J.G., and London, I.M. (1982). *Proc. Nat. Acad. Sci. USA 79*, 7092.

Exton, J.H., Cherrington, A.D., Blackmore, P.F., Dehaye, J. P., Strickland, W.G., Jordan, J.E. and Chrisman, T.D. (1981). *In* "Cold Spring Harbor Conferences on Cell Proliferation, Volume 8, Protein Phosphorylation" (O.M. Rosen, and E.G. Krebs, eds.), p. 513.

Foulkes, J.G., and Cohen, P. (1979). *Eur. J. Biochem. 97*, 251.

Foulkes, J.G., and Cohen, P. (1980). *Eur. J. Biochem. 105*, 195.

Foulkes, J.G., Ernst, V., and Levin, D.H. (1983a). *J. Biol. Chem.*, in press.

Foulkes, J.G., Erickson, E., and Erickson, R.L. (1983b). *J. Biol. Chem. 258*, 431.

Foulkes, J.G., Cohen, P., Strada, S.J., Everson, W.V., and Jefferson, L.S. (1983c). *J. Biol. Chem. 257*, 12493.

Friedman, D.L., and Larner, J. (1963). *Biochemistry 2*, 669.

Goldberg, N.D., Villar-Palasi, C., Sasko, H., and Larner, J. (1967). *Biochim. Biophys. Acta 148*, 665.

Goris, J., Defreyn, G., and Merlevede, W. (1979). *FEBS Lett. 99*, 279.

Goris, J., Dopere, F., Vandenheede, J.R., and Merlevede, W. (1980). *FEBS Lett. 117*, 117.

Hardie, D.G., and Guy, P.S. (1980). *Eur. J. Biochem. 110*, 167.

Hemmings, B.A., Yellowlees, D., Kernohan, J.C., and Cohen, P. (1981). *Eur. J. Biochem. 119*, 443.

Hemmings, B.A., Aitken, A., Cohen, P., Rymond, M., and Hofmann, F. (1982a). *Eur. J. Biochem. 127*, 473.

Hemmings, B.A., Resink, T., and Cohen, P. (1982b). *FEBS Lett., 150,* 319.
Huang, F.L., and Glinsmann, W.H. (1976). *Eur. J. Biochem. 70,* 419.
Humble, E., (1980). *Biochim. Biophys. Acta 626,* 179.
Humble, E., Dahlqvist-Edberg, U., Ekman, P., Netzel, E., Ragnarsson, U., and Engstrom, L. (1979). *Biochem. Biophys. Res. Commun. 90,* 1064.
Ingebritsen, T.S., and Gibson, D.M. (1980). *In* "Molecular Aspects of Cell Regulation", Vol. 1 (P. Cohen, ed.) p. 63. Elsevier-Biomedical, Amsterdam.
Ingebritsen, T.S., and Cohen, P. (1983a). *Science,* in press.
Ingebritsen, T.S., and Cohen, P. (1983b). *Eur. J. Biochem.,* in press.
Ingebritsen, T.S., Foulkes, J.G., and Cohen, P. (1983a). *Eur. J. Biochem.,* in press.
Ingebritsen, T.S., Blair, J., Guy, P.S., Witters, L., and Hardie, D.G. (1983b). *Eur. J. Biochem.,* in press.
Ingebritsen, T.S., Stewart, A.A., and Cohen, P. (1983c). *Eur. J. Biochem.,* in press.
Kasuga, M., Zick, Y., Blithe, D.L., Crettaz, M., and Kahn, C. R. (1982). *Nature 298,* 667.
Khatra, B.S., Chiasson, J.L., Shikama, H., Exton, J.H., and Soderling, T.R. (1980). *FEBS Lett. 114,* 253.
Klee, C.B., and Krinks, M.H. (1978). *Biochemistry 17,* 120.
Klee, C.B., Crouch, T.H., and Krinks, M.A. (1979). *Proc. Natl. Acad. Sci. USA 76,* 6270.
Klee, C.B. *et al,* this volume.
Larner, J. (1972). *Diabetes 21,* (Suppl. 2), 428.
Larner, J. Galasko, G., Cheng, K., DePaoli-Roach, A.A., Huang, L., Daggy, P., and Kellogg, P. (1979). *Science 206,* 1408.
Nimmo, H.G., and Cohen, P. (1974). *FEBS Lett. 47,* 162.
Nimmo, H.G., and Cohen, P. (1977). *Adv. Cyclic Nucleotide Res. 8,* 145.
Nimmo, G.A., and Cohen, P. (1978). *Eur. J. Biochem. 87,* 341.
Nishimura, J., Huang, J.S., and Deuel, T.F. (1982). *Proc. Natl. Acad. Sci. USA 79,* 4303.
Nishizuka, Y., *et al.,* this volume.
Parker, P.J., Embi, N., Caudwell, F.B., and Cohen, P. (1982a). *Eur. J. Biochem. 124,* 47.
Parker, P.J., Caudwell, F.B., and Cohen, P. (1982b). *Eur. J. Biochem., 130,* 227.
Pato, M., Adelstein, R.S., Crouch, D., Safer, B., Ingebritsen, T.S., and Cohen, P. (1983). *Eur. J. Biochem.,* in press.
Picton, C., Aitken, A., Bilham, T., and Cohen, P. (1982). *Eur. J. Biochem. 124,* 37.

Pratt, M.L., Maher, J.F., and Roche, T.E. (1982). *Eur. J. Biochem.* *125*, 349.

Reed, L.J., Pettit, F.A., Bleile, D.H. and Wu, T.L. (1981). *In* "Metabolic Interconversions of Enzymes 1980" (H. Holzer, ed.), p. 124. Springer-Verlag, Heidelberg.

Schlender, K.K., Wei, S.H., and Villar-Palasi, C. (1969). *Biochim. Biophys. Acta* *191*, 272.

Sefton, B.M., Hunter, T., Beeman, K., and Eckhart, W. (1980). *Cell* *20*, 807.

Soderling, T.R., Hickenbottom, J.P., Reimann, E.M., Hunkeler, F.L., Walsh, D.A., and Krebs, E.G. (1970). *J. Biol. Chem.* *245*, 6317.

Stewart, A.A., Hemmings, B.A., Cohen, P., Goris, J., and Merlevede, W. (1981). *Eur. J. Biochem.* *115*, 197.

Stewart, A.A., Ingebritsen, T.S., Manalan, A., Klee, C.B., and Cohen, P. (1982). *FEBS Lett.* *137*, 80.

Stewart, A.A., Ingebritsen, T.S., and Cohen, P. (1983). *Eur. J. Biochem.*, in press.

Takai, Y., Kishimoto, A., Kawahara, Y., Minakuchi, R., Sano, K., Kikkawa, U., Mori, T., Yu, B., Kaibuchi, K., and Nishizuka, Y. (1981). *Adv. Cyclic Nucleotide Res.* *14*, 301.

Van Shaftingen, E., Davies, D.R., and Hers, H.G. (1982). *Eur. J. Biochem.* *124*, 143.

Villar-Palasi, C., and Larner, J. (1960). *Arch. Biochem. Biophys.* *94*, 436.

Villar-Palasi, C., and Larner, J. (1961). *Biochim. Biophys. Acta* *39*, 171.

Wallace, R.W., Tallant, E.A., and Cheung, W.Y. (1980). *Biochemistry* *19*, 183.

Walsh, D.A., Perkins, J.P., and Krebs, E.G. (1968). *J. Biol. Chem.* *243*, 3763.

Wang, J.H., and Desai, R. (1977). *J. Biol. Chem.* *252*, 4175.

Wettenhall, R.E.H., and Cohen, P. (1982). *FEBS Lett.* *140*, 263.

Woodgett, J.R., Tonks, N.K., and Cohen, P. (1982). *FEBS Lett.* *148*, 5.

Wretborn, M., Humble, E., Ragnarsson, U., and Engstrom, L. (1980). *Biochem. Biophys. Res. Commun.* *93*, 403.

THE ROUS SARCOMA VIRUS TRANSFORMING GENE PRODUCT: *IN VITRO* PHOSPHORYLATION WITH ATP AND GTP AS PHOSPHATE DONORS[1,2]

Yosef Graziani
Eleanor Erikson
Raymond L. Erikson

Department of Pathology
University of Colorado School of Medicine
Denver, Colorado

SUMMARY

This report describes quantitative results of *in vitro* phosphorylation of the Rous sarcoma virus transforming gene product, $pp60^{src}$, using ATP or GTP as phosphate donors. The phosphorylated products of approximate M_r = 60,000 that were produced with ATP or GTP were shown to be the same protein molecule since they could be immunoprecipitated with antibody raised against $p60^{src}$ produced by bacterial recombinants. Structural analysis revealed that the use of GTP resulted in phosphorylation of a tyrosine residue in the COOH-terminal region of $pp60^{src}$, apparently the same site which contains the tyrosine phosphorylated in infected cells. In contrast, the use of ATP resulted in phosphorylation of several additional tyrosine residues in the NH_2 terminal region of the molecule. In thermolability studies the t/2 values for the phosphorylation of $pp60^{src}$ in preparations from wild-type virus-infected chicken cells were 5.1 min for both ATP and

[1]*This work was supported by grants CA-15823 and CA-21117 from the National Institutes of Health, by an Award from the American Business Cancer Research Foundation, and indirectly, by a gift from R.J. Reynolds Industries, Inc., to the Department of Pathology.*

[2]*A list of abbreviations used, is included at the end of this chapter.*

ISBN 0-12-387560-9

GTP, whereas the t/2 values for the phosphorylation of $pp60^{src}$ in preparations from temperature-sensitive transformation mutant-infected cells were 1.1 min for both phosphate donors. The k_m values for the phosphorylation of $pp60^{src}$ by ATP and GTP were similar (10-36 μM and 25-36 μM, respectively) and the Vmax values were different (5-7 and 1.5-1.7 nmoles · min^{-1} · mg $pp60^{src-1}$, respectively).

INTRODUCTION

The protein product encoded by the Rous sarcoma virus (RSV) transforming *(src)* gene was identified originally in avian and mammalian cells by Brugge and Erikson in 1977. This product, designated $pp60^{src}$ is a phosphoprotein of M_r = 60,000 (Brugge and Erikson, 1977; Brugge *et al.*, 1978; Purchio *et al.*, 1978; Levinson *et al.*, 1978). It was found that $pp60^{src}$ can phosphorylate specifically intracellular and extracellular protein substrates on tyrosine residues (Collett and Erikson, 1978; Erikson *et al.*, 1979a,b; Hunter and Sefton, 1980; Collett *et al.*, 1980; Erikson and Erikson, 1980; Levinson *et al.*, 1980; Gilmer and Erikson, 1981). The phosphotransferase activity of $pp60^{src}$ is cyclic AMP-independent (Erikson *et al.*, 1979a,b), and studies in this laboratory have shown that during phosphotransferase reactions $pp60^{src}$ itself becomes phosphorylated (Erikson *et al.*, 1979a, b). In contrast to the cAMP- and cGMP-dependent protein kinases, which use only ATP as the phosphate donor, many cyclic AMP-independent protein kinases may use either ATP or GTP as phosphate donor (Rubin and Rosen, 1975). Relatively few studies have been published quantitating the nature of $pp60^{src}$ phosphotransferase activity with GTP as the phosphate donor. It was reported that GTP can be used as substrate for the phosphorylation of TBR IgG by $pp60^{src}$ (Levinson *et al.*, 1978; Richert *et al.*, 1979); however, this reaction is unlikely to show conventional behavior because the enzyme is sequestered in an immune complex. Furthermore, that GTP phosphorylating activity was recently related to GTP-ADP transphosphorylating activity found in the IgG preparations used in those studies (Richert *et al.*, 1979, 1982).

We describe in this paper a more complete analysis of the protein kinase activity of $pp60^{src}$ with ATP or GTP as phosphate donors and present immunological and genetic evidence that both the ATP- and GTP-dependent autophosphorylating activities are encoded by the Rous sarcoma virus transforming gene.

MATERIALS AND METHODS

[γ-^{32}P]ATP and [γ-^{32}P]GTP (5000-9000 Ci/mmol) were prepared as described by Johnson and Walseth (1979). Purification of pp60^{src} from RSV-transformed cells by immunoaffinity chromatography was done as described previously (Erikson *et al.*, 1979a,b; Gilmer and Erikson, 1981; Levinson *et al.*, 1980). Approximate quantitation of the amount of pp60^{src} in the immunoaffinity-purified preparations was obtained by the silver-stained protein determination (Oakley *et al.*, 1980) and will be described elsewhere (Graziani *et al.*, submitted for publication). The phosphotransferase activity of pp60^{src} was measured in 32 µl of 10 mM KPi buffer pH = 7.1, 10 mM $MgCl_2$, 1 mM EDTA and 1 mM β-mercaptoethanol at various concentrations of [γ-^{32}P]ATP or [γ-^{32}P]GTP (5-200 µM) in the presence of approximately $2.5 \cdot 10^7$ cpm of the radiolabeled trinucleotide. The reactions were incubated at 30° for 15 min and terminated by the addition of 8 µl of 5-times concentrated electrophoresis sample buffer and by heating at 95°C for 1 min. Phosphoproteins were resolved by electrophoresis through SDS-polyacrylamide slab gels and the radiolabeled proteins were visualized by autoradiography as described previously (Collett *et al.*, 1979).

RESULTS

Kinetic Parameters of Phosphorylation of pp60^{src}

The phosphorylation of pp60^{src} was determined with ATP or GTP as substrates. As shown in Figure 1, the phosphorylation of pp60^{src} with either ATP or GTP was time-dependent, the reaction was faster at between 0-15 min and reached maximal levels 15-20 min after the reaction was started. The maximal phosphorylation of pp60^{src} by ATP was 4-5 times higher than that by GTP under similar conditions. Under the same conditions, the phosphorylation of α-casein by pp60^{src} with either ATP or GTP was found to be linear with incubation times up to 30 min (data not shown). The time-dependent phosphorylation permitted us to estimate the kinetic parameters of the phosphorylation of pp60^{src} and of α-casein.

The apparent Km and Vmax values for both ATP and GTP were calculated from the curves demonstrated in Figure 2. The apparent Km for ATP of pp60^{src} phosphorylation ranged between 10-36 µM and that for GTP ranged between 25-36 µM under the

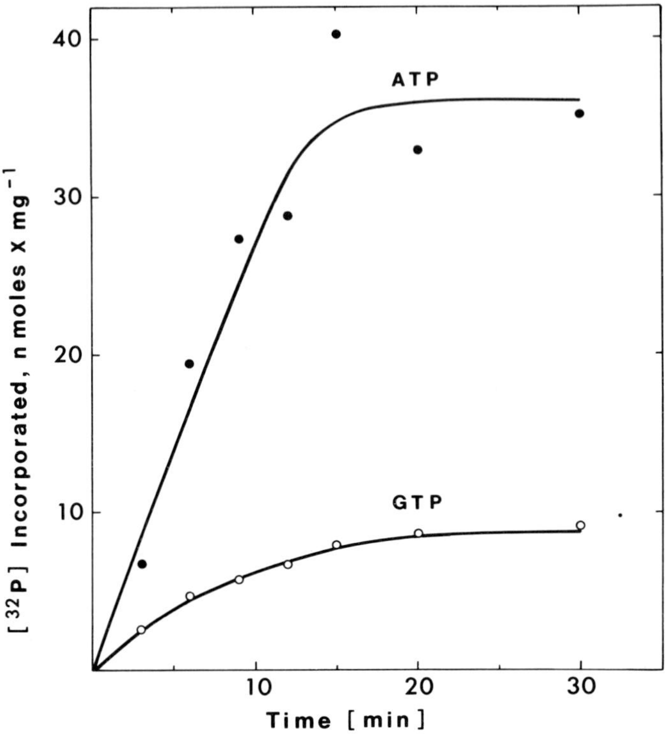

FIGURE 1. Time-dependence of the phosphorylation of pp60^{src} using [γ-^{32}P]ATP or [γ-^{32}P]GTP as phosphate donors. An immunoaffinity-purified preparation of pp60^{src} was incubated in the presence of 20 μM [γ-^{32}P]ATP (○) or 20 μM [γ-^{32}P]GTP (o). The reaction mixtures (300 μl) contained 52.5 ng of pp60^{src}. At the times indicated the reaction was terminated by transferring aliquots of 32 μl to tubes containing 8 μl of 5-times concentrated electrophoresis sample buffer and by heating at 95°C for 1 min. The samples were subjected to 10% polyacrylamide gel electrophoresis, and the phosphorylated bands of pp60^{src} were localized by autoradiography of the dried gels, excised, and the radioactivity was determined.

same conditions. The phosphorylation of α-casein by pp60^{src} gave apparent Km values of 15-40 μM and 150-170 μM for ATP and GTP as phosphorylating substrates, respectively.

The apparent Vmax values for phosphorylation of pp60^{src} with ATP or GTP as substrates were in the range of 5-7.0 and 1.5-1.7 nmoles · min^{-1} · mg pp60^{src}, respectively (Figure 2). Under the same conditions, the Vmax values for phosphoryla-

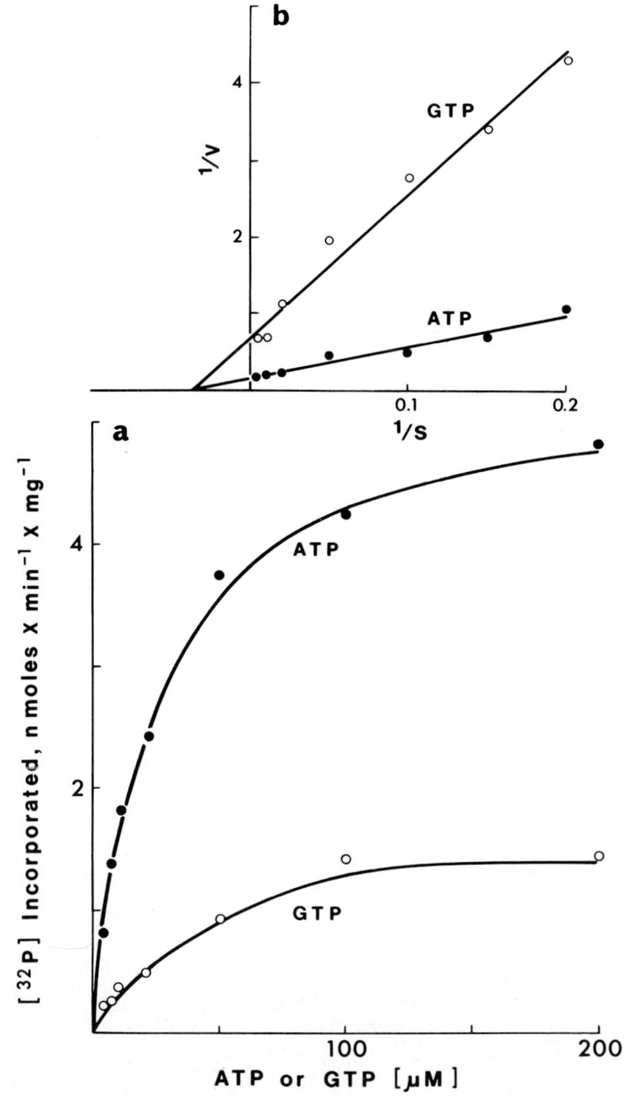

FIGURE 2. Comparison of kinetic parameters of the phosphorylation of pp60^{src} with [γ-^{32}P]ATP or [γ-^{32}P]GTP. a. pp60^{src} was incubated with various concentrations (5-200 μM) of [γ-^{32}P]ATP (•) or [γ-^{32}P]GTP (o). The amount of radiolabel was constant ($\simeq 2.5 \cdot 10^7$ cpm per assay) for both ATP and GTP. b. Double reciprocal curves for estimation of Km and Vmax values of the phosphorylation of pp60^{src} by various concentrations of [γ-^{32}P]ATP or [γ-^{32}P]GTP.

tion of α-casein with ATP or GTP as substrates were 80-210 and 34-42 nmoles · min^{-1} · mg^{-1} $pp60^{src}$, respectively.

Under maximal phosphorylation conditions, the number of phosphorylated sites per molecule of $pp60^{src}$ are in the range of 1.0-1.5 sites with GTP and 5-6 sites with ATP as phosphate donor.

The Effect of Yeast Hexokinase on the Phosphorylation of $pp60^{src}$ in the Presence of [γ-^{32}P]ATP or [γ-^{32}P]GTP

Even though the [γ-^{32}P]GTP prepared in our laboratory was free from [γ-^{32}P]ATP as determined by chromatography on PEI cellulose sheets with 0.75 M KPi, pH = 3.5, or with 1.5 N NH_4 formate, pH = 3.5 (Johnson and Walseth, 1979; Gonzales and Geel, 1975), it was possible that a transphosphorylation reaction from [γ-^{32}P]GTP to ADP or an exchange reaction with ATP, which may exist in the $pp60^{src}$ preparations as bound or free nucleotides, could result in the generation of [γ-^{32}P] ATP, which would then be the actual substrate for the phosphorylations by [γ-^{32}P]GTP that we observed. Experiments to exclude such a possibility are described in Figures 3 and 4. These experiments were deemed crucial because it has been reported (Richert *et al.*, 1982; Blithe *et al.*, 1982) that a M_r = 54,000 protein, allegedly a cleavage product of $pp60^{src}$ does not have GTP phosphorylating activity, and because a previous observation that $pp60^{src}$ could use GTP to phosphorylate TBR IgG (Richert *et al.*, 1979) was found to be due to contaminating GTP-ADP phosphotransferase activity in the IgG preparation. Careful titration of commercial yeast hexokinase under the $pp60^{src}$ phosphorylation reaction buffer conditions, which also contained 5 mM D-glucose, revealed that with 10 mU of yeast hexokinase, 95% of the [γ-^{32}P]ATP at a concentration of 20 μM was converted to radiolabeled glucose-6-phosphate. On the other hand, with [γ-^{32}P]GTP (20 μM) as substrate and under the same conditions only 3% of the radiolabel was found in gluose-6-phosphate (Figure 3). The phosphorylation of $pp60^{src}$ in the presence of either [γ-^{32}P]ATP or [γ-^{32}P]GTP was then examined under the same conditions, with 10 mU of hexokinase and 5 mM D-glucose. As illustrated in Figure 4, the phosphorylation of $pp60^{src}$ with [γ-^{32}P]ATP was completely abolished in the presence of yeast hexokinase and 5 mM D-glucose, whereas the phosphorylation of $pp60^{src}$ by [γ-^{32}P]GTP in the presence of yeast hexokinase and glucose was similar to the control. In addition, the possibility that the radiolabeled phosphate found in $pp60^{src}$ was due to binding of ADP or ATP was investigated by phosphorylation of

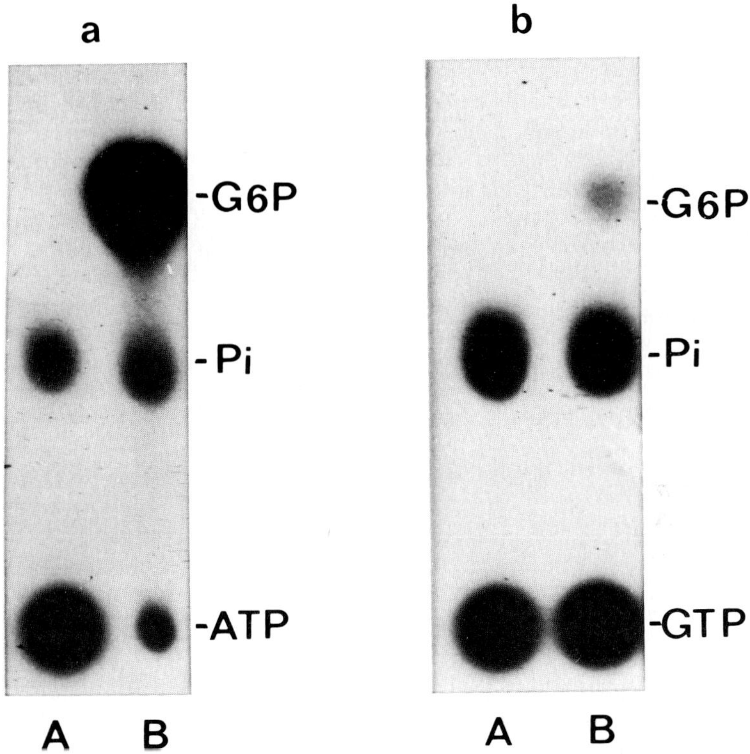

FIGURE 3. Effect of yeast hexokinase on [γ-^{32}P]ATP and [γ-^{32}P]GTP consumption. The activity of yeast hexokinase on phosphorylation of D-glucose, by 20 μM [γ-^{32}P]ATP (panel a) or 20 μM [γ-^{32}P]GTP (panel b), was determined under conditions similar to those described for phosphotransferase reactions. In addition, the reaction mixtures contained 5 mM D-glucose. Samples were incubated without hexokinase (A tracks) or in the presence of 10 mU of yeast hexokinase (B tracks) for 30 min at 30°. The reactions were terminated by applying 2 μl to PEI cellulose sheets which were then developed with 0.5 M ammonium formate (Gonzales and Geel, 1975). G6P = glucose-6-phosphate. PC = inorganic phosphate.

pp60^{src} with [γ-^{32}P]ATP (20 μM with 2.5 · 10^7 cpm per sample). No radiolabel was found in pp60^{src} after this treatment.

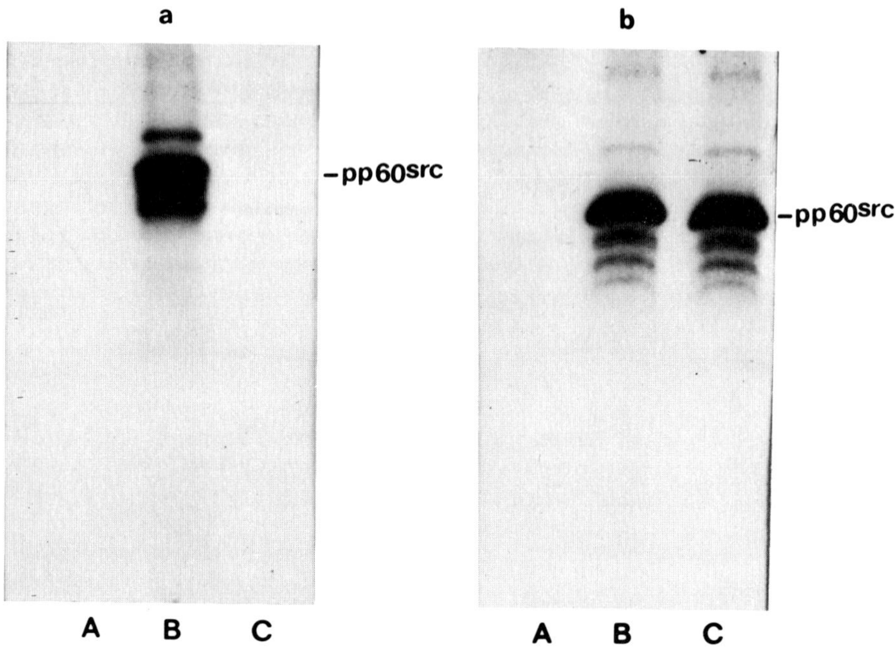

FIGURE 4. The effect of yeast hexokinase on [γ-^{32}P]ATP and [γ-^{32}P]GTP phosphorylation of pp60^{src}. a. Phosphotransferase reactions were carried out in the presence of 20 μM [γ-^{32}P]ATP and 5 mM D-glucose. Track A, plus 10 mU of yeast hexokinase; track B, plus approximately 3 ng of pp60^{src}; track C, plus yeast hexokinase and pp60^{src}. b. Phosphotransferase reactions were carried out in the presence of 20 μM [γ-^{32}P]GTP and 5 mM D-glucose. A, B and C tracks are similar to those described in panel a, except that 10 ng of pp60^{src} was added to the appropriate samples. The samples were subjected to polyacrylamide gel electrophoresis (12.5% acrylamide, 0.1% bisacrylamide) for 3.5 hr.

Immunoprecipitation and Structural Analysis of the Phosphorylated M_r = 60,000 Proteins

In order to determine whether the M_r = 60,000 proteins phosphorylated with either ATP or GTP were indeed pp60^{src}, these proteins were subjected to immunoprecipitation with antibody raised against p60^{src} produced by bacterial recombinants (anti-p60 serum). This antibody is able to immunoprecipitate $pp60^{src}$ even after it has been denatured with SDS (Gilmer and Erikson, 1983). Phosphotransferase reactions were carried out in the presence of either [γ-^{32}P]ATP or [γ-^{32}P]GTP, the products were resolved by SDS-polyacrylamide gel electrophoresis, the phosphorylated proteins that migrated in the M_r = 60,000 region of the gel were eluted from the gel, and then analyzed by immunoprecipitation and polyacrylamide gel electrophoresis. As shown in Figure 5, both the ATP- and GTP-phosphorylated M_r = 60,000 proteins were specifically immunoprecipitated with anti-p60 serum.

Previous studies have shown that $pp60^{src}$ is hydrolyzed by *S. aureus* V8 protease to two major peptides indentified as the N-terminal and C-terminal portions of the molecule (Collett *et al.*, 1979). Partial proteolysis of the immunoprecipitated proteins revealed that both proteins had a similar structure (Figure 5, panel b). The V8 digestion products of the immunoprecipitated M_r = 60,000 phosphoproteins were identical to those produced from $pp60^{src}$. However, the majority of the radiolabel in GTP-phosphorylated $pp60^{src}$ was in the C-terminal region, whereas in the case of ATP-phosphorylated $pp60^{src}$, radiolabel was in both termini, the N-terminal region containing about 3- to 4-fold more radiolabel than the C-terminal region. Phosphoamino acid analysis revealed phosphotyrosine (Figure 6).

$pp60^{src}$ phosphorylated in the presence of [γ-^{32}P]ATP or [γ-^{32}P]GTP was further characterized by two-dimensional tryptic fingerprint separation (Figure 7). A similar peptide, which is negatively charged, appeared in both preparations (panels A and B). These peptides comigrated when a mixture of both preparations was fractionated; therefore, this peptide can be considered as an identical site that is phosphorylated by either ATP or GTP. In addition, this peptide comigrates with the phosphotyrosine-containing peptide from the C-terminal fragment of $pp60^{src}$ prepared from ^{32}P-radiolabeled cells (data not shown). Figure 7 also shows that the [γ-^{32}P]ATP-phosphorylated protein contained several phosphopeptides that are close to their isoelectric point at pH = 6.5 (panel A) and that remain at the origin of the electrophoretic separation. The [γ-^{32}P]GTP-phosphorylated preparation yielded very few phosphopeptides in this region of the fingerprint.

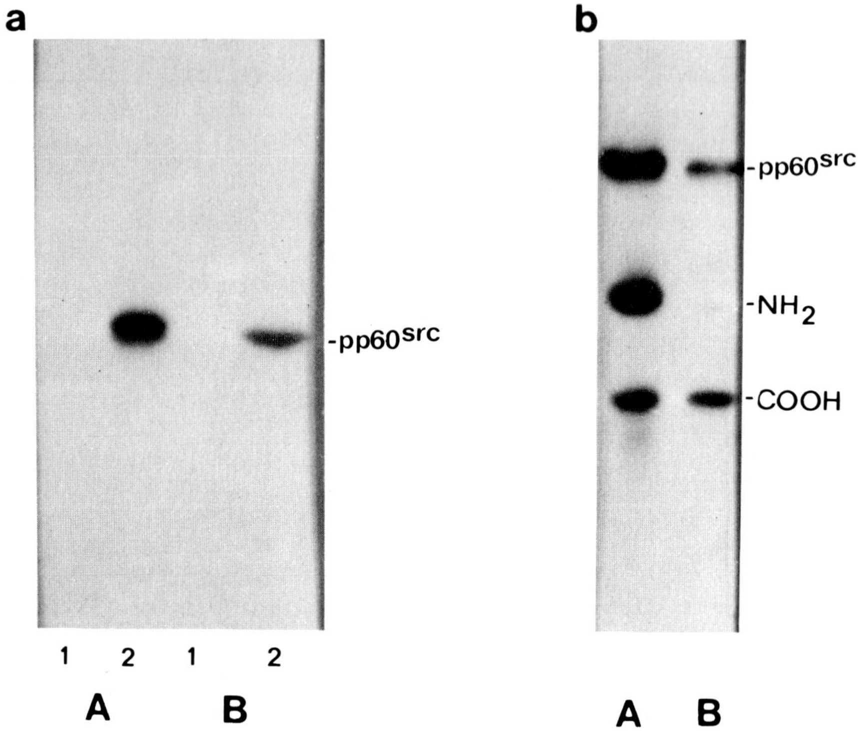

FIGURE 5. Characterization of the M_r = 60,000 proteins phosphorylated by ATP or GTP by immunoprecipitation, and by partial protease digestion. A preparation of $pp60^{src}$ was phosphorylated with 20 μM [γ-^{32}P]ATP or 20 μM [γ-^{32}P]GTP, the products of the reaction were resolved by 10% polyacrylamide gel electrophoresis and localized by autoradiography. The M_r = 60,000 regions of the gel were excised and the phosphoproteins eluted with 0.1% SDS. The eluates were adjusted to the conditions of RIPA buffer (Brugge and Erikson, 1977), and samples were incubated with 2 μl of either preimmune or anti-p60 serum. The immune complexes were collected with the use of protein A-containing S. aureus *and analyzed by polyacrylamide gel electrophoresis. Panel a: A tracks, [γ-^{32}P] ATP-labeled $pp60^{src}$; B tracks, [γ-^{32}P]GTP-labeled $pp60^{src}$; 1, with preimmune serum; 2, with anti-p60 serum. Panel b: The remainder of the eluates were immunoprecipitated with 6 μl of anti-p60 serum, resolved by polyacrylamide gel electrophoresis, and the M_r = 60,000 phosphoproteins were then subjected to partial proteolysis with 0.05 ng* S. aureus *V8 protease*

ATP- and GTP-dependent Kinase Activity of pp60^{src} Prepared from Cells Infected with a Temperature-sensitive Strain of RSV

It has been shown previously that the protein kinase activity of pp60^{src} prepared from cells infected with the temperature-sensitive transformation mutant of RSV, NY68, is more thermolabile than that of pp60^{src} prepared from wild-type virus-infected cells (Erikson *et al.*, 1979a,b; Levinson *et al.*, 1980). Increased thermolability was also found for the autophosphorylating activity of the temperature-sensitive as compared to the wild-type pp60^{src} preparation when ATP was the phosphate donor (Erikson and Purchio, 1982). Furthermore, the specific activities of the temperature-sensitive enzyme are usually about 10-15% that of the wild-type enzyme for both protein kinase and autophosphorylation activities (Erikson and Erikson, unpublished data). In order to answer the question whether the ATP- and GTP-dependent protein kinase and autophosphorylation activities of pp60^{src} are related to the same enzyme, we compared these activities in temperature-sensitive and wild-type pp60^{src} preparations.

The specific activities of ATP- and GTP-dependent α-casein in phosphorylation and autophosphorylation of pp60^{src} are summarized in Table I. The specific activities for the temperature-sensitive enzyme were about 6-8 fold lower than those of the wild-type enzyme with either phosphate donor. These observations indicate that the ATP- and GTP-dependent protein kinase activities are related to the product of the *src* gene and are not a contaminating enzyme from the host cells.

Figure 8 illustrates the thermolability of ATP- and GTP-dependent phosphorylation of pp60^{src} in temperature-sensitive and wild-type preparations. For both ATP and GTP the temperature-sensitive preparation was considerably more thermolabile than the wild-type preparation (t/2 = 1.1 min and 5.1 min, respectively).

The data described above show that the GTP-utilizing enzyme activity that phosphorylates pp60^{src} has the same thermolability as the ATP-utilizing activity that phosphoryl-

(Figure 5 continued)
during re-electrophoresis as described by Cleveland et al. *(1977). Track A, [γ-^{32}P]ATP-labeled pp60^{src}; track B [γ-^{32}P] GTP-labeled pp60^{src}.*

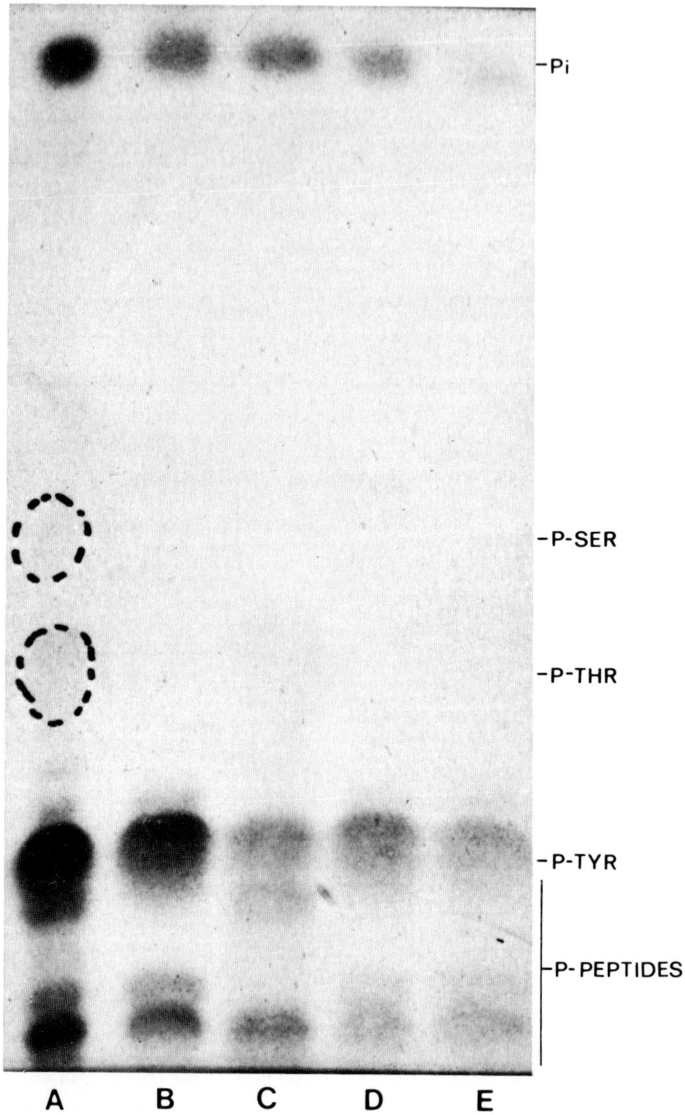

FIGURE 6. Phosphoamino acid analysis of $pp60^{src}$ phosphorylated with either [γ-^{32}P]ATP or [γ-^{32}P]GTP. ATP- and GTP-phosphorylated $pp60^{src}$, and the NH_2 and COOH terminal regions were prepared as described in the legend to Figure 5. After elution from the preparative gels, the radiolabeled proteins were precipitated with TCA, hydrolyzed at 100°C for 3 hr in 6 N HCl, and the phosphoamino acids were resolved by electrophoresis at pH 3.5. Authentic phosphoserine, phospho-

ates $pp60^{src}$, and, moreover, for both ATP and GTP, this activity is more thermolabile in preparations from temperature-sensitive mutant-infected cells than in preparations from wild-type virus-infected cells.

DISCUSSION

By using ATP or GTP as phosphate donors for the phosphorylation of $pp60^{src}$, we found that the phosphorylation by ATP was 4-5 times higher than that by GTP (Figure 1). The trinucleotide concentration used in this study was 20 μM, which is in the range of the apparent Km_{ATP} for IgG, α-casein and tubulin (Levinson *et al.*, 1980; Collett *et al.*, 1980; Richert *et al.*, 1979, 1982). These observations raised two major questions: (1) Are the ATP-and GTP-phosphorylated products of M_r = 60,000 the same protein, and are they indeed $pp60^{src}$? (2) Are the enzyme(s) which use ATP and GTP as phosphate donors for this phosphorylation activity the same enzyme?

The immunological studies demonstrated in Figure 5 show that the M_r = 60,000 proteins which were phosphorylated in the presence of ATP or GTP share the same protein structure since both are recognized by antibody raised against $p60^{src}$ produced in *E. coli*. Moreover, the V8 protease cleavage products of the immunoprecipitated proteins have the same electrophoretical mobility. However, the distribution of phosphorylated sites within $pp60^{src}$ differed when phosphorylation was carried out by ATP or by GTP. GTP-phosphorylated $pp60^{src}$ contained radiolabel mostly in the C-terminal region, in a tryptic phosphopeptide that comigrates with the tyrosine-containing peptide from the carboxy-terminus of $pp60^{src}$ radiolabeled *in vivo*. The sequence of this peptide has recently been resolved (Smart *et al.*, 1981) and the phosphorylation of the tyrosine residue in this peptide confirmed. ATP-phosphorylated $pp60^{src}$, on the other hand, contained radio-

(Fig. 6 continued)
threonine and phosphotyrosine were added to the samples. Track A, ATP-phosphorylated $pp60^{src}$*; B, GTP-phosphorylated* $pp60^{src}$*; C,* NH_2 *terminal region of ATP-phosphorylated* $pp60^{src}$*; D, COOH terminal region of ATP-phosphorylated* $pp60^{src}$*; E, COOH terminal region of GTP-phosphorylated* $pp60^{src}$*.*

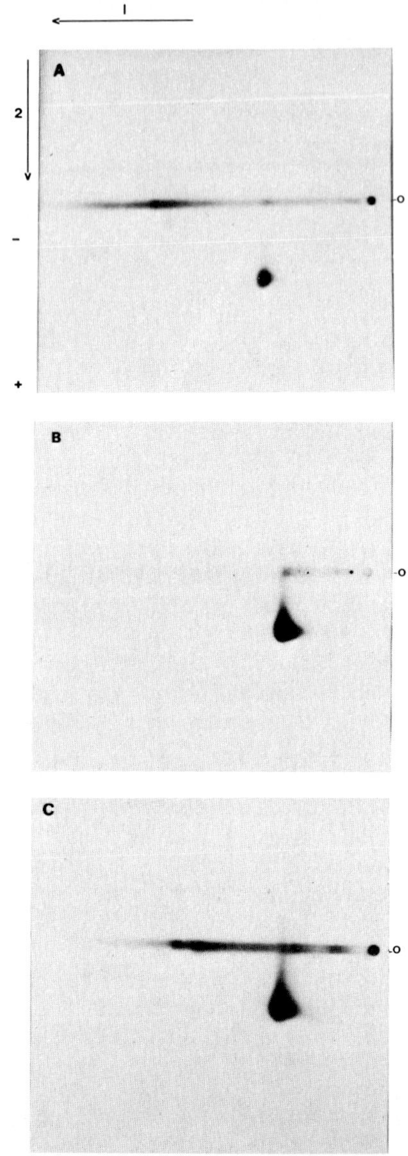

FIGURE 7. Two-dimensional tryptic fingerprints of pp60^{src} radiolabeled with [γ-^{32}P]ATP or with [γ-^{32}P]GTP. pp60^{src} phosphorylated with either ATP or GTP was localized in a preparative gel, eluted from the gels, precipitated, and digested with trypsin as described previously (Erikson et al.*, 1979a). Fractionation of the phosphopeptides involved ascending chromatography in the first dimension. O, origin. An equal number of cpm (approximately 4000 cpm) were analyzed in each case. A, [γ-^{32}P]ATP-labeled pp60^{src}; B, [γ-^{32}P] GTP-labeled pp60^{src}; C, mixture of A and B.*

Table I. Specific Activity of ATP- and GTP-dependent temperature-sensitive (ts) and wild-type (wt) pp60^{src}

		Specific activity[a] nmol · min^{-1} · mg^{-1} pp60src	
		ts enzyme	wt enzyme
phosphorylation of pp60^{src}	ATP	0.50	3.50
	GTP	0.15	0.80
phosphorylation of α-casein	ATP	34	210
	GTP	1.8	14

[a]*The phosphorylation of pp60^{src} and of α-casein was determined as described in the Materials and Methods. α-casein was present at a concentration of 1 mg · ml^{-1}. Each reaction contained 2 μl of pp60^{src} prepared from either temperature-sensitive mutant- or wild-type virus-infected cells.*

label in the same C-terminal peptide and also at several sites in the N-terminal region. Previous analyses (Erikson *et al.*, 1979a) demonstrated radiolabel mainly in the C-terminal peptide of ATP-phosphorylated pp60^{src}. However, in those experiments, the concentration of ATP was 1 μM or less. In the present study we used 20 μM ATP, which is in the range of the apparent Km_{ATP}, for this reaction.

Since thc apparent Km values for the ATP- and GTP-phosphorylation of pp60^{src} were similar (Figure 2) and are in the range of the apparent Km_{ATP} values reported in the literature, the differences between the phosphorylation of pp60^{src} by these two substrates cannot be due to differences in affinity between the catalytic site of the enzyme and the two triphosphates. Furthermore, the similarity of the apparent Km values for pp60^{src} phosphorylation by both substrates may indicate similarity of at least part of the enzymatic sites which are involved in the phosphorylation activity of the pp60^{src} protein.

The question whether the enzymatic activity that phosphorylates pp60^{src} with ATP or with GTP as phosphate donors is enzyme activity of the same protein was investigated by using preparations of pp60^{src} extracted from cells infected from cells infected with the temperature-sensitive transformation mutant of RSV-NY68. As indicated in Table I, the temperature

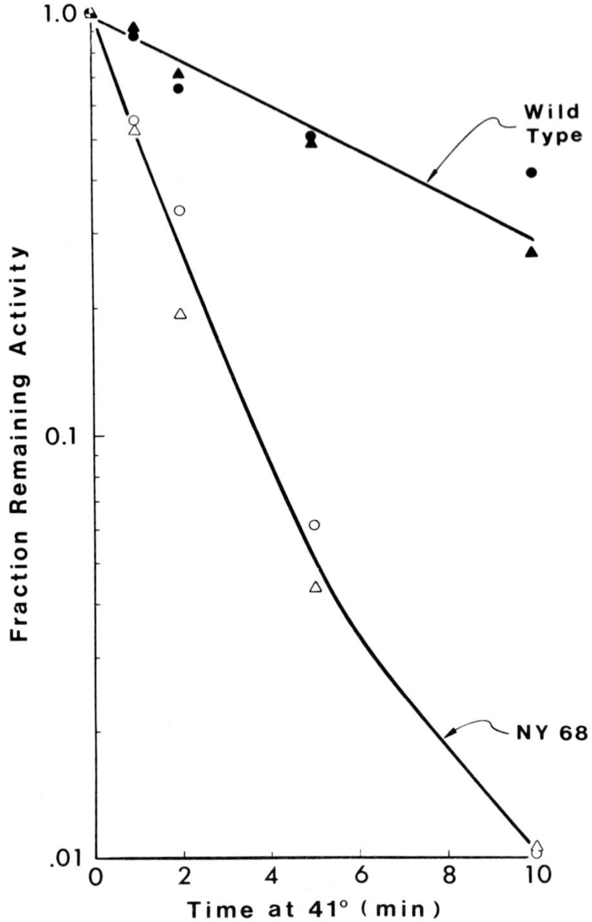

FIGURE 8. Thermolability of the phosphorylation of pp60^{src} with ATP or GTP as phosphate donors. The phosphorylation of pp60^{src} by ATP and GTP in temperature-sensitive and wild-type enzyme preparations was measured as described in Materials and Methods. Three hundred µl of these preparations were incubated at 41° in buffer which contained 10 mM KPi, pH = 7.1, 1 mM EDTA, 1 mM 2-mercaptoethanol, 25% v/v of glycerol and 100 µg/ml BSA. Samples of 10 µl were transferred to precooled tubes at 4° at the times indicated in the figure and the phosphorylation of pp60^{src} by ATP and GTP was determined with 2 µl of the wild-type enzyme and 10 µl of the temperature-sensitive enzyme. The control values were 4789 cpm for the ATP-dependent phosphorylation of pp60^{src} in the

sensitive enzyme had about 6- to 8-fold less activity for both α-casein phosphorylation and pp60^{src} phosphorylation, by either ATP or GTP as substrate, than the wild-type enzyme. Therefore, we may anticipate that both activities are related to the product of the *src* gene and not to a contaminating cellular activity. The similarity in the thermolability of the temperature-sensitive enzyme for ATP- and GTP-phosphorylation of pp60^{src} (t/s = 1.1 min, Figure 8) and of the wild-type enzyme for both substrates (t/2 = 5.1 min) further indicates that this activity is related to the same enzyme.

The reported ability of GTP to be a phosphate donor for pp60^{src} enzymatic activity (Richert *et al.*, 1979) was recently shown to be due to GTP-ADP transferase activity in the IgG used in those studies (Richert *et al.*, 1982; Blithe *et al.*, 1982). In the case of the M_r = 54,000 protein, a protease cleavage product of pp60^{src}, GTP does not serve as phosphate donor for either α-casein or glycerol phosphorylation (Richert *et al.*, 1982; Blithe *et al.*, 1982). Our studies of pp60^{src} phosphorylation in the presence of hexokinase (Figures 3 and 4) exclude the possibility that GTP-ADP phosphotransferase activity is the reason for the GTP-dependent phosphorylating activity of pp60^{src}. The inability of pp54 to use GTP as a phosphate donor is questionable and may be related to the deficiency of the M_r = 6,000 peptide which is cleaved from the original pp60^{src} protein by proteolytic enzymes. The recently predicted sequence of the pp60^{src} protein (Takeya *et al.*, 1982) indicates that the probable location of the major phosphorylated serine residue of pp60^{src} is residue 17. The pp^{54} (Richert *et al.*, 1982; Blithe *et al.*, 1982) and the pp^{52} (Levinson *et al.*, 1980) proteins lack this part of the pp60^{src} molecule and, therefore, they can hardly represent the total physiological properties of pp60^{src}.

The biochemical, immunological and genetic evidence presented in this communication shows that the ATP- and GTP-dependent phosphorylation activities are intrinsic properties of pp60^{src}.

(Figure 8 continued

temperature-sensitive enzyme and 4074 cpm in the wild-type enzyme. The control values for the GTP-dependent phosphorylation of pp60^{src} in the temperature-sensitive and wild-type enzymes were 4680 and 3475 cpm per sample, respectively. ● o, [γ-^{32}P]ATP; △▲, [γ-^{32}P]GTP.

ACKNOWLEDGEMENT

Yosef Graziani is a Visiting Assistant Professor from the Biology Department, Ben-Gurion University of the Negev, Beer-Sheva, Israel.

ABBREVIATIONS

pp^{src} -	The protein product encoded by Rous sarcoma transforming gene. A phosphoprotein of M_r = 60,000.
p^{src} -	Rous sarcoma virus transforming protein.
RSV -	Rous sarcoma virus.
src -	RSV transforming gene.
TBR IgG -	Tumor bearing rabbit immunoglobulin G.
SDS -	Sodium dodecyl sulfate.
RSV, NY68 -	A temperature-sensitive transformation mutant of RSV.
KPi -	Potassium phosphate.
PEI cellulose sheets -	Polyethylimine cellulose sheets.
mU -	Milliunits (yeast hexokinase).
V8 -	*S. aureus* V8 protease.
pp^{54} -	Phosphoprotein of M_r - 54,000.
RIPA buffer -	150 mM NaCl, 1% sodium deoxycholate, 1% Triton X-100, 0.1% SDS, 10 mM Tris-HCl, pH 7.2, containing 1% Trasylol* (a general inhibitor of proteases) and centrifuged at 90,000 rpm for 30 min.

*FBA pharmaceuticals; isolated from animal tissues.

REFERENCES

Blithe, D.L., Richert, N.D., and Pastan, I.H. (1982). *J. Biol. Chem. 257*, 7135.
Brugge, J.S., and Erikson, R.L. (1977). *Nature 269*, 346.
Brugge, J.S., Erikson, E., Collett, M.S., and Erikson, R.L. (1978). *J. Virol. 26*, 773.
Cleveland, D.W., Fisher, S.G., Kirschner, M.W., and Laemmli, U.K. (1977). *J. Biol. Chem. 252*, 1102.
Collett, M.S., and Erikson, R.L. (1978). *Proc. Natl. Acad. Sci. USA 75*, 2021.
Collett, M.S., Erikson, E., and Erikson, R.L. (1979). *J. Virol. 29*, 770.
Collett, M.S., Purchio, A.F., and Erikson, R.L. (1980). *Nature 285*, 167.
Erikson, R.L., Collett, M.S., Erikson, E., Purchio, A.F., and Brugge, J.S. (1979a). *Cold Spring Harbor Symp. Quant. Biol. 44*, 907.
Erikson, R.L., Collett, M.S., Erikson, E., and Purchio, A.F. (1979b). *Proc. Natl. Acad. Sci. USA 76*, 6260.
Erikson, E., and Erikson, R.L. (1980). *Cell 21*, 829.
Erikson, R.L., and Purchio, A.F. (1982). *In* "Advances in Viral Oncology" (G. Klein, ed.), Vol. 1, p. 43. Raven Press, New York.
Gilmer, T.M., and Erikson, R.L. (1981). *Nature 294*, 771.
Gilmer, T.M., and Erikson, R.L. (1983). *J. Virol. 45*, 462.
Gonzales, L.W., and Geel, S.E. (1975). *Anal. Biochem. 63*, 400.
Hunter, T., and Sefton, B.M. (1980). *Proc. Natl. Acad. Sci. USA 77*, 1311.
Johnson, R.A., and Walseth, T.F. (1979). *Adv. Cyclic Nucleotide Res. 10*, 1135.
Levinson, A.D., Opperman, H., Levintow, L., Varmus, H.E., and Bishop, J.M. (1978). *Cell 15*, 561.
Levinson, A.D., Oppermann, H., Varmus, H.E., and Bishop, J.M. (1980). *J. Biol. Chem. 255*, 11973.
Oakley, B.R., Kirsch, D.R., and Morris, N.R. (1980). *Anal. Biochem. 105*, 361.
Purchio, A.F., Erikson, E., Brugge, J.S., and Erikson, R.L. (1978). *Proc. Natl. Acad. Sci. USA 75*, 1567.
Richert, N.D., Davies, P.J.A., Jay, G., and Pastan, I.H. (1979). *J. Virol. 31*, 696.
Richert, N.D., Blithe, D.L., and Pastan, I. (1982). *J. Biol. Chem. 257*, 7143.
Rubin, C.S., and Rosen, O.M. (1975). *Annu. Rev. Biochem. 44*, 831.
Smart, J.E., Oppermann, H., Czernilofsky, A.P., Purchio,

A.F., Erikson, R.L., and Bishop, J.M. (1981). *Proc. Natl. Acad. Sci. USA 78*, 6013.

Takeya, T., Feldman, R.A., and Hanafusa, H. (1982). *J. Virol. 44*, 1.

DISCUSSION

Ingebritsen (U. Colorado): What happens if you phosphorylate with ATP and GTP together?

Graziani: We did a competition experiment with ATP and GTP. When we phosphorylate with ATP, addition of about a 10-fold excess of GTP reduced only about 15-20% the phosphorylation. When we phosphorylate with GTP, and add cold ATP, the phosphorylation was almost totally abolished.

Cohen: If you look at the time course of the reaction with ATP, is the C-terminal fragment phosphorylated preferentially or do they both go up at the same rate?

Graziani: It seems that they are both going up at the same rate, but we are not completely sure. When we use very low concentrations of ATP, 1 μM or less, we also observe much less phosphorylation at the N-terminal region.

Cohen: The reason you see only phosphorylation of the C-terminal region with GTP may be because it is a rather poor nucleotide compared with ATP, and it is difficult to get as much phosphate in. Suppose you did the experiment with GTP using about 20 times higher concentrations of the *src* kinase, and then you repeated the experiment with GTP; you may then be able to force it into the N-terminal site with GTP. In other words, do you just have to get beyond a certain level of phosphorylation to start seeing the N-terminal phosphorylation?

Graziani: We did these experiments in different concentrations of enzyme with 10-fold differences, and we found that both ATP and GTP phosphorylating activities were linear. We also were not able to determine much phosphorylation on the N-side of $pp60^{src}$ by using ten-fold higher concentrations of GTP. A small amount of phosphorylation is there but we were not able to show anything comparable to the ATP phosphorylation: therefore we think that the phosphorylation of the N-terminal region is much more specific for ATP. Another thing that we did do was use another donor dATP. It is known this enzyme user dATP as phosphate donor for its transphosphorylating activity. With this donor we were able to obtain similar observations.

Cohen: Could you review for us what is known about the *in vivo* substrates for the pp60^{src} other than the auto-phosreaction?

Graziani: In our laboratory, the *in vivo* phosphorylation with this enzyme is a protein of 34K which was found several years ago. According to other laboratories, additional proteins can be phosphorylated, including vinculin, but we don't have confirmation for such phosphorylations.

Ochoa (Roche Inst.): In the case of the cellular oncogenes, their gene products do not seem to have any protein kinase activity. This is an interesting difference between oncogenes of acute transforming retroviruses and those of mammalian tumor cells.

Graziani: The protein kinase activity of the sarcoma virus gene product is there. It phosphorylates two proteins: one is 34K and the other is pp60^{src} itself. We know that it is phosphorylated. That means we know that the protein kinase activity in the cells is there. The question is: what are the other protein substrates, if there are other substrates, for this kinase activity? We don't know that answer. In our research we can commit ourselves for the *src* gene product only.

Allfrey: Have you ever tried pyrophosphate as a phosphate donor?

Graziani: Yes, we tried to compete with the phosphorylation by ATP and GTP with pyrophosphate and we did not succeed at all. That means that pyrophosphate is not a substrate for this reaction. The main substrates that are phosphate donors are dATP and ATP; ADP is a strong inhibitor of the reaction.

CALMODULIN MODULATED PROTEIN PHOSPHORYLATION

Claude B. Klee
Dianne L. Newton

Laboratory of Biochemistry
National Cancer Institute
National Institutes of Health
Bethesda, Maryland

INTRODUCTION

During the last decade, it has become clear that the state of phosphorylation of cellular proteins is controlled by the complex interplay of protein kinases and phosphatases and by specific inhibitors and activators of these enzymes. Whereas the central role of cAMP-dependent protein kinase has been known and studied extensively since the discovery of cAMP (Rall *et al.*, 1957) and of the kinase by Walsh *et al.* (1968), the role of Ca^{2+} in regulating protein phosphorylation has been recognized only recently (as reviewed by Schulman *et al.*, 1980; Schulman, 1982; Cohen, 1982) and is still poorly understood. In contrast to cAMP, which activates a single kinase with broad substrate specificity, Ca^{2+} regulates the activity of several kinases, some of which have narrow substrate specificities. These kinases have been difficult to purify and their characterization is only beginning. A Ca^{2+}-dependent, membrane bound kinase discovered by Takai *et al.* (1979) will be described in this volume. Several others are intracellular or membrane-associated and their activity is under the control of the Ca^{2+} calmodulin complex; they are listed in Table I. Many are believed to be responsible for the Ca^{2+}-dependent phosphorylation of several proteins but few have been purified and characterized. The recent discovery by Stewart *et al.* (1982, 1983) of a Ca^{2+} calmodulin-dependent protein phosphatase adds another role for Ca^{2+} in the regulation of protein phosphorylation.

POSTTRANSLATIONAL COVALENT
MODIFICATIONS OF PROTEINS

ISBN 0-12-387560-9

TABLE I. Calmodulin-Regulated Kinases

Enzyme	Subunits M_r	Subunit Structure	Ref.[a]
Myosin kinase	90,000–150,000	monomer	1–3
Light chains			
Phosphorylase kinase	145,000	oligmomer	
Phosphorylase	130,000	$(\alpha\beta\gamma\delta)_4$	2–6
Glycogen synthase	40,000		
	16,500		
Calmodulin-dependent kinases[b]			
a. Glycogen synthase*	53,000	oligomer	
Light chains	51,000	(M_r > 250,000)	7–9
b. Phospholamban			
c. 60,000-dalton protein (skeletal muscle endoplasmic reticulum)			
d. Acetylcholine receptor			
e. Spectrin-like proteins			10
f. Tubulin*	55,000[c]		2
	65,000		11–13
g. Tryptophan hydroxylase*			14
Tyrosine hydroxylase			
Casein			2
h. Synapsin (Protein I)*	50,000[c]		2,15,16
i. Light chains*			
Tubulin and MAPs		oligomer	
Myelin basic protein	49,000	(M_r 500,000–600,000)	17
Casein			
Brain proteins			

[a]*For review of the literature see: 1, Adelstein and Eisenberg, 1980; 2, Klee and Vanaman, 1982; 3, Cohen, 1982; 4, DePaoli-Roach* et al., *1979; 5, Chrisman* et al., *1982; 6, Soderling* et al., *1979; 7, Payne and Soderling, 1980; 8, Ahmad* et al., *1982; 9, Woodgett* et al., *1982; 10, Campbell and MacLennan, 1982; 11, Alper* et al., *1980; 12, Sobue* et al., *1982; 13, Nelson* et al., *1982; 14, Goldenring* et al., *1982; 15, Schulman* et al., *1980; 16, Kennedy* et al., *1982; 17, Fukunaga* et al., *1982.*

[b]*Different calmodulin-dependent kinases have been purified on the basis of their ability to phosphorylate specific substrates. As indicated by the asterisks, several of these kinases exhibit similar structures. Since two such kinases (a and i) appear to be multifunctional, the substrate specificity of the other kinases needs to be ascertained.*

[c]*Major component*

**Kinases with similar structures.*

Multiple Roles of Calmodulin in the Phosphorylation of Brain Proteins

A. Two major enzymes under the control of calmodulin in brain are the Ca^{2+}-dependent cyclic nucleotide phosphodiesterase (Cheung, 1970; Kakiuchi and Yamasaki, 1970) and adenylate cyclase (Brostrom *et al.*, 1975; Cheung *et al.*, 1975). These enzymes determine the level of cAMP and thereby confer upon calmodulin a major role in controlling cAMP-dependent protein phosphorylation. In addition, calmodulin may also regulate Ca^{2+} levels by acting at the synaptic plasma membrane Ca^{2+}-dependent ATPase (Kuo *et al.*, 1979; Sobue *et al.*, 1979) and thereby regulate the activity of Ca^{2+}-modulated kinases.

B. During the last few years several Ca^{2+}- and calmodulin-dependent kinases have been identified and characterized in brain extracts. Protein I or synapsin kinase (Schulman and Greengard, 1978; Kennedy and Greengard, 1981), tryptophan and tyrosine hydroxylase kinase (Yamauchi *et al.*, 1981) and tubulin kinase (Goldenring *et al.*, 1982) are involved in neurotransmitter secretion and synthesis. Myosin kinase (Hathaway *et al.*, 1981; Fukunaga *et al.*, 1982), fodrin (calspectin) kinase (Sobue *et al.*, 1982) and phosphorylase kinase (Yamauchi and Fujisawa, 1980) have also been reported to be under the control of calmodulin in brain extracts but these enzymes remain to be fully characterized.

C. A fraction of cAMP-dependent protein kinase has also been shown to interact directly with calmodulin in a Ca^{2+}-dependent fashion. This interaction is mediated by the regulatory subunit of the enzyme and is specific for the type II enzyme derived from brain tissue. It is not observed with type I or type II enzymes from heart (Hathaway *et al.*, 1981; Klee *et al.*, 1982). Brain type II regulatory subunit is known to be antigenically and chemically different from the type II subunits of the enzymes isolated from other tissues (Erlichman *et al.*, 1980). The characterization of the brain cAMP-dependent kinase is covered in this volume by Rubin (Chapter 5). Although calmodulin has no significant effect on the activity of the dissociated catalytic subunit of the enzyme, crosslinking studies with the holoenzyme after autophosphorylation of the regulatory subunit indicate that calmodulin interferes with the dissociation of the regulatory and catalytic subunits. Brain protein kinase has been purified by affinity chromatography on calmodulin-Sepharose. The enzyme is eluted from the columns as a complex with calcineurin. Its molecular weight, determined by crosslinking with dimethylsuberimidate, is at least 350,000 compatible with a 2:2 complex of the $R_2 \cdot C_2$ holoenzyme and two calcineurin molecules. Addition of 8-Br-cAMP results in a dissociation of the regulatory and catalytic subunits which is prevented by addition of calmodulin in the presence of Ca^{2+} but not in its absence (Krinks and Klee, unpublished observations). The presence of calcineurin is not needed for interaction with calmodulin which is still observed after removal of calcineurin. It is likely that the complex between calcineurin and cAMP-dependent protein kinase is due to the fact that the phosphorylated regulatory subunit of protein kinase is a substrate for calcineurin.

D. Calcineurin, a major calmodulin-binding protein in brain tissue (Klee and Krinks, 1978; Klee *et al*., 1979; Wallace *et al*., 1980; Sharma *et al*., 1979), has recently been shown to be a Ca^{2+}- calmodulin-regulated protein phosphatase (Stewart *et al*., 1982). Like some calmodulin-dependent kinases, calcineurin has a relatively narrow substrate specificity. It catalyzes the dephosphorylation of the -subunit of phosphorylase.kinase, inhibitor-1 of protein phosphatase, myosin light chains and the regulatory subunit of brain cAMP-dependent protein kinase (Stewart *et al*., 1982; Klee *et al*., manuscript in preparation). The enzymatic activity is entirely dependent on Ca^{2+}, which by itself can activate the enzyme partially. Full activation is achieved only in the presence of calmodulin. This dual regulation by Ca^{2+}, acting both directly at the enzyme level and indirectly through a Ca^{2+}-dependent calmodulin interaction, has been taken as further evidence that calcineurin is indeed a phosphatase because it is composed of two subunits, a large 61,000 M_r polypeptide, calcineurin A, believed to be the catalytic subunit and a small Ca^{2+}-binding subunit with M_r 19,000, calcineurin B (Klee *et al*., 1979). The amino acid sequence of the small subunit has recently been determined and revealed the existence of four Ca^{2+}-binding domains (P. Cohen, Chapter 2) similar to those defined by Kretsinger (1980) for all other intracellular Ca^{2+}-binding proteins.

E. In addition to its interaction with the enzymes responsible for the phosphorylation or dephosphorylation of brain soluble proteins calmodulin interacts with several of the substrates of these enzymes and can therefore modify their abilities to act as substrates. Some of the calmodulin-binding proteins involved with or subject to phosphorylation are listed in Table II.

TABLE II. *Calmodulin-binding Proteins in Brain Cytosol*[a]

	Cyclic nucleotide[b] phosphodiesterase	Myosin[b] kinase	cAMP-dependent protein kinase	Protein phosphatase (calcineurin)	Fodrin[b]
mg/Kg	5–10	< 2	10–15	30–50	$1-4x10^3$
subunit M_r	59,000	140,000	55,000	61,000	235,000
			40,000	19,000	230,000
subunit	A_2	A	$R_2 \cdot C_2$	A•B	
	$(CaM \cdot A)_2$	A•CaM	$(CaM \cdot R)_2C_2$	CaM•A•B	
K_{diss}(CaM)	$>4x10^{-9}M$	$>4x10^{-9}M$	$10^{-9}M$	$4x10^{-9}M$	$>10^{-6}M$

[a]From Manalan and Klee, 1983.

[b]Cyclic nucleotide phosphodiesterase, myosin kinase and fodrin are all substrates of cAMP-dependent protein kinase.

Effect of Calmodulin on Phosphorylation

In crude brain extracts additional Ca^{2+} results in decreased phosphorylation. Among other possibilities, this inhibition could be the result of a Ca^{2+}-stimulation of endogenous ATPase. Endogenous calmodulin prevents the demonstration of a calmodulin-dependence for this inhibition. Crude preparations of calmodulin-binding proteins are useful model systems because they contain not only calmodulin-regulated kinases and phosphatases but also the substrates for these enzymes which probably copurify with them during affinity chromatography (Table II). In addition, several calmodulin binding proteins are phosphorylated by cAMP-dependent protein kinase and the effect of calmodulin on their phosphorylation can be studied as well.

As shown in Figure 1, calmodulin-binding proteins prepared by affinity chromatography on calmodulin Sepharose contain cAMP-dependent protein kinase and several of its substrates. A major phosphorylated polypeptide, M_r of 60,000, whose phosphorylation is greatly stimulated upon addition of 8-Br-cAMP is probably the Ca^{2+}-dependent form of phosphodiesterase since the purified enzyme has been shown to be a substrate for cAMP-dependent kinase (Sharma *et al.*, 1980). Other polypeptides which are preferentially phosphorylated in the presence of 8-Br-cAMP include fodrin, and the regulatory subunit of protein kinase which can be labeled with 8-azido cAMP. Several other phosphorylated proteins with M_r of 150,000, 125,000, 94-80,000, 48,000 and 41,000 remain to be identified. They may include synapsin (Huttner and Greengard, 1979) and myosin kinase (Adelstein *et al.*, 1978; Westwood and Perry, 1981; Edelman and Krebs, 1982), known substrates of cAMP-dependent kinase.

Ca^{2+}- and calmodulin-stimulated kinase(s) and their substrates are also present. In the presence of calmodulin and Ca^{2+}, four additional phosphorylated bands are detected with M_r of 51,000, 45,000, 34,000 and 31,000. These appear to be substrates which are specifically labeled by calmodulin dependent kinase(s). Thus, most of the substrates for cAMP- and Ca^{2+} calmodulin-regulated kinases in this preparation are different proteins. When calmodulin-binding proteins are first incubated in the absence of ATP, and in the presence of Ca^{2+} and calmodulin, to allow dephosphorylation of the potential substrates by endogenous calmodulin stimulated protein phosphatase, subsequent phosphorylation by cAMP-dependent kinase or by calmodulin-regulated kinases produce different labeling patterns (Figure 2). With the exception of the 60,000-dalton protein which is phosphorylated both in the presence of Ca^{2+} and cAMP, the other polypeptides are selectively labeled by one or the other kinase. cAMP-dependent kinase phosphorylates high molecular weight proteins, such as fodrin, 170,000 and 150,000-dalton polypeptides whereas the Ca^{2+}-dependent kinase(s) phosphorylate 70,000, 51,000, 45,000, 34,000 and 31,000 M_r polypeptides.

In addition to the phosphorylation of these specific proteins calmodulin and Ca^{2+}, but not Ca^{2+} alone, exert a pronounced inhibitory effect on cAMP-dependent phosphorylation (Figure 1). This inhibition does not appear to be the result of Ca^{2+}-dependent proteolysis because it is not observed in

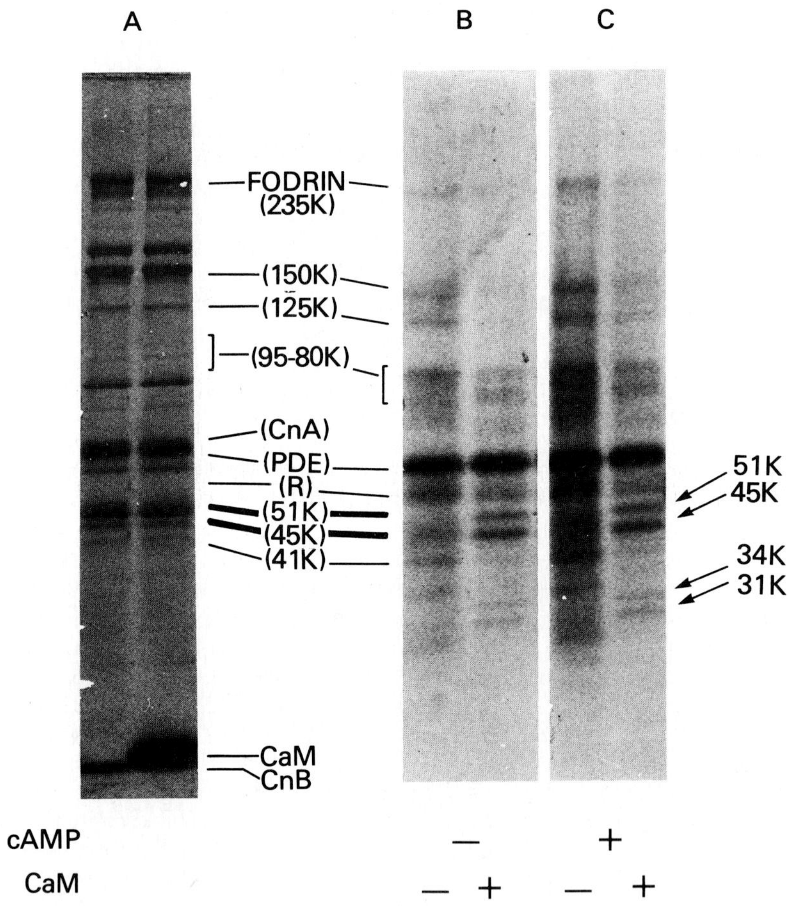

Figure 1

the presence of Ca^{2+} alone and there is no evidence of proteolysis of the polypeptides detected by Coomassie blue staining (Figure 1). Inhibition of cAMP-stimulated phosphorylation is also seen in Figure 2 when the proteins are preincubated with Ca^{2+} and calmodulin, despite the fact that EGTA was present during the phosphorylation step. The inhibition may be the result of incomplete dissociation of calmodulin from those proteins with a high affinity for calmodulin such as cyclic nucleotide phosphodiesterase. The phosphorylation of fodrin, which interacts with calmodulin weakly, (K_{diss}: 10^{-6} M) is not significantly affected. The inhibition of protein phosphorylation could therefore simply be the result of the interaction of calmodulin with the protein substrate, thus rendering the substrate less susceptible to phosphorylation. It has been reported that interaction of phosphorylase kinase with calmodulin inhibits the phosphorylation of the β-subunit by cAMP dependent protein kinase (Cox and Edstrom, 1982).

Alternatively, stimulation of an endogenous calmodulin-stimulated protein phosphatase, such as calcineurin, might be responsible for this inhibitory effect of phosphate incorporation. Calcineurin catalyzes the dephosphorylation of the regulatory subunit of protein kinase. cAMP which dissociates

FIGURE 1. Effect of Calmodulin on the Phosphorylation of Brain Calmodulin Binding Proteins. Calmodulin binding proteins (6 μg in 40 mM Tris-HCl, pH 7.5, 0.4 M NaCl, 0.5 mM dithiothreitol and 2 mM EGTA) were diluted in an incubation mixture containing 10 mM $MgCl_2$ and 80 μM [γ-^{32}P]ATP and incubated at 25° for 10 min in the presence or absence of 10 μM calmodulin and 3 mM $CaCl_2$. Where present, 8 Br-cAMP was 20 μM. The incubation was stopped by addition of a 1% solution of SDS, containing 0.5 M dithiothreitol, and 5 M urea. After boiling for 1 min the samples were subjected to SDS polyacrylamide gel electrophoresis in a 5-15% gradient of acrylamide. Coomassie blue pattern, A; autoradiograms (B and C) obtained under phosphorylation conditions indicated at bottom of figure. Fodrin, calcineurin A (CnA), phosphodiesterase (PDE), regulatory subunit of cAMP-dependent protein kinase (R), calcineurin B (CnB), and calmodulin (CaM) are identified on the gel as well as the M_r of other calmodulin-binding proteins. The arrows on the right indicate the polypeptides specifically phosphorylated in the presence of calmodulin.

the R and C subunits, increases the rate of dephosphorylation (Krinks *et al.*, in preparation). Indeed, when the extent of autophosphorylation of the R subunit is measured in the presence of calcineurin, a pronounced inhibition is observed of cAMP (Figure 3). As shown in Figure 4, not only the R subunit but other phosphorylated calmodulin-binding proteins appear to be dephosphorylated in a Ca^{2+}-calmodulin-stimulated process. This appears to be the case whether the proteins have been phosphorylated by the cAMP- or the calmodulin-dependent kinases.

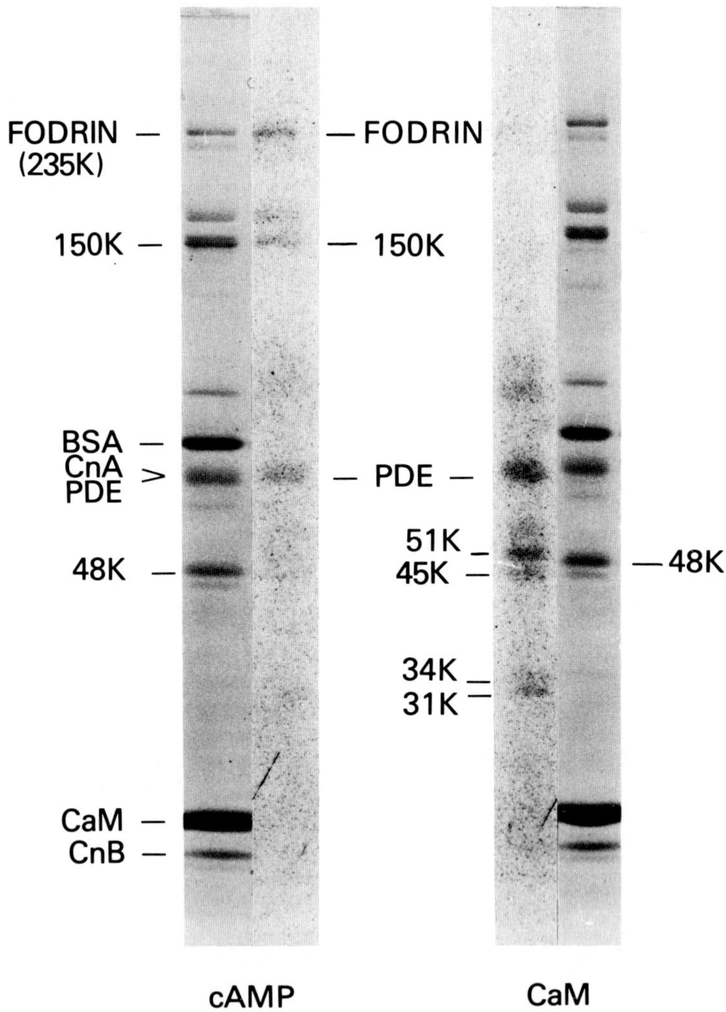

Figure 2

CONCLUSION

Calmodulin-binding proteins, and proteins associated with them during calmodulin affinity chromatography are good substrates for cAMP-dependent and Ca^{2+}-calmodulin regulated kinases. Although several proteins, such as the 60,000 M_r protein mentioned above, synapsin (Huttner and Greengard, 1979), tryptophan hydroxylase kinase (Yamauchi *et al.*, 1981) have been reported to be substrates for both kinases, most appear to be specifically labeled by one or the other kinase. These proteins constitute an interesting model system with which to analyze the role of calmodulin in protein phosphorylation. In the present system addition of calmodulin and Ca^{2+} not only results in the increased phosphorylation of specific proteins phosphorylated by Ca^{2+}-dependent kinases as previously reported (Table I), but also produces a concomitantly decreased phosphorylation of proteins which are phosphorylated by cAMP-dependent protein kinase. The mechanism of this inhibition is not fully understood. It could be the result of the presence of calcineurin, the recently discovered calmodulin-stimulated protein phosphatase (Stewart *et al.*, 1982), which dephosphorylates proteins phosphorylated

FIGURE 2. cAMP and Calmodulin-dependent Phosphorylation of Calmodulin Binding Proteins. Calmodulin-binding proteins (as described in the legend to Figure 1) were incubated at room temperature for 30 min in the presence of 10 mM $MgCl_2$*, 3 mM* $CaCl_2$*, and* 1.5×10^{-5} *M calmodulin to allow dephosphorylation. Phosphorylation was then started by addition of* $[\gamma\text{-}^{32}P]$ *ATP to a final concentration of 70 µM and proceeded for 20 min at room temperature. Left, phosphorylation was carried out in the presence of 17 µM 8-Br-cAMP, and 2 µg of the catalytic unit of cAMP-dependent protein kinase. EGTA (6 mM) was added to prevent calmodulin-dependent phosphorylation. Right, protein kinase inhibitor was added to inhibit cAMP-dependent phosphorylation. The phosphorylated proteins were analyzed as described in legend to Figure 1. The Coomassie blue stained gels are shown on the right and left sides of the autoradiograms. The* M_r *of the major phosphorylated proteins are indicated in the middle. Bovine serum albumin (BSA) was added with the catalytic unit of protein kinase or the protein kinase inhibitor.*

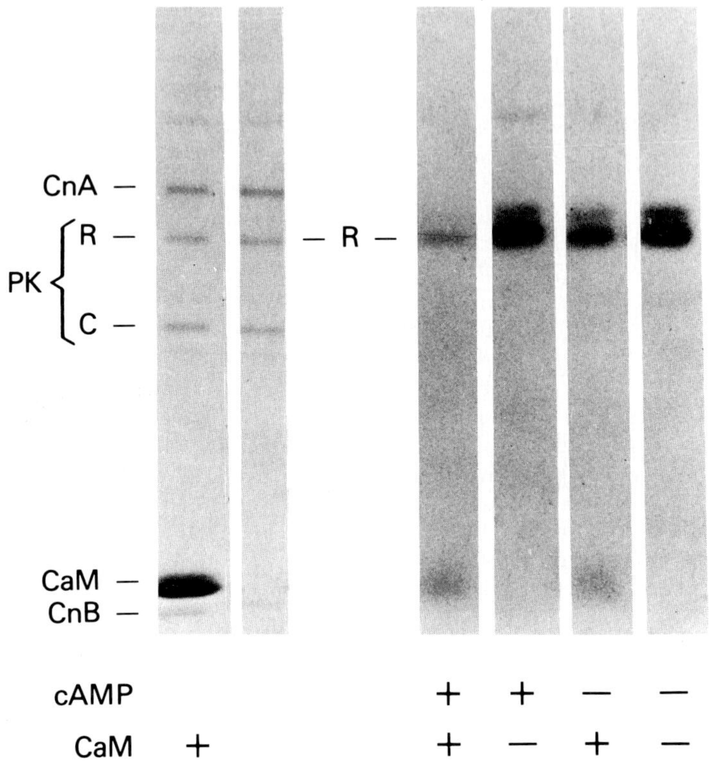

FIGURE 3. Inhibition of the Autophosphorylation of the Regulatory Subunit of cAMP-dependent Protein Kinase by Calmodulin. Equimolar amounts of cAMP-dependent protein kinase (PK), regulatory subunit (R), catalytic subunit (C), and calcineurin (CnA and CnB) in 40 mM Tris-HCl, pH 7.5, 1 mM $MgCl_2$, 2 mM EGTA, 0.5 mM DTT and 0.4 M KCl were incubated in the presence and absence of either 17 μM 8-Br-cAMP or 2.6 mM $CaCl_2$ and 11.6 μM calmodulin (CaM) as noted in the figure. Autophosphorylation was initiated by the addition of 70 μM (γ-^{32}P)ATP and was allowed to proceed for 10 min at room temperature in the presence of 10 mM $MgCl_2$ and 70 mM NaF. Phosphorylation was analyzed as described in legend to Figure 1. The two gels on the left represent the Coomassie blue-stained protein pattern whereas the four gels on the right depict the autoradiographic patterns.

by cAMP-dependent protein kinases such as the α-subunit of phosphorylase kinase, the inhibitor-I of protein phosphatase, the regulatory subunit of protein kinase and also proteins phosphorylated by calmodulin-dependent kinase such as myosin light chains.

An alternative mechanism for the calmodulin-induced inhibition of cAMP-dependent phosphorylation is a decreased affinity of the substrate-calmodulin complex for the kinase. This possibility should be considered since these proteins have been selectively fractionated on the basis of their ability to interact with calmodulin. It has previously been reported that interaction of calmodulin with histones prevents their dephosphorylation by protein phosphatase(s) (Khandelwel *et al.*, 1980; Wolff *et al.*, 1981). A similar mechanism could be invoked for inhibition of phosphorylation.

Calmodulin may also directly inhibit cAMP-dependent protein kinase because, in brain extracts, this kinase interacts with calmodulin (Hathawey *et al.*, 1981). Although this interaction does not significantly alter the activity of the purified enzyme in the presence of saturating levels of cAMP, in a less purified preparation, calmodulin could inhibit the enzyme activation by several other mechanisms.

Regardless of the mechanimm by which calmodulin inhibits cAMP-dependent phosphorylation, the importance of this inhibition should not be underestimated. It can explain in some tissues the antagonistic effect of the two second messengers, Ca^{2+} and cAMP. On the one hand, calmodulin-stimulated phosphodiesterase reduces cyclic nucleotide levels and inhibits cAMP-dependent phosphorylation. On the other hand, stimulation of the Ca^{2+}-calmodulin-dependent kinase maintains high levels of Ca^{2+}-dependent phosphorylation.

The coupled effects of cAMP and Ca^{2+} on the state of phosphorylation of cellular proteins is schematically presented in Figure 5. First, whereas cAMP can inhibit Ca^{2+} stimulation by stimulating Ca^{2+} uptake, Ca^{2+} can inhibit the cAMP signal by stimulating cyclic nucleotide degradation. Second, cAMP-dependent phosphorylation of calmodulin binding proteins modulates the Ca^{2+} regulation of these enzymes. A synergistic effect has been observed with phosphorylase kinase whereas an antagonistic effect is observed with myosin kinase. Interaction of the substrates of cAMP-dependent protein kinase with calmodulin, or their phosphorylation by calmodulin-dependent kinase may in turn modify their ability to be phosphorylated by cAMP-dependent kinase. Third, the presence of a calmodulin stimulated protein phosphatase in some tissues may be responsible for the negative control of the cAMP signal by Ca^{2+}. This complex interplay of the two ubiquitous signals, Ca^{2+} and cAMP, will depend upon the presence in a given tissue of the target proteins and the ki-

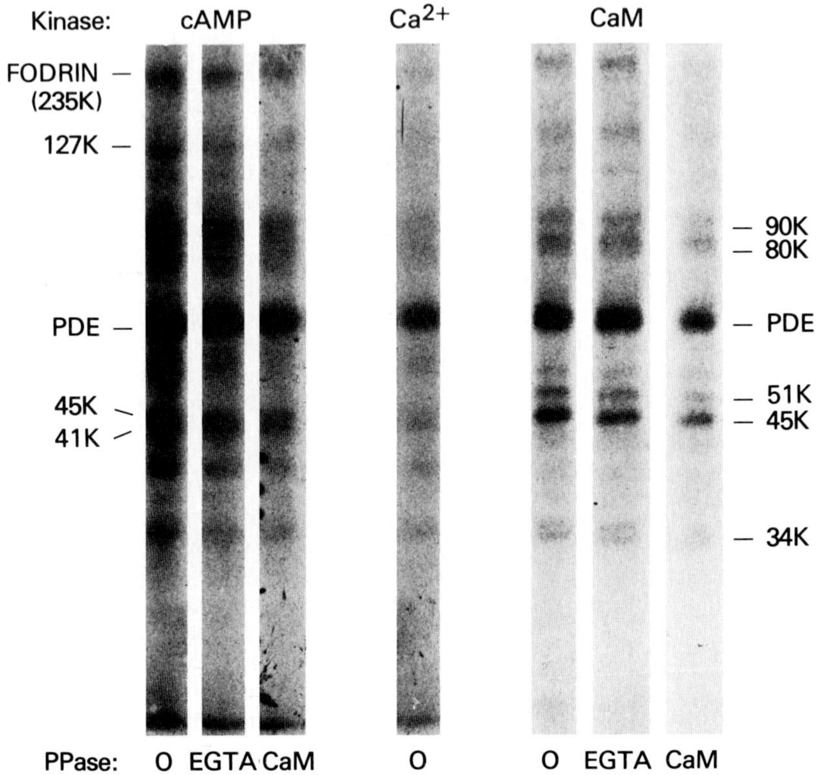

FIGURE 4. Ca^{2+}- and Calmodulin-Stimulated Dephosphorylation of Brain Proteins. Calmodulin-binding proteins were phosphorylated with 125 μM (γ-^{32}P)ATP for 20 minutes as described in the legend to Figure 1. The following phosphorylation conditions were set up to examine the kinases indicated at the top of the figure; 30 μM 8-Br-cAMP and 2 μg catalytic subunit (cAMP), 4.5 mM Ca^{2+} and protein kinase inhibitor (Ca^{2+}), and 4.5 mM Ca^{2+}, 2 x 10^{-5} M CaM and protein kinase inhibitor (CaM). Aliquots of the three incubation

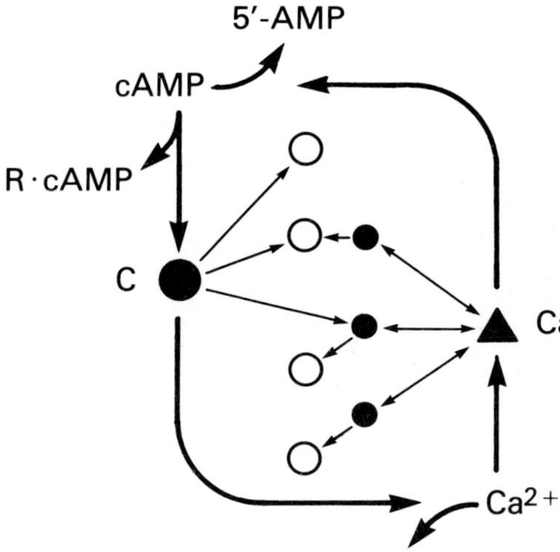

FIGURE 5. Schematic Representation of Calmodulin Modulation of Protein Phosphorylation. R and C are the regulatory and catalytic subunits, respectively, of cAMP-dependent protein kinase; CaM is calmodulin; the open circles are proteins which can be phosphorylated by either cAMP- or calmodulin-dependent kinases and dephosphorylated by calcineurin, a calmodulin-stimulated protein phosphatase. The small filled circles are calmodulin-binding proteins and include protein kinases and calcineurin.

(Figure 4 continued)

mixtures were stopped by addition of the denaturing solution (0). Cold ATP (1 mM) was added to the fraction remaining and protein kinase inhibitor was added to the first group. Phosphatase activity (PPase) was then examined by incubating the phosphorylated proteins for an additional 20 minutes at room temperature in the presence of 0.7 mM EGTA (EGTA) or 1.4 x 10^{-5} *M calmodulin and 3 mM* Ca^{2+} *(CaM) as shown at the bottom of the figure. The proteins were then analyzed by gel electrophoresis. Autoradiograms of the gels are shown in the figure.*

REFERENCES

Adelstein, R.S., and Eisenberg, E. (1980). *Annu. Rev. Biochem.* *49*, 921.

Adelstein, R.S., Conti, M.A., Hathaway, D.R., and Klee, C.B. (1978). *J. Biol. Chem.* *253*, 8347.

Ahmad, Z., DePaoli-Roach, A.A., Roach, P.J. (1982). *J. Biol. Chem.* *257*, 8348.

Alper, S.L., Palfrey, H.C., DeRiemer, S.A., and Greengard, P. (1980). *J. Biol. Chem.* *255*, 11029.

Brostrom, C.O., Huang, Y.C., Breckenridge, B.M., and Wolff, D.J. (1975). *Proc. Natl. Acad. Sci. USA* *72*, 64.

Campbell, K.P., and MacLennan, D.H. (1982). *J. Biol. Chem.* *257*, 1238.

Cheung, W.Y. (1970). *Biochem. Biophys. Res. Commun.* *38*, 533.

Cheung, W.Y., Bradham, L.S., Lynch, T.J., Lin, Y.M., and Tallant, E.A. (1975). *Biochem. Biophys. Res. Commun.* *66*, 1055.

Chrisman, T.D., Jordan, J.E., and Exton, J.H. (1982). *J. Biol. Chem.* *257*, 10798.

Cohen, P. (1982). *Nature* *296*, 613.

Cox, D.E., and Edstrom, R.D. (1982). *J. Biol. Chem.* *257*, 12728.

DePaoli-Roach, A.A., Roach, P.J., and Larner, J. (1979). *J. Biol. Chem.* *254*, 4212.

Edelman, A.M., and Krebs, E.G. (1982). *FEBS Lett.* *138*, 293.

Erlichman, J., Sarkar, D., Fleischer, N., and Rubin, C.S. (1980). *J. Biol. Chem.* *255*, 8179.

Fukunaga, K., Yamamoto, H., Matsui, K., Higashi, K., and Miyamoto, E. (1982). *J. Neurochem.* *39*, 1607.

Goldenring, J.R., Gonzalez, B., and DeLorenzo, R.J. (1982). *Biochem. Biophys. Res. Commun.* *108*, 421.

Hathaway, D.R., Adelstein, R.S., and Klee, C.B. (1981). *J. Biol. Chem.* *256*, 8183.

Huttner, W.B., and Greengard, P. (1979). *Proc. Natl. Acad. Sci. USA* *76*, 5402.

Kakiuchi, S., and Yamazaki, R. (1970). *Biochem. Biophys. Res. Commun.* *41*, 1104.

Kennedy, M.B., and Greengard, P. (1981). *Proc. Natl. Acad. Sci. USA* *78*, 1293.

Kennedy, M.B., McGuinness, T., and Greengard, P. (1982). *12th Ann. Meeting, Society for Neurosciences 8 (Abstract)*, 281.

Khandelwal, R.L., Kotello, M.C., and Wang, J.H. (1980). *Arch. Biochem. Biophys.* *203*, 244.

Klee, C.B., and Krinks, M.H. (1978). *Biochemistry* *17*, 120.

Klee, C.B., and Vanaman, T.C. (1982). *Adv. Prot. Chem. 35,* 213.

Klee, C.B., Crouch, T.H., and Krinks, M.H. (1979). *Proc. Natl. Acad. Sci. USA 76,* 6270.

Klee, C.B., Krinks, M.H., Hathaway, D.R., and Flockhart, D.A. (1982). *In* "Calmodulin and Intracellular Ca^{2+} Receptors" (S. Kakiuchi, H. Hidaka, and A. Means, eds.), p. 303. Plenum Publishing Corporation.

Kretsinger, R.H. (1980). *CRC Crit. Rev. Biochem. 8,* 119.

Kuo, C.-H., Ichida, S., Matsuda, T., Kakiuchi, S., and Yoshida, H. (1979). *Life Sci. 25,* 235.

Manalan, A.S., and Klee, C.B. (1983). *Chemica Scripta 21,* 139.

Nelson, M.J., Daleke, D.L., Huestis, W.H. (1982). *Biochim. Biophys. Acta 686,* 182.

Payne, M.E., and Soderling, T.R. (1980). *J. Biol. Chem. 255,* 8054.

Rall, T.W., Sutherland, E.W., and Berthet, J. (1957). *J. Biol. Chem. 224,* 463.

Schulman, H. (1982). *In* "Handbook of Experimental Pharmacology," Vol. 58/I, p. 425. Springer-Verlag, Berlin, Heidelberg, New York.

Schulman, H., and Greengard, P. (1978). *Nature (London) 271,* 478.

Schulman, H., Huttner, W.B., and Greengard, P. (1980). *In* "Calcium and Cell Function" (W.Y. Cheung, ed.), Vol. 1, p. 219. Academic Press, New York.

Sharma, R.K., Desai, R., Waisman, D.M., and Wang, J.H. (1979). *J. Biol. Chem. 254,* 4276.

Sharma, R.K., Wang, T.H., Wirch, E., and Wang, J.H. (1980). *J. Biol. Chem. 255,* 5916.

Sobue, K., Ichida, S. Yoshida, H., Yamazaki, R., and Kakiuchi, S. (1979). *FEBS Lett. 99,* 199.

Sobue, K., Kanda, K., Yamagami, K., and Kakiuchi, S. (1982). *Biomed. Res. 3,* 561.

Soderling, T.R., Srivastava, A.K., Bass, M.A., and Khatra, B.S. (1979). *Proc. Natl. Acad. Sci. USA 76,* 2536.

Stewart, A.A., and Cohen, P. (1983). *Eur. J. Biochem.* (submitted).

Stewart, A.A., Ingrebritsen, T.S., Manalan, A.S., Klee, C.B., and Cohen, P. (1982). *FEBS Lett. 137,* 80.

Takai, Y., Kishimoto, A., Iwasa, Y., Kawahara, Y., Mori, T., and Nishizuka, Y. (1979). *J. Biol. Chem. 254,* 3692.

Wallace, R.W., Tallant, E.A., and Cheung, W.Y. (1980). *Biochemistry 19,* 1831.

Walsh, D.A., Perkins, J.P., and Krebs, E.G. (1968). *J. Biol. Chem. 243,* 3763.

Westwood, S.A., and Perry, S.V. (1981). *Biochem. J. 197*, 185.
Woodgett, J.R., Tonks, N.K., and Cohen, P. (1982). *FEBS Lett. 148*, 5.
Wolff, D.J., Ross, J.M., Thompson, P.N., Brostrom. M.A., and Brostrom, C.O. (1981). *J. Biol. Chem. 256*, 1846.
Yamauchi, T., and Fujisawa, H. (1980). *FEBS Lett. 116*, 141.
Yamauchi, T., Nakata, H., and Fujisawa, H. (1981). *J. Biol. Chem. 256*, 5404.

DISCUSSION

Ingebritsen (U. Colorado): In the dephosphorylation of the calmodulin binding proteins, in the presence of calcium and calmodulin, can you actually be sure that the dephosphorylation is due to the stimulation of the calcium-calmodulin dependent protein phosphatase as opposed to interacting with the proteins and making them good substrates for some other protein phosphatase in the preparation?

Klee: The same question can be raised for the kinases. The substrate could bind calmodulin, undergo a conformational change and become a better substrate. I cannot rule this out.

Olson: Can you get the effects of calmodulin without calcium?

Klee: All the effects which we observed were Ca^{2+}-dependent and prevented by EGTA.

Olson: So the calmodulin by itself is not an effector?

Klee: In this particular system, no. However, calmodulin does not always require calcium. There is at least one enzyme which has been shown to be stimulated by calmodulin in the absence of calcium: that is adenylate cyclase from *Bordetella pertussis*. The only difference is that more calmodulin is required in the absence of calcium than in its presence. There are possibly other systems in which the effect of calmodulin does not require calcium, but we have not detected them.

Cohen: When you are eluting calmodulin binding proteins from calmodulin Sepharose, are you now using an EGTA gradient routinely? This is not often used in the literature. Is this a good way of getting further fractionation?

Klee: The fractionation is slightly better with an EGTA gradient, but the gradient has to be designed correctly; otherwise one obtains only a very sharp decrease in calcium concentration.

Cohen: If the calcium calmodulin-dependent phosphatase is reversing many of the phosphorylations that are catalyzed by cAMP-dependent kinase, then this could be a nice mechanism in which to explain systems in which the calcium and cAMP second messenger systems act antagonistically; however, I don't know of any situation in the brain where they do act antagonistically.

Klee: I cannot answer your question in brain. Brain is a heterogenous tissue; we will have to work with isolated cells to be able to answer the question. But clearly the response of a given tissue will depend on the calmodulin-binding enzymes which are present in this tissue. For example, other tissues which have less calcineurin than brain does, have less phosphatase activity and will exert less negative control.

Cohen: We obviously have a similar situation with the calcium and calmodulin dependent phosphatase, as one has, for example, with the calmodulin stimulated phosphodiesterase. They are both systems in which Ca^{2+} and calmodulin are turning off the cAMP second messenger system. But exactly why that is important physiologically, I don't think we really understand at the moment.

Klee: It may be premature to look for a physiological significance since in vivo other factors may be present. As soon as we break the cells, we are looking at a completely artificial system.

Nishizuka: Is there any possibility that calcium-dependent proteases are involved in the apparent disappearance of phosphoproteins?

Klee: This is a valid point. We shall have to check it further. We had no evidence of proteolysis by Coomassie blue staining determinations and we assumed that it was not really a problem, but I shall have to look at the data more carefully.

Weiner: In reference to the heterogeneity of the tissue, and the likelihood of course, that some enzymes may be localized in some cells and not in others, I wonder whether you have looked at different regions of the brain? For example, I noticed that your phosphodiesterase was migrating in the SDS gels to about the same place as tyrosine hydroxylase migrates. Have you looked at a region such as the striatum which is rich in tyrosine hydroxylase and seen any difference there?

Klee: I have not done a study of the regional distribution in brain tissue. Your point is correct - most of these calmodulin-regulated enzymes are eluted together from the DEAE cellulose column. Based on this criterium, they are all very similar.

Weiner: Is calcineurin present in glial cells or only in neural cells?

Klee: Calcineurin has been localized in brain by Cheung and his colleagues and found to be concentrated at the postsynaptic regions.

Weiner: So it is neural and glial?

Klee: That does not mean that it is not present in glial cells. It just means that it has not been detected there.

CHARACTERIZATION OF NEURAL-SPECIFIC cAMP-DEPENDENT PROTEIN KINASE[1]

Charles S. Rubin
Dwijen Sarkar
Jill C. Stein
Jack Erlichman

Departments of Molecular Pharmacology,
Medicine and Biochemistry
Albert Einstein College of Medicine
Bronx, New York

INTRODUCTION

Cyclic AMP (cAMP)-dependent protein kinases were discovered by Walsh, Perkins and Krebs in 1968 (Walsh *et al.*, 1968). In the succeeding decade a substantial amount of evidence was accumulated in support of the concept (Kuo and Greengard, 1969) that most (if not all) the actions of cAMP in mammalian cells are mediated through the activation of two forms of cAMP-dependent protein kinase, designated protein kinases I and II (for reviews see Rubin and Rosen, 1975; Nimmo and Cohen, 1977; Greengard, 1978; Krebs and Beavo, 1979).

The structural and functional properties of prototypic protein kinase I from skeletal muscle and protein kinase II from heart have been studied extensively in several laboratories (reviewed in Nimmo and Cohen, 1977). Both enzymes have M_rs of 180,000 and are asymmetric tetramers containing two catalytic (C) and two regulatory (R) subunits. The binding of 2 molecules of cAMP by each R subunit results in the dissociation of two active C monomers and an R_2-$cAMP_4$ complex.

[1]*These investigations were supported by Grants GM-22792, AM-27736 and American Cancer Society Grant NP 377.*

ISBN 0-12-387560-9

Since the properties of the C subunits of protein kinases I and II are virtually identical (Bechtel *et al.*, 1977) the features that distinguish the two enzymes reside in their R subunits. RI and RII clearly differ in their subunit M_rs, isoelectric points, their abilities to be autophosphorylated by C or participate in the non-covalent binding of MgATP in the holoenzyme and their antigenic determinants (Table I).

An intriguing paradox is that the total concentration of cAMP-dependent protein kinases in many mammalian tissues is invariant (~0.3 μM, Hofmann *et al.*, 1977) but the relative proportions contributed by the type I and type II enzymes vary greatly from tissue to tissue and among species (Corbin *et al.*, 1975, Corbin and Keely, 1977). A possible conclusion from such observations is that protein kinases I and II are functionally interchangeable.

TABLE I. Properties of the Regulatory Subunits of Protein Kinases I and II

	RI	RII
Autophosphorylation by C[a]	-	+
MgATP binding	35nM = Kd	-
M_r[b]	48,000	55,000
Re-association with C (0°, -MgATP)	Slow	Fast
Dissociated from C by NaCl or histone	+	-
Isoelectric point	5.2 - 5.5	4.8 - 5.1
Elution from DEAE cellulose [a]	50mM NaCl	150mM NaCl
Reaction with anti-RI IgG	+	-
Reaction with anti-RII IgG	-	+

[a]*Properties observed when R subunits are associated with C subunits in the holoenzyme.*

[b]*Apparent M_r determined by electrophoresis in SDS-polyacrylamide gels.*

IDENTIFICATION OF TWO SUBCLASSES OF TYPE II PROTEIN KINASES: THE NEURAL SPECIFIC AND NON-NEURAL ENZYMES

In a series of studies initiated several years ago on cAMP-dependent protein kinases in bovine brain, we discovered that there are multiple forms of type II protein kinase (Rubin *et al.*, 1979). In subsequent investigations we have obtained evidence that suggests that one subclass of type II kinases may be structurally and functionally specialized for roles in neural and neuroendocrine cells, i.e.; brain protein kinase II may not be interchangeable with protein kinases I and II from non-neural cells.

The principal form of cAMP-dependent protein kinase in cerebral cortex is a type II enzyme and its subcellular distribution is atypical (Table II). Fifty percent of the enzyme activity is tightly associated with the particulate fraction and the bulk of this activity co-purifies with synaptic plasma membranes (Maeno *et al.*, 1971, Rubin *et al.*, 1979). In contrast, more than 90% of the type II kinases from liver, heart and skeletal muscle are found in the cytosol.

The physicochemical properties of the cytosolic and detergent-solubilized type II enzymes from cerebral cortex are very similar to those of cytosolic heart protein kinase II (Rubin *et al.*, 1979). However, when either of the brain enzymes was compared with heart protein kinase II by assessing their abilities to react with high-titer polyclonal antibodies, produced against affinity-purified cerebral cortex RII (Erlichman *et al.*, 1980), marked differences were observed (Figure 1). IgGs in one microliter of serum complexed more than 60% of the brain enzyme, but less than 15% of an equal amount of heart protein kinase II. At a ten-fold higher concentration of antibodies only 50% of the heart enzyme was precipitated while three microliters of serum was sufficient to bind > 90% of cerebral cortex protein kinase II (Figure 1). These observations and more detailed, indirect immunoprecipitation studies revealed that anti-brain RII IgGs sequestered the cerebral cortex enzyme with a 20 to 25-fold higher affinity than heart enzyme (Erlichman *et al.*, (1980) and raised the possibility that there might be two or more subclasses of type II protein kinases.

A modification of a competitive binding radioimmunoassay procedure was employed to examine this possibility (Erlichman et al., 1980). The abilities of partially-purified type II protein kinases to inhibit the binding of a highly purified tracer, ^{125}I-brain RII, by anti-brain RII antibodies were determined. Since the concentrations of the protein kinases

TABLE II. *Distribution of Soluble and Membrane-Associated cAMP-Dependent Protein Kinase Activity in Bovine Tissues*

Tissue	Cytosol		Membranes		% total activity in cytosol
	Specific[a] activity	Units[b]	Specific[a] activity	Units[b]	
Cerebral cortex	0.78	131	3.60	133	50
Liver	0.25	292	0.80	22	93
Muscle	0.70	284	0.44	7	98
Heart	0.79	280	0.27	7	98

[a]Nanomoles of ^{32}P transferred from [γ-^{32}P] ATP to protamine per min per mg of protein.
[b]Nanomoles of ^{32}P transferred to protamine per 5g wet weight.

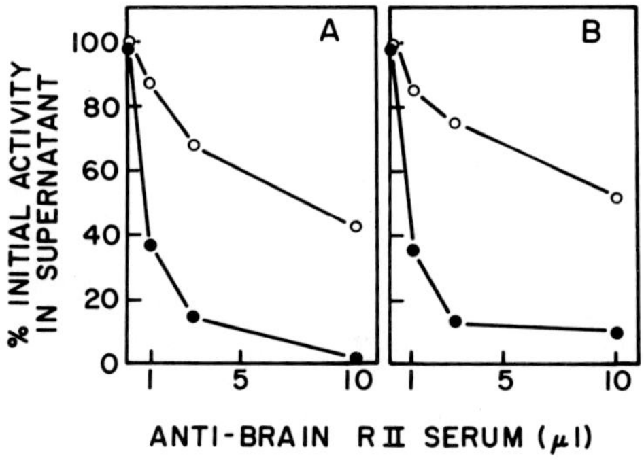

FIGURE 1. Immunoprecipitation of cAMP-binding and catalytic activities of type-II protein kinases by anti-brain RII serum. Brain protein kinase (50 pmoles) and heart protein kinase (52 pmoles) were incubated with varying amounts of rabbit anti-brain RII serum for 16 hr and subsequently precipitated by the addition of anti-rabbit IgG serum. Immunoprecipitates were isolated by centrifugation, and aliquots of the supernatant were assayed for cAMP-binding activity (A) and phosphotransferase activity (B). Data are presented as a percentage of original activity remaining in the supernatant fraction. (●) Cerebral cortex cytosolic kinase; (○) heart kinase.

were independently estimated by the equilibrium binding of [^{3}H]cAMP at saturation the competitive binding assays could be used to compare directly the affinty of the antibodies for the various type II enzymes. Type II protein kinases from liver, heart, skeletal muscle and kidney exhibited very similar competitive inhibition profiles and were less that 5% as potent as the bovine cerebral cortex enzyme (Figure 2). The non-neural kinases apparently constitute a highly related subset of type II enzymes that are immunologically distinct from cerebral cortex protein kinase II.

To determine whether the cerebral cortex enzyme was unique or related to other kinases, type II protein kinases from several regions of the brain and from neuroendocrine tissues were compared with the cerebral cortex enzyme in additional radioimmunoassays (Figure 3). Type II protein kinases from bovine cerebellum, midbrain, medulla and brainstem (Figure 3A), as well as from pineal gland, posterior pituitary, anterior pituitary and adrenal medulla (Figure 3B) generated competitive inhibition curves that were congruent with the curve obtained using cerebral cortex protein kinase II. Thus

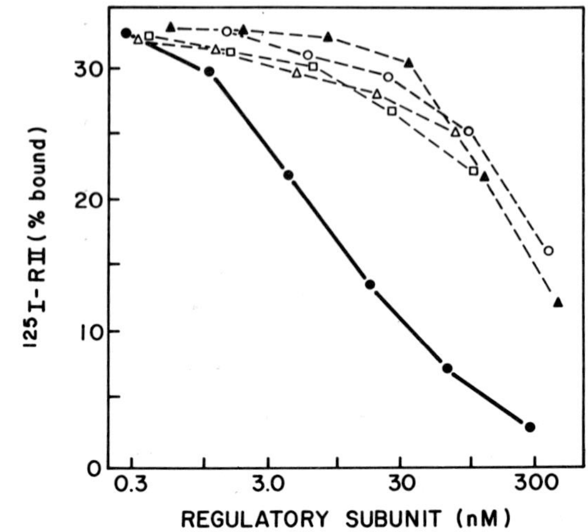

FIGURE 2. Radioimmunoassays for type-II protein kinases from brain and nonneural tissues. Type-II enzymes were prepared from cerebral cortex (●), muscle (○), heart (△), kidney (□), and liver (▲). Competitive binding immunoassays were carried out as described in Erlichman et al. *(1980) using anti-cerebral cortex RII serum and* ^{125}*I-cerebral cortex RII as tracer. Results are expressed as a percentage of added* ^{125}*I-RII bound to anti-RII IgG. Under standard conditions 35% of the tracer was bound by the antibodies in the absence of competing non-radioactive ligand.*

neural and neuroendocrine cells contain a second, distinct subset of neural-specific type II protein kinases.

The marked antigenic differences between neural and nonneural protein kinases have been highly conserved in several mammalian species, including humans (Rubin *et al.*, (1981). In addition a preliminary survey suggests that many of the cells and tissues that contain the neural specific type II kinase are members of the APUD series (amine precursor uptake and decarboxylation endocrine cells) (Pearse, 1968; Pearse, 1969; Pearse and Takor-Takor, 1979). Therefore cells containing neural-specific protein kinase II may exhibit several common features including: 1) similar ultrastructural properties; 2) a capacity to accumulate and metabolize biogenic amines; 3) the ability to synthesize and secrete structurally-related peptide hormones and neurotransmitters and 4) an embryological ancestry that can be traced back to the neural crest (for a review on the properties of APUD cells see Pearse

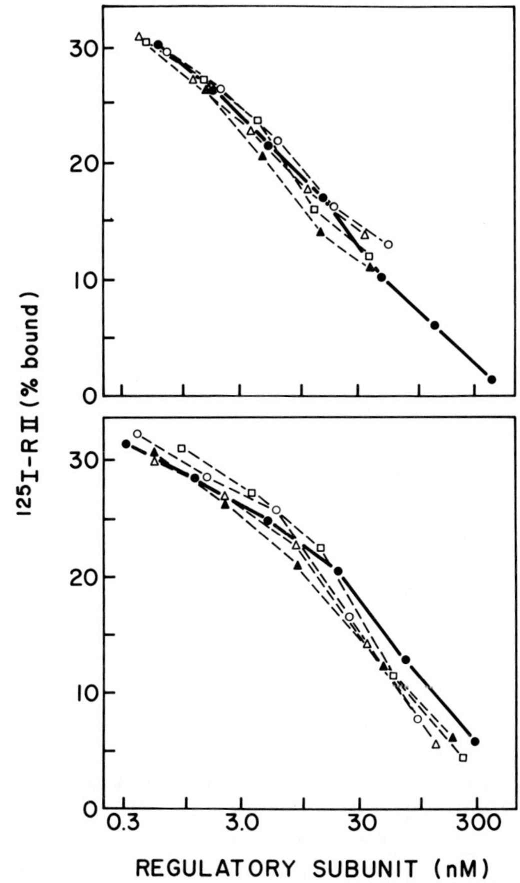

FIGURE 3. Radioimmunoassays of type-II protein kinases from neural tissues. Competitive binding radioimmunoassays were performed using type-II kinases from: (A) cerebral cortex (●), cerebellum (△), medulla (□), midbrain (▲), and brain stem (○); (B) cerebral cortex (●), anterior pituitary (○), posterior pituitary (△), pineal (□), and adrenal medulla (▲). Radioimmunoassays were performed and the resulting data were expressed as indicated in Fig. 2.

and Takor-Takor, 1979). The expression of a unique subclass of type II protein kinases in neural and neuroendocrine cells suggests the possibility that the structural and functional properties of this enzyme may be specifically adapted to the regulation of metabolism and function in the central nervous and neuroendocrine systems.

STRUCTURAL PROPERTIES OF BRAIN RII SUBUNITS

The immunological disparity between neural and non-neural protein kinases implies significant structural differences between the two enzymes. Large quantities of heart RII can be readily purified to homogeneity by conventional and affinity chromatography procedures and the complete amino acid sequence of this polypeptide was recently determined (Takio *et al*, 1982). In contrast, no sequence data are available on brain RII and the isolation of homogenous samples of brain RII has proved to be quite difficult because of the high concentration of proteases in brain and the specific, high and moderate affinity binding of a variety of brain proteins to the neural cAMP-binding subunit (see below). Nevertheless small quantities of highly purified brain RII have been isolated and several novel structural properties of brain RII have been documented.

Dephosphorylated heart RII exhibits an apparent M_r of 55,000 upon electrophoresis in SDS-polyacrylamide gels. After its phosphorylation via incubation with C (catalytic subunit) and MgATP heart RII displays a relative mobility corresponding to M_r of 57,000 (Figure 4B). Although the structural basis of this alteration in mobility is unknown, it serves as a useful analytical probe. Phosphorylation of brain RII under identical conditions results in the appearance of two species with M_rs of 55,000 and 57,000 (Figure 4A). Both polypeptides appear to be R subunits since they are phosphorylated by C (catalytic subunit), covalently labeled with the photoaffinity analog 8-N_3[^{32}P]-cAMP and specifically immunoprecipitated by anti-RII serum (Figure 4, Rubin *et al.*, 1979). These results raise the possibility that there are multiple species of RII subunits in cerebral cortex type II protein kinases. Lohmann *et al.*, (1980) have also observed two forms of RII subunits in bovine brain.

The limited supply of homogeneous brain RII prompted us to study functional domains that can be covalently labeled with ^{32}P. These include the autophosphorylation site and one of the intrachain cAMP binding sites (Kerlavage and Taylor, 1982). The autophosphorylation sites in highly purified heart and brain RII were labeled by incubation with C (catalytic subunit) and [γ-^{32}P]ATP. After peroxyformic acid oxidation and complete digestion with trypsin, ^{32}P-labeled phosphopeptides were resolved by 2-dimensional mapping on thin layers

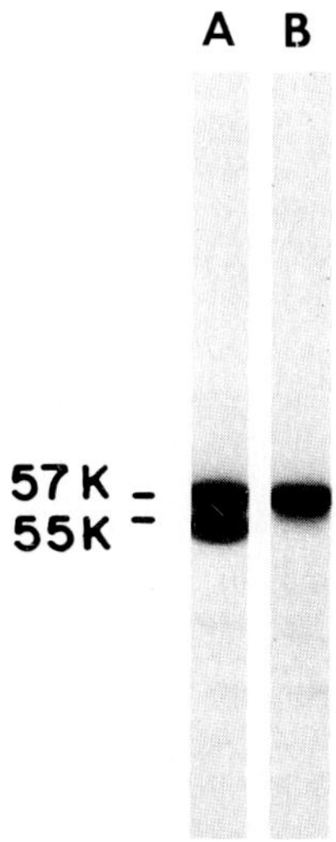

FIGURE 4. Comparison of phosphorylated R subunits obtained from neural and non-neural protein kinases. Type II protein kinases from bovine cerebral cortex and skeletal muscle were autophosphorylated by pre-incubation for 15 min at 0° in 25 mM 2-(N-morpholino)ethanesulfonic acid buffer (MES) pH 6.8, containing 10 mM $MgCl_2$ and 0.5 mM ATP. After the addition 0.1 M NaF and 0.1 M EDTA to inhibit protein phosphatases the samples were photoaffinity-labeled with 8-N_3-[^{32}P] cAMP and immunoprecipitated with specific anti-R II serum as described by Rubin et al. *(1979). Immunoprecipitates were then analyzed by electrophoresis on 0.1% SDS-10% polyacrylamide gels followed by autoradiography as described previously (Rubin* et al. *1979). In the autoradiogram presented, lane A received 0.2 pmol cerebral cortex RII; lane B received 0.2 pmol skeletal muscle RII.*

of cellulose and were detected by autoradiography. All of the ^{32}P incorporated into heart RII appeared in a single phosphopeptide fragment (Figure 5, center). Primary sequence data indicate that this tryptic peptide contains 20 amino acids corresponding to residues 94 to 113 (Takio *et al.*, 1982). Tryptic digests of brain RII contained two labeled phosphopeptides that displayed similar electrophoretic mobilities, but were resolved in the chromatographic dimension (Figure 5, left). The minor fragment appeared to correspond to the labeled peptide derived from the heart autophosphorylation site, while the predominant brain phosphopeptide seemed to be unique (Figure 5, center). This was confirmed

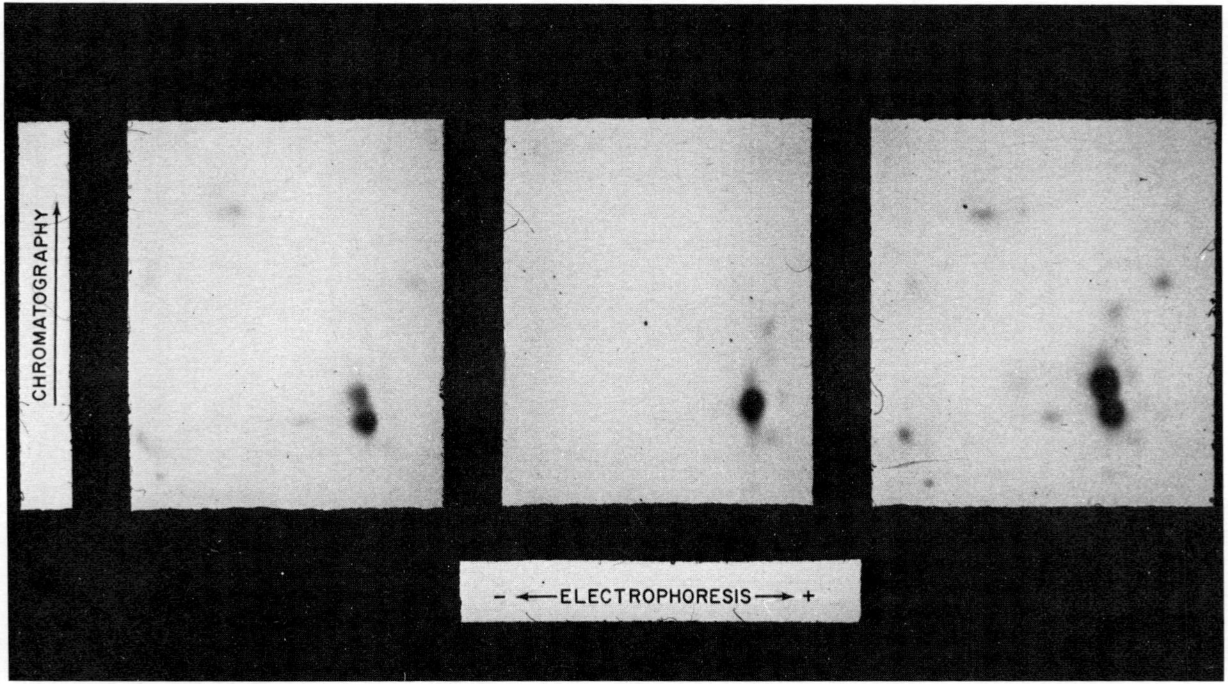

FIGURE 5

when mixtures of the two tryptic digests were analyzed. Only two fragments were observed and the phosphopeptide with the higher chromatographic mobility was enriched (Figure 5, right). Thus, the heart phosphopeptide and the minor fragment from brain RII co-migrate on 2-dimensional maps. The latter may be derived either from contaminating non-neural protein kinase II or from a third form of RII present in brain. The principal fragment containing the autophosphorylation site in brain RII is closely related to, but distinct from the heart RII phosphopeptide.

Similar analyses of tryptic digests of heart and brain RII subunits that were covalently labeled with 8-N_3[^{32}P]cAMP have disclosed significant structural differences between the cAMP-binding domains in the two subclasses of protein kinase II (Figure 6). The multiplicity of ^{32}P-labeled tryptic fragments observed in these experiments could result from the reaction of the photogenerated nitrene intermediates with 2 or more distinct regions of primary sequence that fold to form the cyclic nucleotide binding site and/or the labeling of two binding sites.

FIGURE 5. Two-dimensional maps of ^{32}P-labeled tryptic peptides derived from phosphorylated brain and heart RII subunits. Highly purified preparations of brain or heart RII (5 μg) were phosphorylated by incubation for 45 min at 0°C in MES buffer, pH 6.8 containing 10 mM $MgCl_2$, 3.3 μM [γ-^{32}P]ATP, 12 μM cAMP and 2 μg homogeneous C subunit. The phosphorylation reaction was terminated by the addition of 10% trichloroacetic acid and protein precipitation was facilitated by the subsequent addition of 0.2 mg albumin. Protein precipitates were freed of trichloroacetic acid by ether extraction and the resulting pellets were dried under vacuum. The samples were then subjected to performic acid oxidation (Hirs, 1967) and digestion with 4 μg trypsin for 4 hr at 37°C (Smyth, 1967). The mixture of tryptic peptides was lyophilized and then dissolved in 20 μl of H_2O. Aliquots (8 μl) were applied to 20 cm x 20 cm plates coated with a thin layer of cellulose (1.2mm) for 2-dimensional mapping. In the first dimension samples were subjected to electrophoresis in 1.2% pyridine/ 1.2% acetic acid (vol/vol), pH 4.7, for 50 min at 750 V. Chromatography in the second dimension was carried out for 4 hr in 1-butanol/pyridine/acetic acid/H_2O (60:40:12:48:). After drying in air ^{32}P-phosphopeptides were visualized by autoradiography. The autoradiograms presented were obtained from brain RII (left), heart RII (Center) and a mixture of brain and heart RII subunits (right).

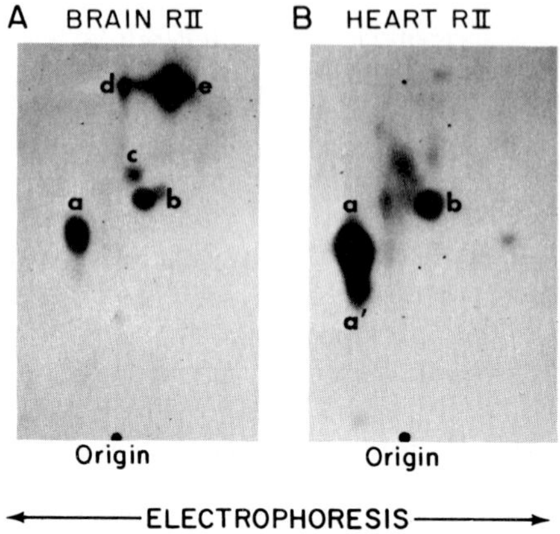

FIGURE 6. Two-dimensional maps of ^{32}P-labeled tryptic peptides derived from photoaffinity-labeled brain and heart RII subunits. Highly purified heart or brain RII (5µg) were covalently labeled with 8-N_3-[^{32}P]cAMP as described previously (Rubin et al., *1979). Tryptic peptide mapping analyses were then performed as described in Fig. 5. In the autoradiogram presented ^{32}P-phosphopeptides labeled a and b were common to both RII subunits; c, d, and e were found only in brain RII, while a' was characteristic of heart RII.*

Differences in structure in domains that would be expected to be highly conserved raise the possibilities that neural and non-neural RII subunits are products of different genes and that larger differences in 3-dimensional structure and primary sequence may be found in other regions of RII.

Monoclonal antibodies are highly sensitive probes for elucidating structural dissimilarities between related proteins. Small amounts (0.2 to 0.4 mg) of highly purified brain RII were denatured with 5M guanidine thiocyanate and injected into BALB/c mice. Responding animals were boosted once and spleen cell/myeloma fusions were performed 3 days later. A panel of hybridomas were identified, subcloned and injected into BALB/c mice to produce ascites tumors secreting high concentrations of IgGs. Immunoglobulins from ascites fluid were purified to homogeneity by affinity chromatography on protein-A Sepharose (Figure 7).

Since the purified brain RII preparations used for immunization contained subunits with apparent M_rs of both 55,000 and 57,000 (See Figure 4) the monoclonal antibodies (MAbs) were screened for their ability to preferentially complex RII

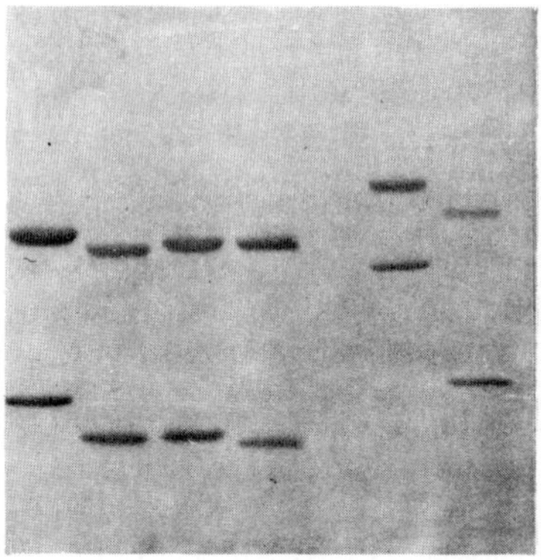

FIGURE 7. Characterization of purified, anti-RII monoclonal antibodies. Ascites fluid (2 ml, 150 mg protein) was diluted with an equal volume of 0.1 M potassium phosphate, pH 8.0, and applied to a column (0.9 cm x 9 cm) of protein A-Sepharose equilibrated in the same buffer. Only IgGs were bound to the column under these conditions. After washing with buffer at pH 8.0 IgG subclasses were eluted according to the sequential procedure of Ey et al. *(1978). IgG_1 was eluted with citrate buffer at pH 6.0; IgG_{2a} with citrate buffer at pH 4.5 and IgG_{2b} with citrate buffer at pH 3.5. Samples were subjected to electrophoresis on a 0.1% SDS-10% polyacrylamide gel and proteins were visualized by staining with Coomassie blue (Rubin* et al.*, 1979). The amounts of monoclonal antibodies (MAbs) applied to the gel were 16 μg MAb-124, 12 μg MAb-107, 11 μg MAb-625 and 11 μg MAb-622. Lane A received 3 μg albumin (M_r = 67,000) and 3 μg fumarase (M_r = 48,000). Lane B received 3 μg pyruvate kinase (M_r = 57,000) and 3 μg carbonic anhydrase (M_r=30,000).*

subunits from neural or non-neural type II protein kinases. MAb-294 immunoprecipitated heart RII over a narrow concentration range, but reacted poorly with brain RII (Figure 8). Conversely, MAb-820 bound a significant fraction of brain RII under the same conditions, but was unreactive with the heart enzyme (Figure 8). Thus, both types of RII subunits contain apparently unique antigenic determinants that are either absent or inaccessible in the other subclass of type II enzyme. In accord with these findings, Mumby and Beavo (1981) have reported that monoclonal antibodies prepared against heart RII bound brain RII with lower affinity.

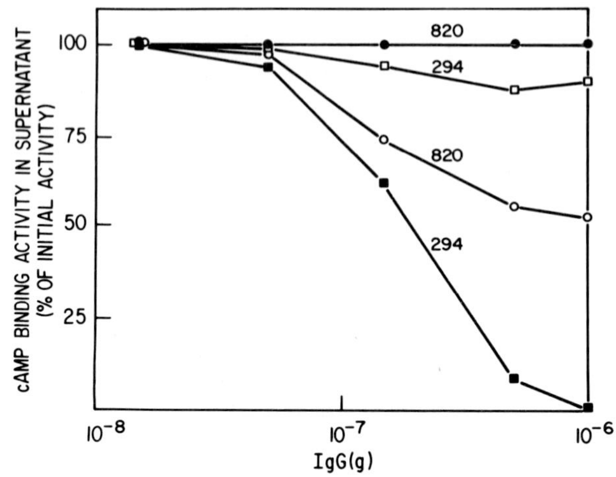

FIGURE 8. Immunoprecipitation of brain and heart RII subunits by monoclonal antibodies. Aliquots containing 3 pmol of partially-purified type II protein kinases from either brain or heart cytosol were preincubated for 30 min at 0°C in 0.1 ml, 50 mM Tris-HCl buffer, pH 8.0, containing 8 mm EDTA, 100 µg albumin and 4 µM [^{3}H] cAMP to achieve the equilibrium binding of the cyclic nucleotide and dissociation of the holoenzyme. Subsequently, the indicated amounts of purified MAbs-294 and 820 were added and the incubations were continued for 90 min at 0°C. Next 40 µl of a 10% suspension of heat-killed formalin-fixed S. aureus was added to adsorb the IgG-RII-[^{3}H] cAMP complexes. The bacteria were pelleted by centrifugation at 2,000 x g for 5 min. The supernatant solution was removed and the pellet was resuspended in 1 ml of ice-cold 50 mM Tris-HCl, pH 8.0, and transferred to a glass fiber filter. The filters were washed with 10 ml of 50 mM Tris-HCl, pH 8.0 and then were dried. Radioactivity on the filters was determined in a scintillation counter. The data are presented as the per cent of cAMP-binding activity remaining in the supernatant fraction. This was calculated by subtracting the amount of cAMP-binding activity on the filters from the total cAMP-binding activity in the sample. The precipitation of RII subunits derived from brain protein kinase II is shown in the open symbols; the precipitation of RII subunits derived from heart protein kinase II is shown in the solid symbols.

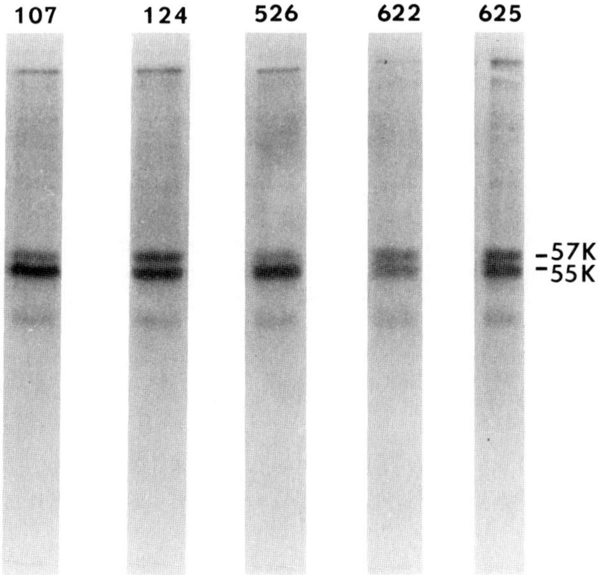

FIGURE 9. Recognition of the 55,000 dalton and 57,000 dalton brain RII subunits by monoclonal antibodies. Samples (0.1 pmol) of brain protein kinase II were subjected to electrophoresis on 0.1% SDS-10% polyacrylamide gels and analysis by the Western blot technique of Burnette (1981). In brief this entailed the electrophoretic transfer of the resolved polypeptides to nitrocellulose, an incubation with 0.2 mg of the indicated monoclonal antibodies, treatment with ^{125}I-protein A (100,00 cpm per lane) and visualization of the RII-IgG-^{125}I-protein A complexes by autoradiography.

Several MAbs, such as 107 and 526, exhibit a high degree of selectivity for the 55,000 dalton brain RII subunit (Figure 9). These MAbs should prove to be especially useful for effecting the separation of the 55,000 and 57,000 dalton RII subunits which co-purify in conventional purification procedures and affinity chromatography on agarose derivatized with cAMP analogs.

MAbs have also provided a new insight on the relationship between membrane-associated and cytosolic type II protein kinases in brain. Previous studies failed to disclose any differences in the physicochemical and immunological properties of the two enzymes (Rubin *et al.*, 1979). However, MAb-

124 recognized and complexed cytosolic protein kinase II in the presence and absence of Triton X-100, but failed to precipitate detergent-solubilized, membrane protein kinase II (Table III). The same result was obtained after Triton X-100 was removed from the membrane-derived kinase by chromatography on DEAE-cellulose. Several other MAbs bound both enzymes avidly and displayed only a slight preference for the cytosolic enzyme (Table III). These results suggest that subtle structural differences may distinguish the cytosolic, neural-specific protein kinase from the membrane-associated type II enzyme. Whether the difference between these two enzymes is related to diverging primary sequences, posttranslational modifications or the binding of a membrane protein to a specific interaction site on RII remains to be determined.

IDENTIFICATION OF A CALMODULIN BINDING PROTEIN THAT CO-PURIFIES WITH BRAIN RII

Brain RII can be readily purified to 60 to 90% homogeneity by conventional and affinity chromatography techniques, but it has proved to be difficult to obtain large quantities of more highly purified brain RII. A major reason is that several brain proteins specifically and tightly bind to RII. Under certain conditions 40% of brain protein kinase II was found associated with calcineurin, a Ca^{2+}-calmodulin activated phosphoprotein phosphatase (Hathaway *et al.*, 1981)

TABLE III. Precipitation of Brain RII Subunits by Monoclonal Antibodies[a]

Antibody	*Source of protein kinase*	*pmol RII precipitated*
124	*brain cytosol*	*1.14*
124	*brain membranes*	*0.12*
526	*brain cytosol*	*1.29*
526	*brain membranes*	*0.93*
107	*brain cytosol*	*1.50*
107	*brain membranes*	*0.99*

[a]*Samples containing 3 pmol of either cytosolic or detergent-solubilized, membrane-associated type II protein kinases were treated with 1 μg of the indicated antibodies and assayed as described in Figure 8.*

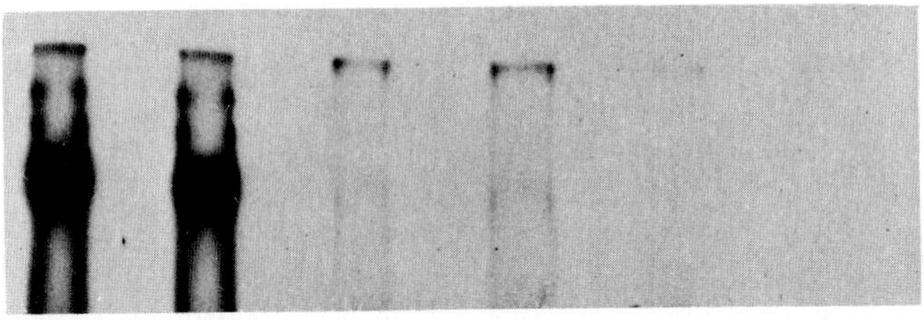

FIGURE 10. Identification of a calmodulin binding protein that co-purifies with brain RII. Samples (10 μg protein) of purified brain RII and heart RII subunits were incubated with 17 nM ^{125}I-calmodulin (6 x 10^5 cmp) in 0.1 ml of 50mM Tris-HCl buffer, pH 7.5, containing 1 mM EGTA and 3.75 mM $CaCl_2$ at 4°. After 30 min the samples were transferred to the wells of a 7.5% polyacrylamide gel prepared using the buffer systems described by Laemmli (1970). SDS and mercaptoethanol were omitted from the sample and all solutions used for electrophoresis. Electrophoresis, staining, destaining and autoradiography were carried out as previously described (Rubin et al., *1979). In the autoradiogram shown lanes 1 and 2 received purified brain RII; lanes 3 and 4, heart RII; lane 5 received brain RII that had been incubated with ^{125}I-calmodulin in the presence of 1 mM EGTA and the absence of calcium; lane 6 received heart RII that had been incubated with ^{125}I-calmodulin in the presence of 1 mM EGTA and the absence of calcium. The upper half on the gel is shown - the lower portion of the gel contained only a very intense band of free ^{125}I-calmodulin at the front.*

while different procedures result in the binding of 30% of protein kinase II to microtubules via a specific interaction with MAP 2 (Theurkauf and Vallee, 1982). We have recently characterized a second calmodulin binding protein that is very tightly associated with brain RII.

Highly purified preparations of cytosolic brain RII obtained by DEAE-cellulose chromatography, gel filtration on ACA-34 and affinity chromatography on 8-(6-)-aminohexyl amino cAMP-Sepharose contain a contaminating polypeptide with a M_r of 75,000. This protein is not removed by prolonged washing of the affinity column with 2M NaCl, 1% Triton X-100, 10mM 5'-AMP, 10 mM adenosine or combinations of these agents, or by subsequent chromatographic procedures.

When highly purified brain and heart RII preparations were pre-incubated with calcium and ^{125}I-calmodulin and then were subjected to polyacrylamide gel electrophoresis under nondenaturing conditions a major calmodulin binding component was observed in the brain RII sample (Figure 10, lanes 1 and 2). Heart RII exhibited no calmodulin binding activity (Figure 10 lanes 3 and 4). Furthermone, the binding of calmodulin to the brain RII sample was calcium dependent (Figure 10 lanes 5 and 6). The calmodulin binding protein had a sedimentation coefficient of 9S (Figure 11), indicating that it was considerably larger than free RII (5S) and that it might be a complex of the 75,000 dalton polypeptide(s) and RII dimer.

This possibility was investigated with two specific MAbs. During the selection of hybridomas secreting anti-brain RII MAbs several IgG-producing "negative" clones were isolated. These were screened for reactivity with the 75,000 dalton protein by the Western blot technique (Figure 12). MAb-918

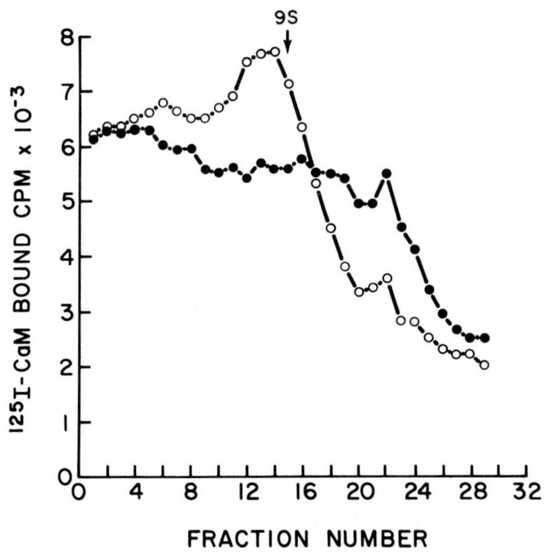

FIGURE 11. Sedimentation of the calmodulin binding component associated with brain RII. A sample of highly purified brain RII (15 μg protein) was sedimented through a glycerol gradient (10 - 40%) containing ^{125}I-CaM (1.2 x 10^5 cpm/ml) as described by Richman and Klee (1978). Radioactivity in fractions from the experimental samples (○-○) and control gradients (●-●) that received no RII was determined in a gamma counter. Fumarase (8.9S), LDH (7.4S) and hemoglobin (4.5S) were included as standards for the estimation of the sedimentation coefficient.

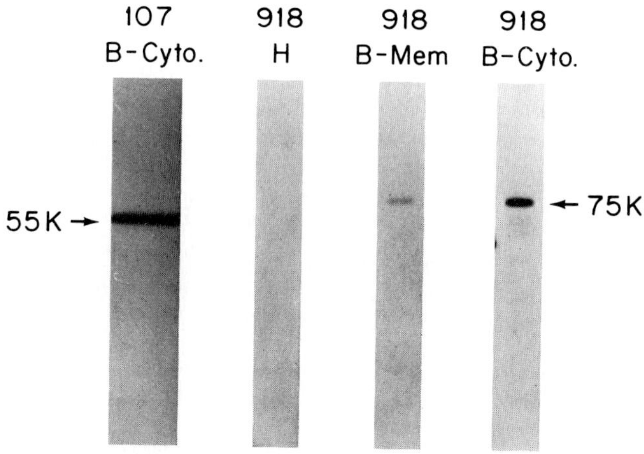

FIGURE 12. Identification of a monoclonal antibody that recognizes the 75,000 dalton protein associated with brain RII. Highly purified preparations (10 μg protein) of cytosolic and membrane-associated brain RII and heart RII were subjected to electrophoresis on 0.1% SDS-10% polyacrylamide gels and the immunoreactivity of the resolved polypeptides with MAb-918 and RII-specific MAb-107 were determined by the Western blot precedure as indicated in Figure 9 and Burnette (1981).

bound to the 75,000 dalton protein in the cytosolic brain RII preparation and identified a lower level of a similar protein in a purified, membrane-associated RII sample (Figure 12). This polypeptide was not observed in heart RII preparations. MAbs directed against brain RII (e.g. MAb-107, Figure 12) only complexed RII and were unreactive with the larger polypeptide. Western blot analyses on proteins transferred from non-denaturing gels revealed that brain RII migrates as 7 distinct species (Figure 13). In contrast, MAb 918 complexed only the two more slowly migrating species that correspond to ^{125}I-calmodulin binding proteins (Figure 13). These observations suggest that the calmodulin binding component in the brain RII preparation is a complex containing one RII dimer and one or two molecules of the 75,000 dalton protein.

The function of this calmodulin binding protein is not currently known. It does not appear to be a cAMP phosphodiesterase, a phosphatase or a protein kinase. It is phosphorylated by both cAMP-dependent and calcium-phospholipid activated protein kinases.

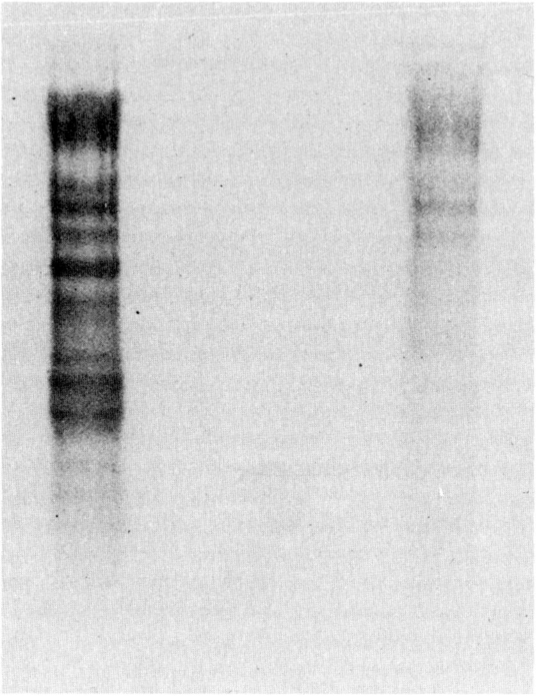

FIGURE 13. Identification of the proteins associated with calmodulin binding activity. Purified brain RII (10 μg protein) was subjected to electrophoresis on non-denaturing gels and Western Blot analyses as described in Figures 9 and 10 using anti-RII MAb 107 (left) and anti-75k MAb 918 (right).

On the basis of these observations and those of others it is possible to speculate that virtually all of the type II protein kinase in brain is associated either with specific structures such as synaptic plasma membranes (Maeno *et al.*, 1971; Rubin *et al.*, 1979), microtubules (Theurkauf and Vallee, 1982) and the Na channel (Costa *et al.*, 1982) important regulatory proteins such as calmodulin binding proteins (Hathaway *et al.*, 1981 and text). It will therefore be of interest to determine the nature of the specific binding sites on brain RII responsible for these interactions and to investigate the role such specific localizations play in regulating protein phosphorylation in brain.

REFERENCES

Bechtel, P.J., Beavo, J.A., and Krebs, E.G. (1977). *J. Biol. Chem. 252,* 2691.

Burnette, W.N. (1981). *Anal. Biochem. 112,* 195.

Corbin, J.D., and Keely, S.L. (1977). *J. Biol. Chem. 252,* 910.

Corbin, J.D., Keely, S.L., and Park, C.R. (1975). *J. Biol. Chem. 250,* 218.

Costa, M.R.C., Casnellie, J.E., and Catterall, W.A. (1982). *J. Biol. Chem. 257,* 7918.

Erlichman, J., Sarkar, D., Fleischer, N., and Rubin, C.S. (1980). *J. Biol. Chem. 255,* 8179.

Ey, P.L., Prowse, S.J., and Jenkin, C.R. (1978). *Immunochemistry 15,* 429.

Greengard, P. (1978). *Science 199,* 146.

Hathaway, D.R., Adelstein, R.S., and Klee, C.B. (1981). *J. Biol. Chem. 256,* 8183.

Hirs, C.H.W. (1967). *Methods Enzymol. 11,* 197.

Hofmann, F., Bectel, P.J., and Krebs, E.G. (1977). *J. Biol. Chem. 252,* 1441.

Kerlavage, A., and Taylor, S.S. (1982). *J. Biol. Chem. 257,* 1749.

Krebs, E.G., and Beavo, J.A., (1979). *Annu. Rev. Biochem. 48,* 923.

Kuo, J.F., and Greengard, P. (1969). *Proc. Nat. Acad. Sci. USA. 64,* 1349.

Laemmli, U.K. (1970). *Nature 227,* 680.

Lohmann, S.M., Walter, U., and Greengard, P. (1980). *J. Biol. Chem. 255,* 9985.

Maeno, H., Johnson, E.M., and Greengard, P. (1971). *J. Biol. Chem 246,* 134.

Mumby, M., and Beavo, J.A. (1981). *Cold Spring Harbor Conferences on Cell Proliferation 8,* 105.

Nimmo, H.G., and Cohen, P. (1977). *Adv. Cyclic Nucleotide Res. 8,* 145.

Pearse, A.G.E. (1968). *Proc. Royal Soc. London B 170,* 72.

Pearse, A.G.E. (1969). *J. Histochem. Cytochem. 17,* 303.

Pearse, A.G.E., and Takor-Takor, T., (1979). *Fed. Proc. 38,* 2288.

Richman, P.G., and Klee, C.B. (1978). *J. Biol. Chem. 252,* 6323.

Rubin, C.S., and Rosen, O.M. (1975). *Annu. Rev. Biochem. 44,* 831.

Rubin, C.S., Rangel-Aldao, R., Sarkar, D., Erlichman, J., and Fleischer, N. (1979). *J. Biol. Chem. 254,* 3797.

Rubin, C.S., Fleischer, N., Sarkar, D., and Erlichman, J. (1981). *Cold Spring Harbor Conferences on Cell Proliferation 8*, 1333.
Smyth, D.G. (1967). *Methods Enzymol. 11*, 214.
Takio, K. Smith, S.B., Krebs, E.G., Walsh, K.A., and Titani, K. (1982). *Proc. Nat. Acad. Sci. USA 79*, 2544.
Theurkauf, W.E., and Valee, R.B. (1982). *J. Biol. Chem. 257*, 3284.
Walsh, D.A., Perkins, J.P., and Krebs, E.G. (1968). *J. Biol. Chem. 243*, 3763.

DISCUSSION

Klee: Have you tested the neuroblastoma-glioma hybrid cells to see what type of regulatory subunit they have?

Rubin: We have not looked.

Klee: I thought there was some evidence from Greengard's laboratory that they contain essentially the type I subunit?

Rubin: Yes. Greengard's lab and Preshad in Houston have shown that if you treat neuroblastoma-glioma hybrids or the parenteral neuroblastoma cells with 8-Br-cAMP or hormones, you get about a 3-fold induction of R1. We have not dealt with that system for the following reason: those cells in the absence of any treatment have free R1, not holo-enzyme. They do not seem to be normal. The Friend cells contain fully associated PKI and PKII. They have a system where the hybrids are making lots of free R1 and they induce more, but that's already abnormal. For that reason we have not looked at those cells. One interesting point along those lines is that when we differentiate the Friend cells, the expression of C and R is perfectly coordinate. When we treat them with 8-Br-cAMP, that coordination is completely eliminated and we only induce RII. C levels stay the same.

Klee: Has anyone been able to induce neuroblastoma-glioma hybrid cells to produce the type II subunit?

Rubin: Not as far as I know.

Klee: Of the two bands, the 55 and the 57K, which one binds calmodulin?

Rubin: 55K is the one that is specifically induced.

Cohen: We have been interested for several years in the idea that calmodulin could, in some situations, be a bifunctional protein interacting with two proteins simultaneously. This came out of some experiments in which we found that the calmodulin which is tightly bound to phosphorylase kinase can activate other calmodulin-dependent enzymes, even when it is not extracted from phosphorylase kinase. I was rather impressed by the interaction of the brain kinase with other calmodulin-binding proteins. Do you think there is a possibility that some of these could represent a similar situation - that calmodulin binds to the RII, and this targets it to a variety of calmodulin binding proteins, a type of double-headed interaction system? Would you or Dr. Klee like to comment on that?

Rubin: I think it's a great idea, but we haven't done any work on that really.

Klee: I think there is really no direct evidence for a termany complex. We did try to detect a complex of phosphorylase kinase with phosphodiesterase at high concentrations of phosphodiesterase and were never able to detect it.

Cohen: Negative results never mean anything! What is the evidence that the 55K and the 57K proteins are really different? Why couldn't the 55 be a differentially phosphorylated form of the 55K? The autophosphorylation site is not the only site on the RII subunit which is phosphorylated, and the 57K could be phosphorylated on sites that are absent in the 55K proteins. This could explain your difference in mobility.

Rubin: That is certainly a possibility but in the absence of phosphorylation they are both 55K. It is this one additional autophosphorylation that shifts it, so I don't think it is due to other phosphorylation sites to begin with. Secondly, if the 57K in brain is the heart-type enzyme (for the sake of simplicity), then it is clear from Dr. Klee's work, for example, that the heart-type enzyme R subunit does not interact with calmodulin or calcineurin. The functional properties, the binding properties, are clearly different. Immunologically we have monoclonal antibodies that seem to recognize only the 55K and not the 57K, so I think there is a reasonable amount of evidence that they are in fact different.

Cohen: Can you cleave the brain enzyme like the heart enzyme into 37K and 20K proteins by limited proteolysis. If so, do either of those fragments bind calmodulin by the overlay technique?

Rubin: You can make those cleavages and get fragments that are approximately those sizes. I don't know the answer to the second question; it will have to be done.

Shenolikar (U. of Texas, Houston): You say that fat cells also produce the brain-type enzyme. Have you looked at differentiation of pre-adipocytes?

Rubin: Yes we have. The reason it came as a great shock to us is that we looked in the 3T3L1 system, on which we have worked, and to our surprise the brain antibodies work very well and it is the neural-type enzyme. We went back to animals and demonstrated it in animals as well. It seems to be that what we call the neural-type enzyme also occurs in fat cells. Now in terms of differentiation, there is a report in the literature which says (and this is only in reference to the PKI, PKII story) that there is a major change in protein kinase I during differentiation and that this may be related to the ability of those cells to become sensitive to lipolytic hormones. We have not been able to reproduce those results in the 3T3L1 system. We see that both kinases are present in the undifferentiated state and the differentiated state. Their relative proportions change, but they are there in both cases and nothing spectacular happens.

PHOSPHOLIPID TURNOVER AS TRANSMEMBRANE SIGNALLING FOR PROTEIN PHOSPHORYLATION AND PLATELET ACTIVATION[1]

Kimihiko Sano
Kozo Kaibuchi
Masahiko Hoshijima
Junji Yamanishi
Ushio Kikkawa
Yoshimi Takai
Yasutomi Nishizuka[2]

Department of Biochemistry
Kobe University School of Medicine
Kobe, Japan
and
Department of Cell Biology
National Institute for Basic Biology
Okazaki, Japan

SUMMARY

When platelets are stimulated by either thrombin, collagen or platelet-activating factor, phosphatidylinositol is rapidly broken down, and diacylglycerol is produced transiently. This diacylglycerol initiates the selective activa-

[1]*This investigation has been supported in part by research grants from the Scientific Research Fund of the Ministry of Education, Science and Culture (1982), the Intractable Diseases Division, Public Health Bureau, the Ministry of Health and Welfare (1981-1982), by a Grant-in-Aid of New Drug Development from the Ministry of Health and Welfare (1979-1982), and by the Yamanouchi Foundation for Research on Metabolic Disorders (1982) and the Mitsuhisa Cancer Research Foundation (1981).*

[2]*The paper was presented by Dr. Nishizuka*

ISBN 0-12-387560-9

tion of a Ca^{2+}-activated, phospholipid-dependent protein kinase. In *in vivo* as well as *in vitro* systems the protein kinase preferentially phosphorylates an endogenous platelet protein having an approximate molecular weight of 40,000 (40K protein). Under appropriate conditions synthetic diacylglycerol when directly added to intact platelets may also induce this protein phosphorylation. It is shown that the phosphorylation of 40K protein while a prerequisite is not the only requirement for the platelet activation measured by the release of serotonin. The simultaneous addition of diacylglycerol and a low concentration of ionophore A23187 results in the efficient release of serotonin, suggesting that the receptor-linked protein phosphorylation and Ca^{2+} mobilization are synergistically required to cause a full physiological cellular response. It is found that cyclic AMP and cyclic GMP do not antagonize each other but both prevent the activation of this protein kinase by inhibiting the signal-dependent phosphatidylinositol breakdown.

INTRODUCTION

Various hormones and neurotransmitters exhibit their physiological effects by interaction with their specific cell surface receptors. Such signal-receptor interaction induces a series of structural and biochemical changes in membranes. It is well known that α-adrenergic and muscarinic cholinergic stimulators, peptide hormones, and many other biologically active substances provoke phosphatidylinositol turnover in their target tissues (for a review, see Michell, 1975). In addition, the stimulation of these receptors generally accompanies immediate mobilization of Ca^{2+} and elevation of cyclic GMP but not of cyclic AMP. The receptor-linked breakdown of phosphatidylinositol is catalyzed by phospholipase C, and the primary products of this reaction are identified as diacylglycerol and inositol phosphate (partly recovered as inositol 1,2-cyclic phosphate) (Dawson *et al.*, 1971). Recent studies in this laboratory (Takai *et al.*, 1979; Kishimoto *et al.*, 1980) have clarified that diacylglycerol plays an essential role in the transmembrane control of cellular events through the activation of a novel protein kinase. The protein kinase absolutely requires Ca^{2+} and phospholipid for its enzymatic activity, and is specifically activated by diacylglycerol that is produced in a signal-dependent manner. The present article will briefly describe a role of this protein kinase in controlling platelet activation and the serotonin release reaction. The inhibition of phosphatidylinositol breakdown by both cyclic AMP and cyclic GMP will also be described. This inhibition constitutes a feedback mechanism that prevents this signal transduction. The protein kinase will be referred to tentatively as C-kinase, and the cyclic AMP-dependent and cyclic GMP-dependent protein kinases will be designated as A-kinase and G-kinase, respectively.

PATHWAY OF SIGNAL TRANSDUCTION

The principal pathway of the signal transduction is schematically outlined in Figure 1. Although C-kinase absolutely requires both Ca^{2+} and phospholipid for catalytic activity, the enzyme is solely dependent on diacylglycerol for its activation at physiologically low concentrations of Ca^{2+}. Kinetic analysis indicates that in the presence of a small quantity of diacylglycerol the apparent affinity of this enzyme for Ca^{2+} as well as for phospholipid is dramatically increased, and the enzyme is fully active without a net increase in the Ca^{2+} concentration (Kishimoto *et al.*, 1980; Kaibuchi *et al.*, 1981). The diacylglycerol active in this role contains at least one unsaturated fatty acid such as arachidonate. It may be recalled that phosphatidylinositol

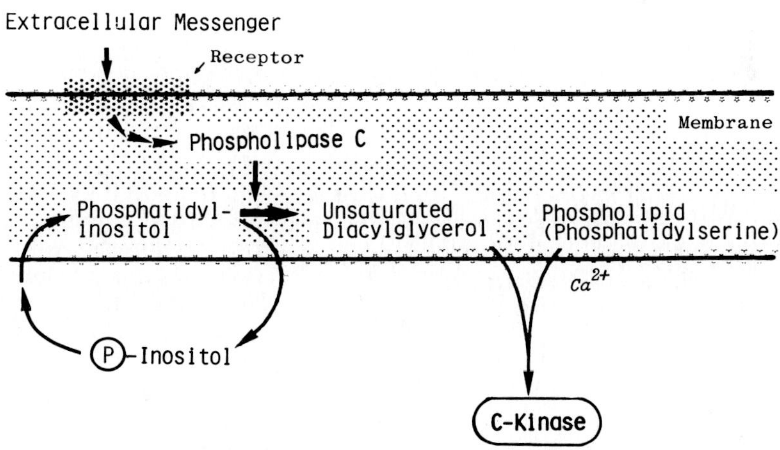

FIGURE 1. A proposed mechanism of signal transduction. (Adapted from Takai et al., *1981b).*

normally present in mammalian tissues contains predominantly 1-stearoyl-2-arachidonyl fatty acid on the *sn*-glycerol backbone (Holub *et al.*, 1970). Among various phospholipids present in membranes, phosphatidylserine is indispensable, but other phospholipids show positive or negative cooperativity for this enzyme activation (Kaibuchi *et al.*, 1981). For instance, phosphatidylethanolamine further stimulates the enzyme, and the full enzymatic activity is obtained at the 10^{-7} *M* range of Ca^{2+}. On the other hand, phosphatidylcholine and sphingomyelin are both inhibitory. It is likely, therefore, that the asymmetric distribution of various phospholipids in the lipid bilayer appears to be favorable for the activation of C-kinase. No other divalent cations can substitute for Ca^{2+}, except Sr^{2+} which is less than 10% as active as Ca^{2+}.

C-Kinase shows higher specific activity in platelets than in any other tissues and/or organs so far examined (Minakuchi *et al.*, 1981). This enzyme appears to lack tissue and species specificity in its physical and kinetic properties. The enzyme has been recently purified to homogeneity from rat brain (Kikkawa *et al.*, 1982). The molecular weight is roughly 77,000 as estimated from its sedimentation coefficient, and no subunit structure is demonstrated. Some of the physical and kinetic properties of C-kinase are summarized in Figure 2. Its polypeptide chain appears to be composed of two functionally different domains. One is a hydrophobic domain which may bind to membranes presumably as directed by extracellular messengers, whereas the other is a hydrophilic domain which carries the catalytically active center.

In *in vitro* systems C-kinase shows a broad substrate specificity. It has been shown that C-kinase and A-kinase often use the same phosphate acceptor proteins. For example, both kinases phosphorylate seryl and threonyl residues in histone, protamine, myelin basic proteins, microtubule-associated proteins, and membrane-bound proteins to variable extents. However, analysis of the sites in these proteins which are phosphorylated *in vitro* suggests that these protein kinases show distinctly different catalytic properties (Iwasa *et al.*, 1980). C-Kinase phosphorylates itself, but the significance of this autophosphorylation is not known. C-Kinase does not react with its own tyrosyl residues nor those in any substrate proteins tested.

RECEPTOR-LINKED ACTIVATION OF PROTEIN PHOSPHORYLATION

Platelets are frequently used as a model system for hormone action, since many extracellular messengers such as

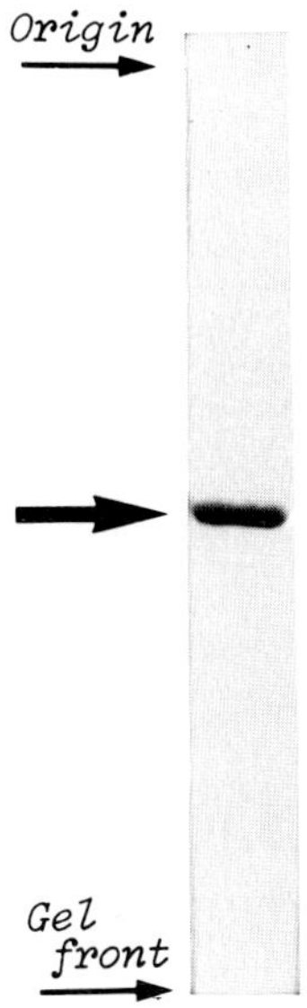

Molecular weight	77,000
Stokes radius	42 Å
S value	5.1 S
Isoelectric point	pH 5.6
Optimum pH	7.5 ∿ 8.0
Optimum Mg^{2+}	5 mM
Nucleotide specificity	ATP
Activators	Ka value
Phosphatidylserine	36 μg/ml
Diacylglycerol	0.5 μg/ml
Ca^{2+}	4×10^{-7}M

FIGURE 2. Physical and kinetic properties of C-kinase. A single polypeptide band indicates a homogeneous preparation of C-kinase purified from rat brain (Kikkawa et al., *1982). The physical and kinetic parameters have been obtained under conditions described elsewhere (Takai* et al., *1979; Kishimoto* et al., *1980).*

thrombin, collagen and platelet-activating factor (PAF) induce platelet aggregation and release of dense body constituents such as ADP and serotonin. When platelets are stimulated, two endogenous proteins with approximate molecular weights of 40,000 (40K protein) and 20,000 (20K protein) are selectively and heavily phosphorylated, and these phosphorylation reactions always accompany the release of serotonin (Lyons *et al.*, 1975; Haslam and Lynham, 1977). 20K protein is myosin light chain, and a calmodulin-dependent specific protein kinase has been proposed to catalyze this reation (Hathaway and Adelstein, 1979). A series of recent studies in this laboratory has shown that the C-kinase outlined above is responsible for 40K protein phosphorylation as schematically shown in Figure 3 (Kawahara *et al.*, 1980; Ieyasu *et al.*, 1982; Sano *et al.*, 1983). 40K protein partially purified from human platelets serves as a preferable substrate

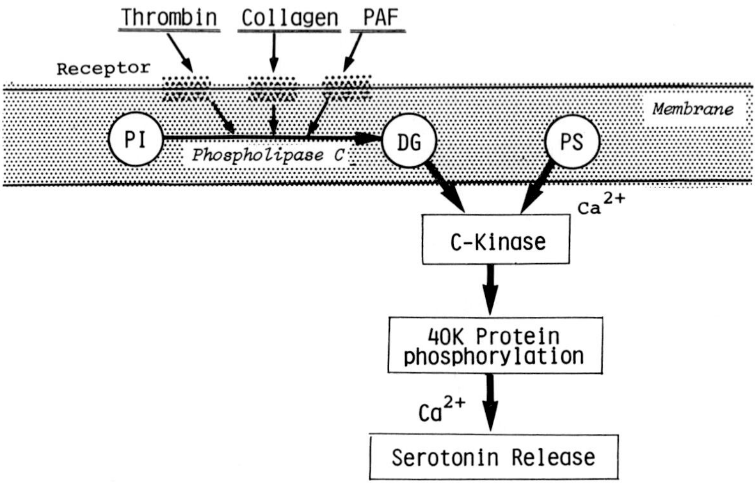

FIGURE 3. Schematic representation of platelet 40K protein phosphorylation and release reaction. PI, phosphatidylinositol; DG, diacylglycerol; and PS, phosphatidylserine.

for the homogeneous preparation of C-kinase, and the reaction is absolutely dependent on the simultaneous presence of Ca^{2+}, phospholipid and diacylglycerol (Sano *et al.*, 1983). The fingerprints of tryptic phosphopeptides prepared from radioactive 40K protein samples, which are phosphorylated in such a purified cell-free system and in intact platelets stimulated by thrombin, collagen and PAF, are all identical (Ieyasu *et al.*, 1982; Sano *et al.*, 1983).

Another set of experiments given in Figure 4 reveals that, upon stimulation of platelets by thrombin or collagen, diacylglycerol is rapidly produced, and that the reaction is immediately followed by 40K protein phosphorylation and serotonin release. Quantitative analysis of various phospholipids indicates that this diacylglycerol is derived most likely from phosphatidylinositol as proposed by Rittenhouse-Simmons (1979) and Bell and Majerus (1980). Essentially similar results have been obtained with PAF instead of thrombin and

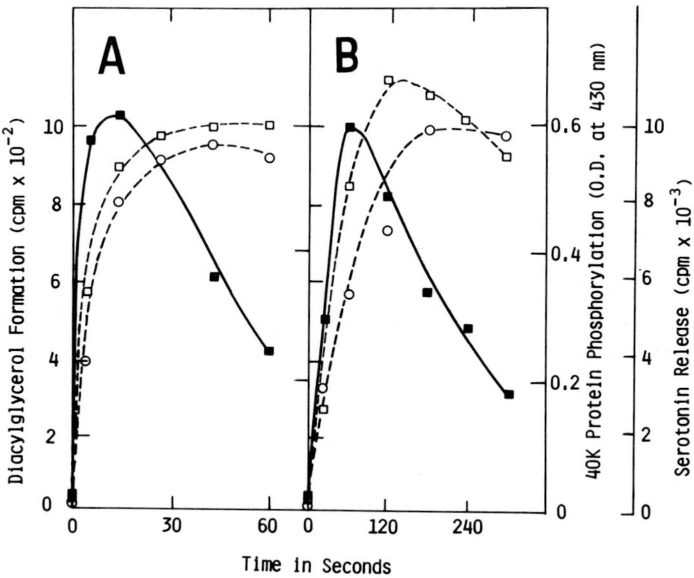

FIGURE 4. Activation of platelets by thrombin and collagen. Human platelets, which were labelled with either [^3H] arachidonic acid, ^{32}Pi or [^{14}C] serotonin, were stimulated by thrombin (0.05 unit/10^8 platelets) or collagen (50 µg/ml) at 37°C for various periods of time as indicated. Diacylglycerol formation, 40K protein phosphorylation and serotonin release were measured as described elsewhere (Sano et al., *1983). A, stimulated by thrombin; and B, stimulated by collagen. ■——■, diacylglycerol formation; □-------□, 40K protein phosphorylation; and ○-----○, serotonin release.*

collagen (Ieyasu *et al.*, 1982). In fact, it is shown that these extracellular messengers greatly enhance the incorporation of radioactive phosphate into phosphatidate as well as into phosphatidylinositol as a result of the enhanced turnover of this particular phospholipid. The diacylglycerol once produced in this way disappears very rapidly, and the amount of it accumulated is usually less than the amount of phosphatidylinositol expended. This poor stoichiometry and rapid disappearance of diacylglycerol are probably due to its conversion to phosphatidate through the action of diacylglycerol kinase, and also due to its further degradation to arachidonate for thromboxane synthesis. Under all conditions so far tested, the receptor-linked formation of diacylglycerol is always associated with 40K protein phosphorylation as well as serotonin release.

Diacylglycerol is normally almost absent from membranes, but is transiently produced from phosphatidylinositol in the signal-dependent manner described above. However, some synthetic diacylglycerols, which are capable of activating C-kinase *in vitro*, directly cause the rapid phosphorylation of 40K protein in intact platelets when exogenously added. For instance, in confirmation of the previous observation (Mori *et al.*, 1982), Figure 5 shows that 1-oleoyl-2-acetyl-glycerol is effective to activate C-kinase in the purified cell-free

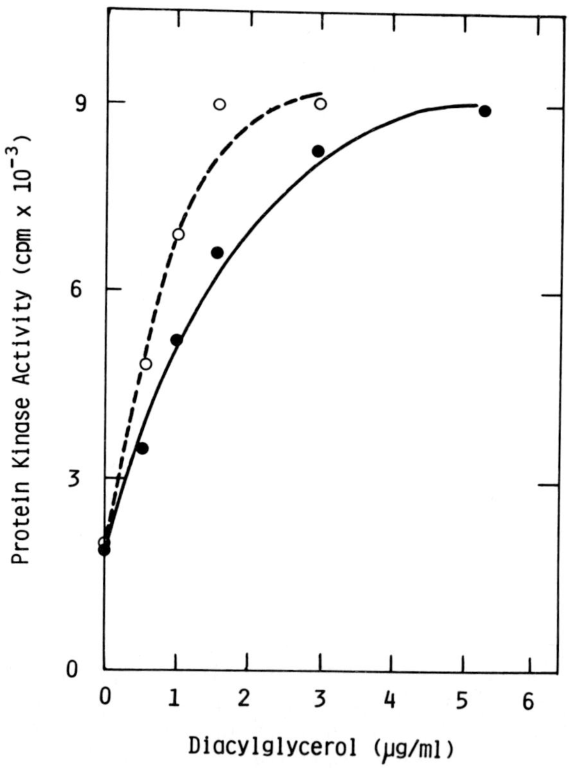

FIGURE 5. Activation of C-kinase by synthetic diacylglycerol in a purified cell-free system. A homogenous preparation of C-kinase was assayed at 4 x 10^{-6} M $CaCl_2$ in the presence of various amounts of either diolein or 1-oleoyl-2-acetyl-glycerol and fixed amounts of phospholipid. Other detailed conditions were described elsewhere (Mori et al.*, 1982). ●——●, 1-oleoyl-2-acetyl-glycerol; and ○----○, diolein.*

system. In this experiment the synthetic diacylglycerol is mixed first with phospholipid in a small volume of chloroform. After removal of this solvent under nitrogen, the lipid mixture has been suspended by sonication in a buffer medium and added to the reaction mixture for enzyme assay. On the other hand, in the experiment with intact platelets the synthetic diacylglycerol has been suspended by sonication in 1% dimethylsulfoxide in an aqueous buffer solution, and directly added to the platelet suspension. It is shown that this diacylglycerol induces rapid phosphorylation of 40K protein in the total absence of phosphatidylinositol breakdown and endogenous diacylglycerol formation (see below). Dimethylsulfoxide alone is rather inhibitory, and there is no indication that this effect of exogenous diacylglycerol is simply due to the damage of cell membranes. The radioactive 40K protein which is phosphorylated in this way has been isolated from platelets and subjected to tryptic digestion, followed by fingerprint analysis. The results indicate that C-kinase is indeed responsible for this protein phosphorylation. Thus, it is suggestive that under appropriate conditions, this synthetic diacylglycerol intercalates into the phospholipid bilayer and activates C-kinase directly. This assumption has been supported in part by the fact that the exogenously added synthetic diacylglycerol is rapidly converted *in situ* to the corresponding phosphatidate, that is 1-oleoyl-2-acetyl-3-phosphoglycerol, presumably through the action of diacylglycerol kinase present in the cytoplasmic side of the membrane as schematically shown in Figure 6.

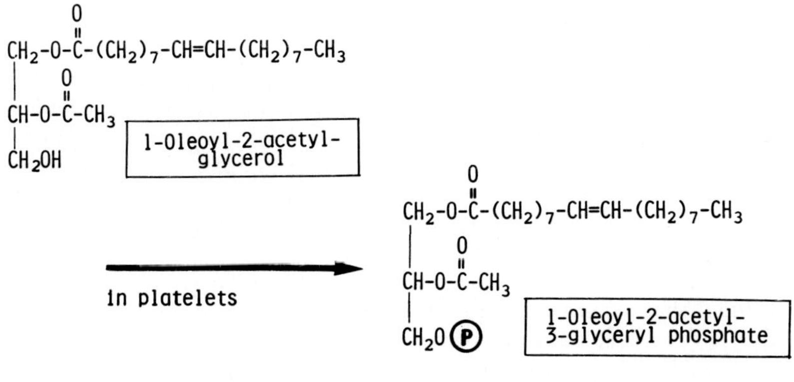

FIGURE 6. Structure of an active synthetic diacylglycerol and its conversion to phosphatidate in intact platelets.

This unique phosphatidate has been isolated by thin layer chromatography, and identified as such by comparison with an authentic sample of the synthetic product. In intact platelets diacylglycerols possessing two long fatty acyl moieties such as diolein are far less effective in the promotion of 40K protein phosphorylation presumably due to their steric properties or for some other reason.

Based on the experimental results mentioned above it may be concluded that, at least in platelets, C-kinase is *ipso facto* activated by extracellular messengers, and that diacylglycerol derived from the receptor-linked breakdown of phosphatidylinositol serves as a signal for the transmembrane control of protein phosphorylation. At present, however, the precise role of 40K protein in the platelet activation remains to be clarified.

SYNERGISTIC ACTIONS OF CALCIUM AND PROTEIN PHOSPHORYLATION

It is plausible that the interaction of natural extracellular messengers such as thrombin, collagen and PAF with their specific platelet receptors immediately mobilizes Ca^{2+} and provokes phosphatidylinositol breakdown. Although the activation of C-kinase absolutely requires both Ca^{2+} and diacylglycerol, extracellular messenger as well as synthetic diacylglycerol directly added to intact platelets causes the full activation of this enzyme in the absence of Ca^{2+} in the medium. This is presumably due to the fact that a small quantity of diacylglycerol dramatically increases the affinity of this enzyme for Ca^{2+} to the 10^{-7} *M* range, and renders the enzyme fully active without a net increase in the cytosolic Ca^{2+} concentration as mentioned above. However, the activation of this protein kinase appears to be a prerequisite but not a complete requirement for the release reaction, and the addition of synthetic diacylglycerol alone does not result in the full response of platelets to cause release of serotonin. As schematically shown in Figure 7, it is theoretically possible to induce Ca^{2+} mobilization and 40K protein phosphorylation independently by the addition of Ca^{2+} ionophore and synthetic diacylglycerol, respectively. In the experiment given in Figure 8, the radioactive platelets are incubated with 1-oleoyl-2-acetyl-glycerol in the presence or absence of a low concentration of ionophore A23187. It is shown that the release of serotonin is markedly enhanced by the addition of ionophore A23187, and that the full physiological response may be observed in the presence of both

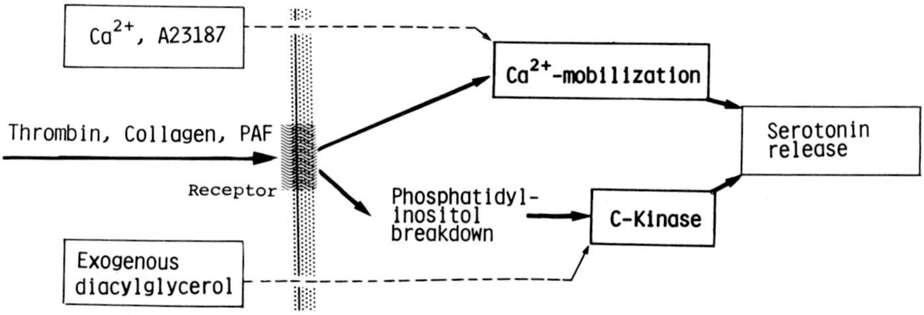

FIGURE 7. The independent but synergistic actions of Ca^{2+} and protein phosphorylation for platelet activation.

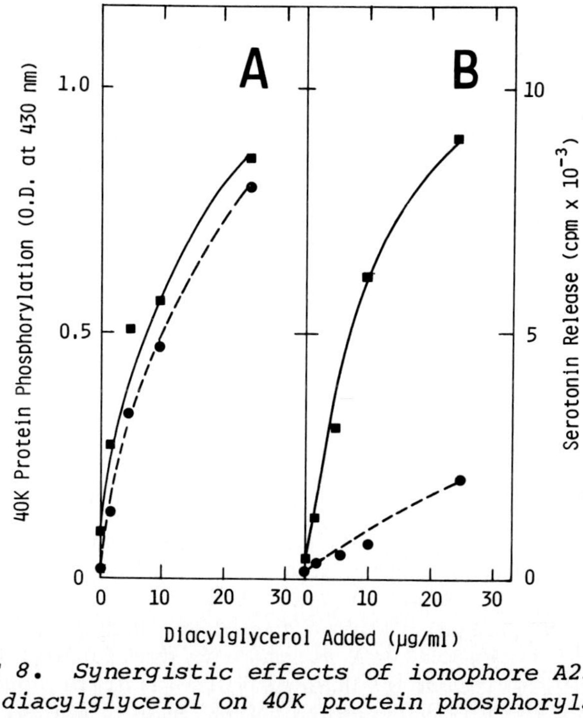

FIGURE 8. Synergistic effects of ionophore A23187 and synthetic diacylglycerol on 40K protein phosphorylation and serotonin release. The radioactive human platelets were stimulated by various concentrations of 1-oleoyl-2-acetyl-glycerol in the presence or absence of 0.4 μM ionophore A23187 as indicated. 40K protein phosphorylation and serotonin release were determined as specified elsewhere (Kawahara et al.*, 1980). A, 40K protein phosphorylation; and B, serotonin release. ■——■, in the presence of synthetic diacylglycerol and ionophore A23187 (0.4 μM); and ●----●, in the presence of synthetic diacylglycerol alone. (Adapted from Kaibuchi* et al.*, 1982).*

diacylglycerol and ionophore A23187. In contrast, 40K protein phosphorylation is not significantly affected by the same treatment, and this protein is almost fully phosphorylated by the exogenous addition of diacylglycerol alone, just as it is by natural extracellular messengers such as thrombin, collagen and PAF. The Ca^{2+} ionophore alone does not induce endogenous diacylglycerol formation and 40K protein phosphorylation, nor cause release of serotonin at the concentration employed in the present experiment. However, ionophore at more than the micromolar range appears to be toxic and causes by itself both 40K protein phosphorylation and serotonin release. These effects are probably due to enhancement of non-specific degradation of phospholipid, and/or to activation of C-kinase in the absence of diacylglycerol by a large increase of Ca^{2+} (Takai *et al.*, 1979; Kishimoto *et al.*, 1980; Kaibuchi *et al.*, 1981). Likewise, synthetic diacylglycerol alone at more than 0.1 mg/ml causes release of a significant amount of serotonin. The exact reason for this enhanced serotonin release is not known, but it is possible that diacylglycerol or phosphatidate therefrom may act as a promoter of membrane fusion or a Ca^{2+} ionophore (Allan and Michell, 1975; Tyson *et al.*, 1976). Perhaps, Ca^{2+} may play diverse roles in this physiological process, but the results briefly presented above seem to suggest that both Ca^{2+} mobilization and protein phosphorylation are indispensable for full cellular response to extracellular messengers.

INHIBITORY ACTION OF CYCLIC NUCLEOTIDES

Although in receptor mechanisms there should be a dramatic variation and heterogeneity from tissue to tissue, most tissues seem to possess at least two major classes of receptors for controlling cellular functions and proliferation. One class of receptors are related to cyclic AMP, and the other generally induce phosphatidylinositol turnover as well as immediate Ca^{2+} mobilization as mentioned above. In addition, the stimulation of the latter class of receptors often produces cyclic GMP (for a review, see Michell, 1975). In *bidirectional control systems* for most tissues such as platelets, the receptors that induce phosphatidylinositol turnover normally promote activation of cellular functions, whereas the receptors that produce cyclic AMP usually antagonize such activation. With human platelets it has been shown that phosphatidylinositol breakdown, diacylglycerol formation, 40K protein phosphorylation, and serotonin release are all inhibited in a parallel manner by prostaglandin E1 as well as by dibutyryl cyclic AMP (Takai *et al.*, 1982a,b). A series of analyses has indicated that cyclic AMP profoundly inhibits phosphatidylinositol turnover presumably through the phosphorylation of a distinct protein having a molecular weight of about 50,000 (50K protein), and thereby counteracts C-kinase activation. The experiment given in Figure 9 shows that the phosphorylation of 40K protein and that of 50K protein are opposingly controlled by the two different receptors mentioned above.

Although cyclic GMP has been repeatedly shown to be increased by stimulation of receptors that induce phosphatidylinositol turnover, its definitive role in biological regulation is not yet well understood. It remains a puzzle that G-kinase shows very similar, if not identical, catalytic properties to those of A-kinase (Hashimoto *et al.*, 1976). Cyclic GMP has been suggested to act as a negative rather than a positive messenger in human platelets and to constitute an immediate feedback control that results in an over-

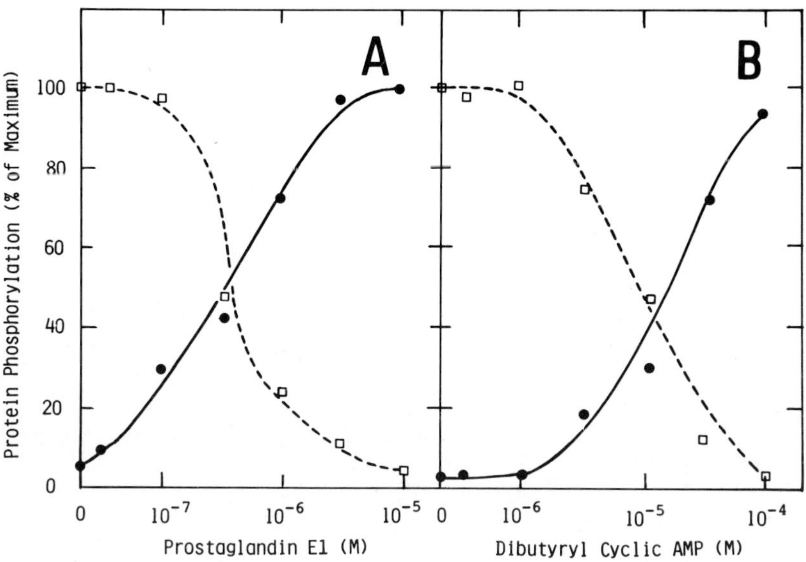

FIGURE 9. Opposing control of platelet 40K and 50K protein phosphorylation by thrombin and cyclic AMP. Detailed conditions are described elsewhere (Takai et al., *1982a,b).* ●——●, *50K protein phosphorylation; and* □----□, *40K protein phosphorylation.*

response (Haslam *et al.*, 1980; Takai *et al.*, 1981a). In In fact, either sodium nitroprusside or 8-bromo-cyclic GMP inhibit the thrombin-induced phosphatidylinositol breakdown, diacylglycerol formation, 40K protein phosphorylation and serotonin release in a parallel manner (Takai *et al.*, 1981a; 1982a,b). Again, the 50K protein mentioned above is inversely phosphorylated in the presence of increasing amounts of sodium nitroprusside or 8-bromo-cyclic GMP.

These results appear to be compatible with the supposition that cyclic AMP and cyclic GMP do not antagonize each other, but similarly inhibit the phospholipid degradation in the activated cell membrane. Although the mechanism of cyclic GMP accumulation is not clear at present, it is possible that arachidonate peroxide and prostaglandin endoperoxides may serve as activators for guanylate cyclase (Hidaka and Asano, 1977; Graff *et al.*, 1978). Such a possible receptor-linked phospholipid metabolic cascade for signal transduction is schematically given in Figure 10.

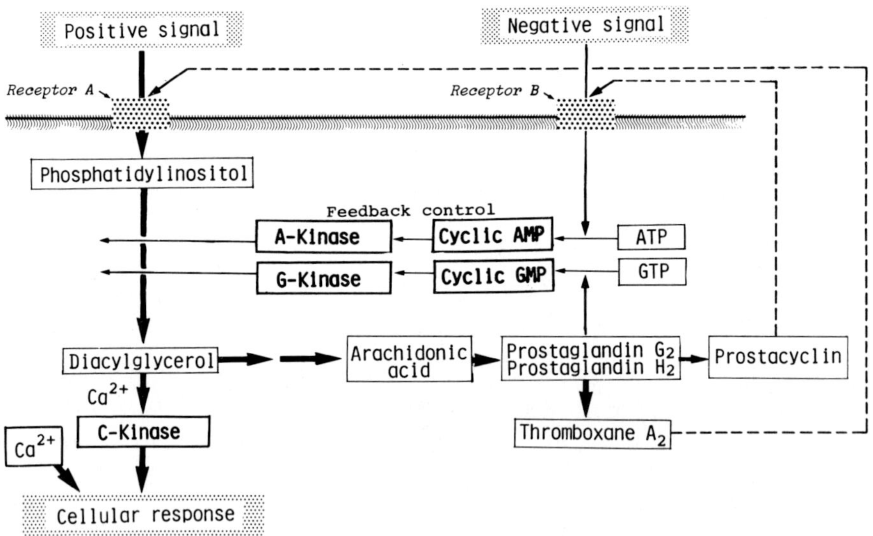

FIGURE 10. Receptor-linked phosphatidylinositol degradation in platelets. (Adapted from Takai et al.*, 1982a).*

CONCLUSION

Although the experimental support described in this article has resulted from studies with a limited number of a specific tissue, it is suggestive that C-kinase plays a role of crucial importance in signal transduction during platelet activation. Michell (1975) has postulated that phosphatidylinositol turnover may be involved in the Ca^{2+}- gate opening, but this attractive hypothesis is still a matter of discussion. Using platelets as a model system, the present studies have attempted to explore a possible relationship between Ca^{2+} mobilization and protein phosphorylation. The experimental results described above seem to suggest that under limited conditions either the increase in Ca^{2+} concentration or the receptor-linked activation of C-kinase alone is not sufficient to cause a physiological response such as release of serotonin. The evidence available at present is not conclusive, but suggests that protein phosphorylation and Ca^{2+} mobilization are independent events but synergistically involved in the transmembrane control of platelet functions.

The present article also shows that in platelets cyclic AMP and cyclic GMP do not antagonize each other, but both serve as potent inhibitors for the receptor-linked phosphatidylinositol breakdown presumably through the action of the respective protein kinases. Thus, it is attractive to suggest that these cyclic nucleotides may play integrated parts in the extracellular and intracellular circuits, leading to feedback control to prevent over-response. However, the molecular basis of their inhibitory actions is not clear at present.

ACKNOWLEDGEMENTS

The authors are grateful to Dr. T. Fujikura for the preparation of 1-oleoyl-2-acetyl-3-phosphoglycerol, and also to Mrs. S. Nishiyama and Miss K. Yamasaki for their skillful secretarial assistance.

REFERENCES

Allan, D., and Michell, R.H. (1975). *Nature 258*, 348.

Bell, R.L., and Majerus, P.W. (1980). *J. Biol. Chem. 255*, 1790.

Dawson, R.M.C., Freinkel, N., Jungalwala, F.B., and Clarke, N. (1971). *Biochem. J. 122*, 605.

Graff, G., Stephenson, J.H., Glass, D.B., Haddox, M.K., and Goldberg, N.D. (1978). *J. Biol. Chem. 253*, 7662.

Hashimoto, E., Takeda, M., Nishizuka, Y., Hamana, K., and Iwai, K. (1976). *J. Biol. Chem. 251*, 6287.

Haslam, R.J. and Lynham, J.A. (1977). *Biochem. Biophys. Res. Commun. 77*, 714.

Haslam, R.J., Salama, S.E., Fox, J.E.B., Lynham, J.A., and Davidson, M.M.L. (1980). *In* Platelets: "Cellular Response Mechanisms and Their Biological Significance" (A. Rotman et al., eds.), p. 213. John Wiley & Sons Ltd.

Hathaway, D.R. and Adelstein, R.S. (1979). *Proc. Natl. Acad. Sci. USA 76*, 1653.

Hidaka, H. and Asano, T. (1977). *Proc. Natl. Acad. Sci. USA 74*, 3657.

Holub, B.J., Kuksis, A., and Thompson, W. (1970). *J. Lipid Res. 11*, 558.

Ieyasu, H., Takai, Y., Kaibuchi, K., Sawamura, M., and Nishizuka, Y. (1982). *Biochem. Biophys. Res. Commun. 108*, 1701.

Iwasa, Y., Takai, Y., Kikkawa, U., and Nishizuka, Y. (1980). *Biochem. Biophys. Res. Commun. 96*, 180.

Kaibuchi, K., Takai, Y., and Nishizuka, Y. (1981). *J. Biol. Chem. 256*, 7146.

Kaibuchi, K., Sano, K., Hoshijima, M., Takai, Y., and Nishizuka, Y. (1982). *Cell Calcium 3*, 323.

Kawahara, Y., Takai, Y., Minakuchi, R., Sano, K., and Nishizuka, Y. (1980). *Biochem. Biophys. Res. Commun. 97*, 309.

Kikkawa, U., Takai, Y., Minakuchi, R., Inohara, S., and Nishizuka, Y. (1982). *J. Biol. Chem. 257*, 13341.

Kishimoto, A., Takai, Y., Mori, T., Kikkawa, U., and Nishizuka, Y. (1980). *J. Biol. Chem. 255*, 2273.

Lyons, R.M., Stanford, N., and Majerus, P.W. (1975). *J. Clin. Invest. 56*, 924.

Michell, R.H. (1975). *Biochim. Biophys. Acta 415*, 81.

Minakuchi, R., Takai, Y., Yu, B., and Nishizuka, Y. (1981). *J. Biochem. 89*, 1651.

Mori, T., Takai, Y., Yu, B., Takahashi, J., Nishizuka, Y., and Fujikura, T. (1982). *J. Biochem. 91*, 427.

Rittenhouse-Simmons, S. (1979). *J. Clin. Invest. 63*, 580.
Sano, K., Takai, Y., Yamanishi, J., and Nishizuka, Y. (1983). *J. Biol. Chem. 258*, 2101.
Takai, Y., Kishimoto, A., Iwasa, Y., Kawahara, Y., Mori, T., and Nishizuka, Y. (1979). *J. Biol. Chem. 254*, 3692.
Takai, Y., Kaibuchi, K., Matsubara, T., and Nishizuka, Y. (1981a). *Biochem. Biophys. Res. Commun. 101*, 61.
Takai, Y., Kishimoto, A., Kawahara, Y., Minakuchi, R., Sano, K., Kikkawa, U., Mori, T., Yu, B., Kaibuchi, K., and Nishizuka, Y. (1981b). *Adv. Cyclic Nucleotide Res. 14*, 301.
Takai, Y., Kaibuchi, K., Sano, K., and Nishizuka, Y. (1982a). *J. Biochem. 91*, 403.
Takai, Y., Minakuchi, R., Kikkawa, U., Sano, K., Kaibuchi, K., Yu, B., Matsubara, T., and Nishizuka, Y. (1982b). *In* "Progress in Brain Research, Characterization and Function of Brain Phosphoproteins" (W.H. Gispen and A. Routtenberg, eds.) 56, p. 287. Elsevier Biomedical Press, North-Holland.
Tyson, C.A., Zande, H.V., and Green, D.E. (1976). *J. Biol. Chem. 251*, 1326.

DISCUSSION

Carroll (OMRF): Platelet activation is also characterized by myosin light chain phosphorylation. Does your synthetic diacylglycerol dependent protein kinase catalyze the myosin light chain phosphorylation as well?

Nishizuka: When platelets are activated by either thrombin, collagen or platelet-activating factor, two proteins having molecular weights of 40K and 20K are rapidly phosphorylated. The 20K protein phosphorylation is catalyzed by myosin light chain kinase, and this reaction is easily demonstrated by the addition of calcium ionophore (A-23187). In fact, at a low concentration of A-23187, only 20K protein is phosphorylated, and 40K protein is not phosphorylated. In contrast, when platelets are activated by diacylglycerol, only 40K protein is phosphorylated and 20K protein is not. Therefore, 40K and 20K proteins may be used as very nice monitors for diacylglycerol production and calcium ion increase, respectively.

Carroll: Hirata has suggested that the 40,000 molecular weight protein, at least in neutrophils, may be a lipomodulin that negatively regulates phospholipases and that this inhibition is relieved by phosphorylation. Do you have any evidence as to whether that is true or not in your system?

Nishizuka: I could not comment on this point, but we feel that perhaps phospholipases are not rate-limiting. In most tissues, the activities of phospholipases C and phospholipase A_2 are apparently abundant in the cytoplasm. It is possible that the rate-limiting step is the structure or susceptibility of membrane phospholipids. But the right answer to your question is in the black box at this time, and we have virtually no experimental evidence to discuss the regulatory mechanisms of phospholipases.

Olson: How does the 40K protein phosphorylation relate to serotonin release?

Nishizuka: I do not know. It is an interesting but difficult question to answer.

Johnson: What is the interrelationship between thrombin and C kinase?

Nishizuka: How do thrombin and other extracellular messengers provoke the PI breakdown and calcium movement? This is still unknown.

Kahn: My question relates to the role that you are proposing for calcium in the activation. You have two slides which suggest that it is not playing a physiological role in the phosphorylation of 40K, but that its role is distal to that, possible helping to fuse membranes to release serotonin. Do you agree with that?

Nishizuka: It is possible. But, the role of calcium should be diverse.

Klee: Are the phorbol esters acting indirectly at the membrane level or directly at the level of the enzyme?

Nishizuka: Phorbol ester alone is unable to activate the enzyme. Perhaps it may be intercalated into the lipid bilayer, and thereby modify the hydrophobic interaction between membrane and enzyme.

Cohen: Did you look at all to see whether C kinase can be directly regulated by cyclic AMP or cyclic GMP dependent protein kinases?

Nishizuka: There is no such indication.

Cox (U. of Oklahoma): You are using 10^{-5} calcium in your *in vitro* experiments? How do you account for the fact that it doesn't seem to require any calcium *in vivo*? How do you account for it still being active in the cytosol *in vivo* without Ca^{2+}?

Nishizuka: In the presence of both phosphatidylserine and phosphatidylethanolamine, the enzyme is fully activated by diacylglycerol at 10^{-7} *M* calcium. In fact, diacylglycerol alone, when added exogenously, induced 40K protein phosphorylation *in vivo* without apparent increase in calcium. However, kinetic studies indicate the C-kinase may also be activated by a large increase in the calcium concentration without diacylglycerol. But, this may not be physiological.

Ochoa (Roche Inst.): Is there self phosphorylation of the C-kinase?

Nishizuka: Yes, mostly on the serine of the molecule.

HISTONE H1 PHOSPHORYLATION: PURIFICATION AND GENERAL PROPERTIES OF A cAMP INDEPENDENT HISTONE H1 KINASE[1]

C. Quirin-Stricker

Laboratoire de Génétique Moléculaire des Eucaryotes du CNRS
Faculté de Médecine
Institut de Chimie
Strasbourg, France

INTRODUCTION

Many proteins in general, and histones in particular, are subject to posttranslational modification. This includes acetylation of ε-NH_2 groups in lysine, methylation of arginine, poly-ADP ribosylation and phosphorylation. It seems increasingly more probable that protein phosphorylation could play an important role in the regulation of cell growth, proliferation and differentiation. The dominant role of chromatin organization in cell proliferation and differentiation is more and more often proposed (Nicolini, 1979). The *in vivo* phosphorylation of histone H1 has already been reported by several groups in rapidly dividing cells (Langan, 1974) and in cells grown in tissue culture (Bradbury *et al.*, 1973, 1974; Lake, 1973; Hohman *et al.*, 1976; Ajiro *et al.*, 1982). Several workers have emphasized a strong correlation between histone H1 phosphorylation and chromatin condensation (Inglis *et al.*, 1976; Gurley *et al.*, 1978; Krystal and Poccia, 1981; Matsumoto *et al.*, 1980). It is chiefly histone H1 phosphorylation that has been proposed as a trigger mechanism of chromatin condensation during mitosis (Bradbury *et al.*, 1973; Hohmann *et al.*, 1975; Matthews, 1979). It has also been

[1]*This work was supported by grants from the Délégation Générale à la Recherche Scientifique (77-7-1749) and I.N.S.E.R.M. (CRL 801.036).*

ISBN 0-12-387560-9

shown that *in vivo* histone H1 phosphorylation becomes significant in S-phase (2 PO_4/molecules) and culminates in mitosis where histone H1 has at least 4-6 PO_4/molecules (Gurley *et al.*, 1978; Ajiro *et al.*, 1981). As our purpose is to understand the function of the mitosis-specific histone H1 phosphorylation and how it is controlled, our main interest has been to isolate, purify, and determine the general characteristics of a histone H1 specific kinase.

Our previous work on the phosphorylation of eukaryotic initiation factors allowed us to study the protein kinases associated with the microsomes in mouse plasmacytomas, i.e. plasma cell tumors. Several years ago we found that one group of microsomal protein kinases appeared to be distinct from the other ribosomal and soluble protein kinases generally described (Schmitt *et al.*, 1977): it phosphorylates histone H1 to a very high degree. Therefore, we proceeded with further purification and characterization of this histone H1 kinase.

Isolation and Purification of a Specific Histone Hl Protein Kinase from Plasmacytoma Microsomes

We used the solid transplantable plasma cell tumor RPC5. The purification procedure employed provided a nearly homogeneous preparation of the enzyme. The procedure has been described (Quirin-Stricker *et al.*, 1981). Briefly, the enzyme was purified from a microsomal fraction by a high salt wash, ammonium sulfate precipitation, chromatography on DEAE-cellulose, hydroxy apatite, phosphocellulose, and Sephadex G-200 columns (Figure 1). A chromatography profile of the histone H1 protein kinase obtained after the last step of purification using gel filtration on a Sephadex G-200 column is shown in Figure 2. The profile corresponds to H1 kinase activity tested in the presence of calf thymus histone H1 and (γ-^{32}P) ATP. It shows a large predominant peak of H1 kinase activity eluting at about 90,000-120,000 daltons followed by 'shoulders' in the range of 50,000-68,000 daltons. Very often two small H1 kinase activity peaks of higher molecular weight precede the major peak. The G-200 filtration step largely eliminated the contaminating protein kinases which phosphorylate casein, protamine, histone H2 and ribosomal proteins. It is worth mentioning that problems encountered during the purification of the enzyme concerned the protection of the kinase from proteolytic activity, its tendency to aggregate, and its partial deactivation when it is in a highly purified state. The purified enzyme was termed 'PKH1'.

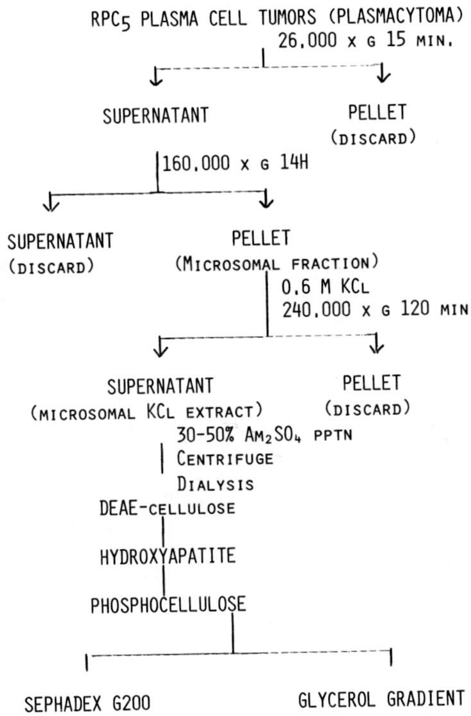

FIGURE 1. Preparation and Purification of Plasmacytoma Histone H1 Kinase.

General Properties of the Specific Histone H1 Kinase

All the properties of the enzyme were determined at pH 7.6 as the maximum activity of the enzyme was observed between 7-8. A summary of the main characteristics of microsomal histone H1 kinase are presented in Table I. Under our conditions, histone H1, phosphorylated *in vitro* to saturation with purified microsomal histone H1 kinase from plasmacytomas contains about 3 to 4 moles of phosphate per mole of histone H1. Langan (1978) found that histone H1 isolated from calf thymus or slow-growing tissues contained less than 0,2 mole of phosphate per mole of histone H1.

Kinetic Parameters. The apparent Km values for ATP and GTP are 58 μM and 1.4 μM respectively. It is worth mentioning that this kinase prefers GTP over ATP, a characteristic

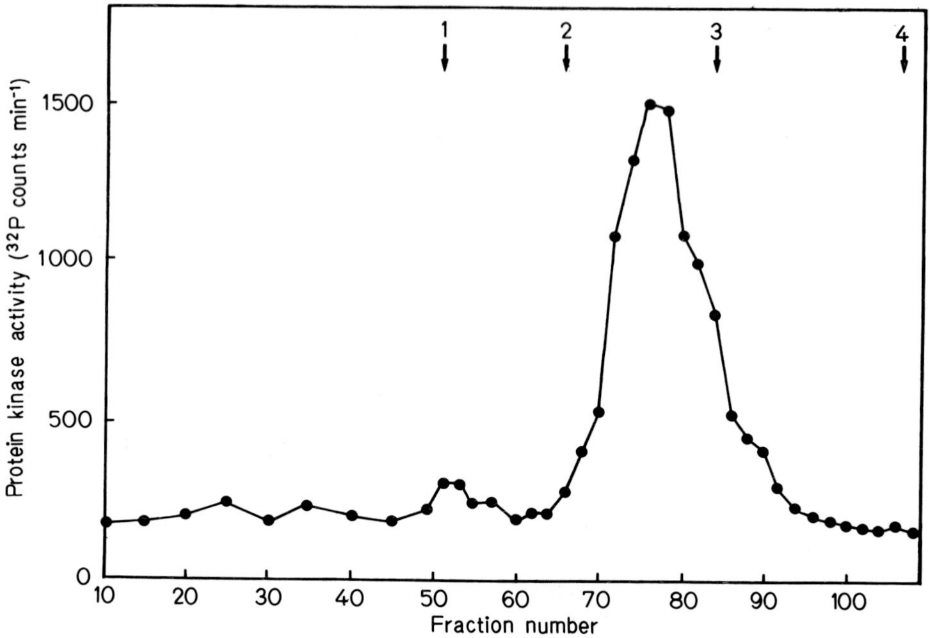

FIGURE 2. Gel filtration on Sephadex G-200 of microsomal histone H1 kinase obtained from phosphocellulose (Figure 1). Activity was assayed with calf thymus histone H1 as substrate and [$\gamma^{32}P$]GTP. Molecular weight standards used to calibrate the column are indicated by arrows: 1 = exclusion volume; 2 = IgG; 3 = serum albumin; 4 = myoglobin.

which is very different from the histone kinases generally described in the literature. The apparent Km for histone H1 is 0.59 μM (calculations of Km for histone H1 assume a molecular weight of 23,000) or 14 μg/ml histone H1. This indicates a high affinity of the enzyme for histone H1.

Metal Ions. Several bivalent ions were studied with regard to their effect on the activity of the histone H1 kinase. Only Mg^{2+} is highly active in the reaction. The optimum concentration for Mg^{2+} is 10-20 mM. Several other bivalent ions were tested at different concentrations in the presence or absence of Mg^{2+}. All other ions tested; Mn^{2+}, Ca^{2+}, Cu^{2+} and F^{2+}, Fe^{3+}; had an inhibitory effect on histone H1 kinase activity. A very interesting observation is that Cu^{2+} was the most inhibitory at low the lowest concentration (0.5 mM).

TABLE I. General Properties of Plasmacytoma "Histone H1 Kinase"

- *Optimum* Mg^{2+}		*10-20 mM*
- *Optimum pH*		*7,5*
- *Phosphate donors*		*ATP*
		GTP
	Km for ATP	*58 μM*
	Km for GTP	*1.4 μM*
	Km for Histone H1	*0.59 μM - 14 μg/ml*
	Mn^{2+} *5 mM*	*90% Inhibition*
	Ca^{2+} *0.5 mM*	*10% "*
- *Different*	*" 5 mM*	*57% "*
Effector	Cu^{2+} *0.5 mM*	*100% "*
Agents:	Fe^{2+} *2 mM*	*100% "*
	Fe^{3+} *2 mM*	*100% "*
	Phosphate 0.2 M	*90% "*
	NaCl 0.2 M	*20% "*
	KCl 0.3 M	*30% "*
	cAMP 1.10^{-5} - 1.10^{-7}*M*	*No stimulation*
	cGMP 2.10^{-5} - 2.10^{-7}*M*	*No stimulation*
- *Heat stable inhibitor protein of PKcAMP*		*No response*
- *Hemin 25 μM*		*20% Inhibition*
50 μM		*50% "*
- *dsRNA 0.1 - 0.4 μg/ml*		*No effect*
- *Heparine 0.5 μg/ml*		*No effect*
- *Spermidine 0.5 - 5 mM*		*No effect*

Histone H1 kinase appeared to be less sensitive towards salt, since high concentrations of NaCl or KCl (>0.3 M) decrease only slightly the activity of the kinase.

These results clearly show that this enzyme is distinct from the cytosolic, ribosomal and most of the nuclear kinase activities reported in the literature (Rubin and Rosen, 1975).

Effects of Cyclic Nucleotide. Histone H1 protein kinase is not dependent on cyclic nucleotides, as it is not stimulated either by cAMP (10^{-5} to -10^{-7} M) or by cGMP (10^{-5} to -10^{-7} M). Unlike the cytosolic and ribosomal cAMP-dependent protein kinases, histone H1 kinase is not inhibited by the heat-stable inhibitor of cAMP-dependent protein kinase.

TABLE II. Substrate Specificity of "Histone H1 Kinase"

Substrate added		*nmol/mg*
Calf thymus whole Histone		*12.1*
	Histone H1	*15.6-34*
	Histone H2b	*0.3*
	Histone H2a	*0*
	Histone H3	*0*
	Histone H4	*0*
Protamine		*0*
Casein mixture		*0*
	eIF_2	*0*
	eIF_3	*0*
	ribosomal proteins	*0*
	actin	*0*
	α-actinin	*0*
	desmin	*0*
	vimentin	*0*

Substrate Specificity. The ability of histone H1 kinase to phosphorylate various substrates was evaluated (Table II). With a purified preparation only histone H1 is phosphorylated. No phosphorylation was obtained with either H2A, H2B, H3, H4, protamine or casein. Eukaryotic initiation factors eIF2, eIF3, ribosomal proteins, as well as several major cytoskeletal proteins such as actin, α-actinin, desmin, and vimentin, were also tested as substrates. No phosphorylation of any of these substrates occurred in the presence of PKH1. These results were obtained by two different methods: by counting the total radioactivity of the substrates after the kinase assay, and by autoradiography of the substrates after electrophoresis on a polyacrylamide gel. The enzyme will however phosphorylate histone H5.

Approximate Molecular Weight

The native enzyme. When the histone H1 kinase was subjected to glycerol gradient centrifugation, it sedimented as several peaks of activity. By molecular sieving on Sephadex G-200 one major peak of histone H1 kinase activity was found

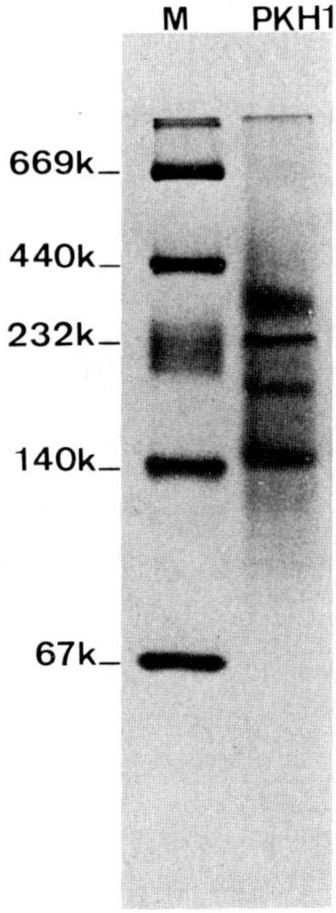

FIGURE 3. Polyacrylamide gel electrophoresis of purified histone H1 kinase under non-denaturing conditions. The numbers on the left indicate the molecular weight of the marker proteins.

in the range of molecular weight 90,000-120,000, with minor peaks of activity at higher and lower molecular weights. These results suggest heterogeneity of the enzyme and indicate that it exists as different molecular forms in our preparation. Further information was obtained by gel electrophoresis on a 5-20% gradient polyacrylamide slab gel under non-denaturing conditions. This showed 3 to 4 major distinct protein bands corresponding to approximate molecular weights of about 290,000, 240,000, 190,000 and 120,000 daltons (Figure 3).

Detection of histone H1 kinase activity on non-denaturing polyacrylamide gels. The assay for determining histone H1 kinase activity on non-denaturing gels was developed from the method described by Hirsch and Rosen (1974) for locating the kinase after electrophoresis. Radioactive histone H1 was present at the position occupied by the kinase and was detected by autoradiography.

One single narrow region of radioactivity was detected corresponding to a protein of about 240,000 daltons. Longer exposure time showed a wider zone of radioactivity.

Under denaturing conditions. The non-denaturing polyacrylamide gel was cut into slices and each protein band was eluted and analysed by electrophoresis on a gradient dodecylsulfate-polyacrylamide gel. All the bands either gave a single protein band of about 58,000 daltons (Figure 4), or a closely spaced doublet of about 56,000-58,000 daltons.

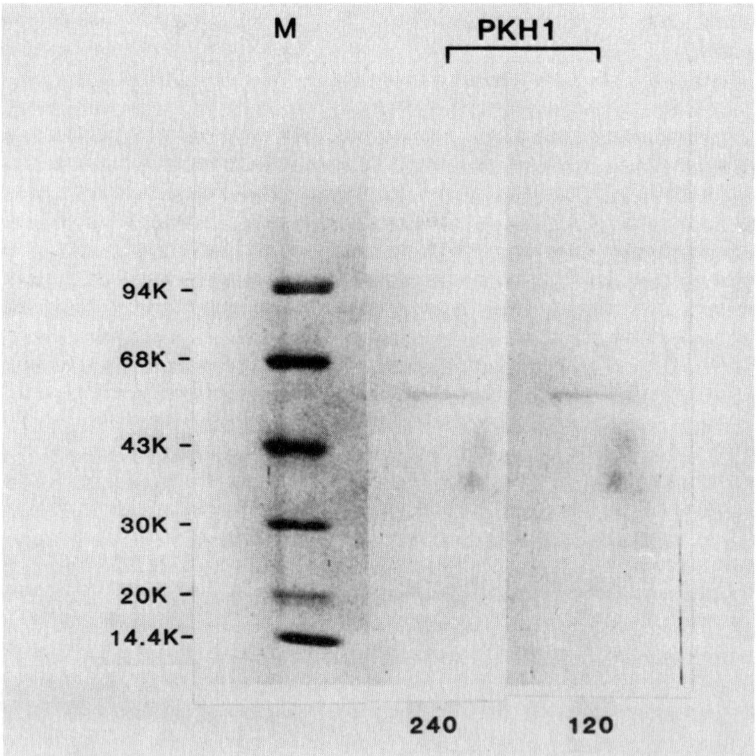

FIGURE 4. After elution of the protein bands 240 K and 120 K from non-denaturing gels, they were electrophoresed on a gradient polyacrylamide slab gel in the presence of sodium dodecylsulfate. The numbers at the bottom indicate the approximate molecular weight of bands 2 and 4 from the non-denaturing gel. The numbers on the left indicate the molecular weight of the marker proteins.

All these data indicate that the histone H1 kinase is an oligomeric enzyme formed probably by four identical subunits of about 58,000 daltons. The molecular weight of the native enzyme might be about 240,000 daltons. Higher molecular weights could correspond to different aggregated forms of the enzyme.

The tetramer form can easily be dissociated into trimers, dimers and monomers. These forms were probably due to incomplete dissociation in the presence of mercaptoethanol. We have shown that all forms, i.e. the tetramer, trimer, dimer and monomer form, display antigen function after analysis using the western blotting technique (to be published).

Determination of Phosphorylation Sites of Chicken Erythrocyte H5 by the Specific Histone H1 Kinase

Due to the close structural and functional relationship between histones H1 and H5, the histone H1 kinase was assayed with chicken erythrocyte histone H5. Sites of phosphorylation of histones H5 have been investigated *in vitro* using the cAMP independent histone H1 kinase from plasma cytoma cells. The sites of phosphorylation in histone H5 are distributed *in vitro* in two distinct regions of the protein: the amino-terminal region (residues 1-21) and mainly the carboxy-terminal region (residues 101-189). No site has been detected in the globular part of the molecule (residues 22-100). This fact has been already mentioned for histone H1 which, like histone H5, has three different domains in its molecule: a short apolar amino-terminal part in random coil (residues 1-34), a globular central part (residues 35-120), and a highly basic carboxy-terminal part in random coil (residue 121-213).

Radioautography of the peptide map of the tryptic hydrolysate of histone H5 phosphorylated with microsomal histone H1 kinase shows five ^{32}P labelled peptides. Only serine residues are found phosphorylated. Structural studies of the phosphorylated tryptic peptides, isolated by ion-exchange chromatography on Chromobeads-P by A. Martinage (1981), allowed us to localize the phosphoserine residues in the sequence of the protein. The sites of phosphorylation of histone H5 are presented in Figure 5.

Five serine residues in histone H5 were found phosphorylated to different extents by the histone H1-specific kinase. Two minor sites (serine-3 and serine-7) were identified in the short hydrophobic sequence (residues 1-11) of the amino-terminal region of histone H5. The three other phosphorylation sites were found in the highly basic carboxy-terminal

AMINO ACID SEQUENCES OF H5 CHICKEN ERYTHROCYTE

Phob

Thr - Glu - Ser (P) - Leu - Bal - Leu - Ser (P) - Pro - Ala - Pro [10] - Ala / Lys - Pro - Lys - Gln/Arg - Val - Lys - Ala - Ser - Arg [20] - Arg

Ser - Ala - Ser - His - Pro - Thr - Tyr - Ser - Glu - Met - Ile - Ala - Ala - Ala - Ile - Arg - Ala - Glu - Lys [40] - Ser - Arg

Gly - Gly - Ser - Ser - Arg - Gln - Ser - Ile - Gln - Lys - Tyr - Ile - Lys - Ser - His - Tyr - Lys - Val [60] - Gly - His - Asn

Ala - Asp - Leu - Gln - Ile - Lys - Leu - Ser - Ile - Arg - Arg - Leu - Leu - Ala - Ala - Gly - Val [80] - Leu - Lys - Gln - Thr

Lys - Gly - Val - Gly - Ala - Ser - Gly - Ser - Phe - Arg - Leu - Ala - Lys - Ser - Asp - Lys [100] - Ala - Lys - Arg - Ser (P) [104] - Pro

Gly - Lys - Lys - Lys - Lys - Ala - Val - Arg - Arg - Ser - Thr - Ser (P) [117] - Pro - Lys - Lys - Ala - Ala - Arg - Pro - Arg - Lys

Ala - Arg - Ser - Pro - Ala - Lys - Lys - Pro - Lys - Ala - Thr - Ala - Arg - Lys [140] - Ala - Arg - Lys - Lys - Ser - Arg - Ala

Ser (P) [148] - Pro - Lys - Lys - Ala - Lys - Lys - Pro - Lys - Thr - Val - Lys - Ala [160] - Lys - Ser - Arg - Lys - Ala - Ser - Lys - Ala

Lys - Lys - Val - Lys - Arg - Ser - Lys - Pro - Arg - Ala - Lys - Ser [180] - Gly - Ala - Arg - Lys - Ser - Pro - Lys - Lys - Lys half of the protein, at serine residues 104, 117 and 148. Serines 117 and 148 are the major sites of phosphorylation of histone H5 for the H1-specific kinase, since about 80% of the total ^{32}P-labelling was incorporated in these residues. By comparison calf thymus histone H1 in the presence of microsomal histone H1 kinase was preferentially phosphorylated at threonine residues (P-Ser/P-Thr ratio: 0.3-0.4).

The sites of growth-associated phosphorylation of histone H1 subfraction RTL-3, as reported by Langan (1978), are given for comparison in Figure 6. The major sites of growth-associated phosphorylation are located at positions 16, 136, 153 and 180 in histone H1.

Both enzymes, the microsomal histone H1 kinase (PKH1) and the chromatin bound growth-associated kinase (HKG), phosphorylate serine and threonine residues located in the highly basic domain of the protein which is thought to interact strongly in chromatin with the phosphate groups of DNA. Both enzymes recognize the same specific amino acid sequence. The major phosphorylated sites are included in a repeat structure where the sequence Pro-Lys-Lys is adjacent to either a serine or threonine residue.

In conclusion, microsomal histone H1 kinase from plasmacytoma cells phosphorylates histone H5 at sites similar to the sites phosphorylated *in vivo* and *in vitro* in histone H1 (Langan, 1977). These data allow us to conclude that both enzymes are similar, or share a common catalytic domain, or are interconvertible forms of the same enzyme.

FIGURE 5. Amino acid sequences of chicken erythrocyte histone H5.

H1-3L [HKG]

Lys - Ser(P) [16] - Pro - Ala - Lys - Lys - Lys - Lys

Ala - Thr(P) [136] - Pro - Lys - Lys - Pro - Lys - Lys

Lys - Thr(P) [153] - Pro - Lys - Lys - Ala - Pro - Lys

Lys - Ser(P) [180] - Pro - Lys - Lys - Ala - Lys - Ala

H5 [PKH1]

Arg - Ser(P) [104] - Pro - Glyc- Lys - Lys - Lys - Lys

Thr - Ser(P) [117] - Pro - Lys - Lys - Ala - Ala - Arg

Ala - Ser(P) [148] - Pro - Lys - Lys - Ala - Lys - Lys

FIGURE 6. Comparative study of histone H1 kinase activity in normal and neo-plastic tissues

In order to compare the activity of histone H1 kinase from normal and neoplastic tissues, we evaluated the kinase activity in a partially purified enzyme extract (0.2 M phosphate eluate from the hydroxylapatite column) of microsomes from plasmacytoma, Morris hepatoma, and normal liver. The protein extract put on the column corresponded to the same wet weight of each tissue. Either [γ^{32}P] ATP or [γ^{32}P] GTP was used as phosphate donor (Figure 7). In the presence of [γ^{32}P] ATP the activity of PKH1 in plasmacytoma and hepatoma is respectively 7-fold and 4-fold higher than in liver. When [^{32}P] GTP is used as substrate, this difference is much higher. In this case, the activity of the histone H1 kinase of

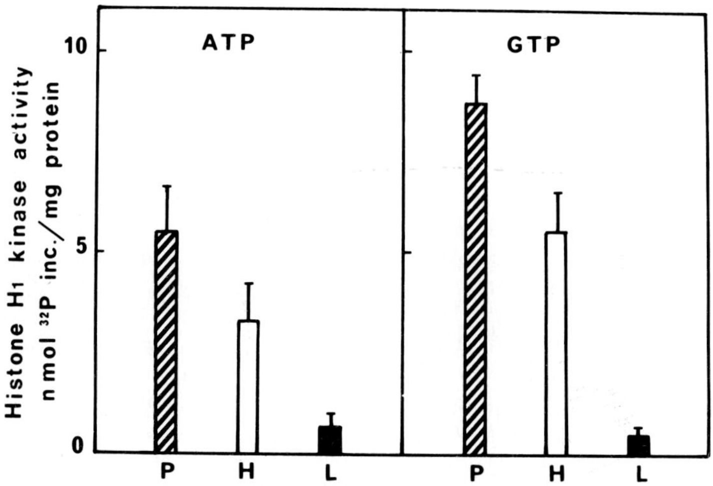

FIGURE 7. Comparative analysis of histone H1 kinase activity of plasmacytoma, Morris hepatoma and normal liver. Histone H1 kinase (0.2 M phosphate hydroxylapatite eluate) from mouse plasmacytoma ( ), Morris hepatoma ( ) and mouse liver ( ) were assayed in the presence of purified histone H1 and [γ^{32}P]ATP or [γ^{32}P]GTP. The protein extract applied onto the hydroxylapatite column corresponded to the same wet weight of each tissue.

plasmacytoma and hepatoma is respectively 17-fold and 11-fold higher than that of liver. Thus, we have shown a good correlation of histone H1 kinase activity and the rate of cell division, since the kinase was predominant in two neoplastic tissues, i.e. plasmacytoma and hepatoma, and negligible in a non-proliferating tissue, i.e. liver.

The low level of kinase activity in liver cannot be due to the presence of phosphatase activities coeluting with the kinase, as we detected no phosphatase activity in the enzyme extract from liver and tumor cells.

Intracellular Localization of Histone H1 Kinase

To eliminate the possibility that microsomal histone H1 kinase is an extraction artefact due to disruption of some nuclei, we investigated the intracellular location of histone H1 kinase. It was examined by using the indirect immunofluorescence technique. Antiserum against purified histone

H1 kinase was obtained from rabbits. Using fluorescein-conjugated goat rabbit IgG, immunofluorescence was performed on randomly growing 3T3 and plasmacytoma cells.

The specificity of anti-histone H1 kinase serum was shown by the double imunodiffusion technique of Ouchterlony and by solid phase radioimmunochemical tests. Plasmacytoma cells exhibited intense fluorescence throughout the entire cytoplasm. In 3T3 cells a diffuse perinuclear fluorescence which extended into the cytoplasm was observed. Nuclear fluorescence was not detected. These results suggest either that the H1 kinase antigenic determinants are not present in the nucleus or are not available in the nuclei due to side chain modification, conformational differences or masking by other non-histone proteins (to be published). I would also like to mention that in randomly growing cells 96% or more of the cells are in the interphase state. Another question is still open: is the nuclear fluorescence visible only in cells at mitosis?

Immunofluorescence studies at the different phases of the cell cycle and on cell populations blocked in either G1, S and mitosis are under investigation.

Most of the reports in the literature on the occurrence and properties of different histone kinases are somewhat confusing due to the presence of different protein kinases in the preparation. The significant aspects of this study lie in the purification of a specific histone H1 kinase from microsomes and the production of specific polyclonal antibodies against this enzyme. Other important observations result from the main properties of microsomal PKH1: when comparing it with nuclear HKG from rapidly dividing cells (Lake, 1973; Langan, 1978; Zeilig and Langan, 1980), it was shown that both enzymes are identical at least in their catalytic domain. The fact that many divalent ions are very strong inhibitors of the purified microsomal histone H1 kinase, in particular Cu^{2+}, leads to the suggestion that during interphase microsomal PKH1 might be associated in the cytoplasm with a metalloprotein which has been dissociated from the enzyme by our extraction procedures using high salt.

I would like to make the following speculation: PKH1 might exist stored in an inactive form in the cytoplasm, its site of synthesis, during the whole cell cycle until it is needed, then it must be activated, i.e. translocated to the nucleus, to cause superphosphorylation of histone H1 at specific sites which might induce the condensation of the chromatin during mitosis.

As yet there is no direct proof for any specific function of histone H1 phosphorylation in mitosis. Therefore to know the real biological function of the specific histone H1 kinase, we plan to introduce the purified histone H1 kinase into mammalian cells to show whether chromatin condensation can be induced by introducing the kinase, and stopped by introducing the specific antibodies against the kinase.

ACKNOWLEDGMENTS

I am very grateful to the following persons who contributed to the results presented in this paper: Drs. M. Schmitt, M. Martinage, P. Sautiere and J. Tiphine.

REFERENCES

Ajiro, K., Borun, T.W., and Cohen, L.H. (1981a). *Biochemistry 20,* 1445.

Ajiro, K., Borun, T.W., Shulman, S.D., McFadden, G.M. and Cohen, L.H. (1981b). *Biochemistry 20,* 1454.

Bradbury, E.M., Inglis, R.J., Matthews, J.R., and Sarner, N. (1973). *Eur. J. Biochem. 33,* 131.

Bradbury, E.M., Inglis, R.J., and Matthews, H.R. (1974). *Nature 247,* 257.

Gurley, L.R., D'Anna, J.A., Barham, S.S., Deaven, L.L., and Tobey, R.A. (1978). *Eur. J. Biochem. 84,* 1.

Hirsch, A., and Rosen, O.M. (1974). *Anal. Biochem. 60,* 389.

Hohmann, P., Tobey, R.A., and Gurley, L.R. (1976). *J. Biol. Chem. 251,* 3685.

Inglis, R.J., Langan, T.A., Matthews, H.R., Hardie, D.G., and Bradbury, E.M. (1976). *Exp. Cell Res. 97,* 418.

Isenberg, I. (1979). *Annu. Rev. Biochem. 48,* 159.

Lake, R.S. (1973). *J. Cell Biol. 58,* 317.

Langan, T.A., and Hohmann, P. (1974). *Fed. Proc. 33,* 1597.

Langan, T.A. (1978). *In* "Methods in Cell Biology", Vol. 19 (G. Stein, J. Stein, and L.J. Kleinsmith, eds.), p. 143. Academic Press, New York.

Martinage, A., Mangeat, P., Laine, B., Couppez, M., Sautiere, P., Marchis-Mouren, G., and Biserte, G. (1980). *FEBS Lett. 118,* 323.

Martinage, A., Quirin-Stricker, C., Champagne, M. and Sautiere, P. (1981). *FEBS Lett. 134,* 103.

Matsumoto, Y., Yasuda, H., Mita, S., Marunouchi, T. and Yamada, M. (1980). *Nature 284,* 181.

Matthews, H. R. (1980). *In* "Protein Phosphorylation and Bioregulation" (G. Thomas, E.J. Podesto, and J.K Bordon,

eds.) (FMI EMBO Workshop, Basel), p. 83. Karger, Basel.
Nicolini, C.A., ed. (1979). "Chromatin Structure and Function". Plenum Press, New York.
Quirin-Stricker, C., and Schmitt, M. (1981). *Eur. J. Biochem. 118,* 165.
Rubin, C.S., and Rosen, O.M. (1975). *Annu. Rev. Biochem. 44,* 831.
Schmitt, M., Quirin-Stricker, C., and Kempf, J. (1982). *Biochimie 64,* 13.
Zeilig, C.E., and Langan, T.A. (1980). *Biochem. Biophys. Res. Commun. 95,* 1372.

DISCUSSION

Berger: Does the kinase work on H1 in chromatin? Do you get the same phosphorylation pattern?

Stricker: In the presence of purified PKH1, we have not succeeded in phosphorylating Histone H1 when incubating intact chromatin or nucleosomes obtained after microsomal digestion. It seems that the sites for phosphorylation are not available for interaction with the enzyme. Histone H1 extracted either from nucleosomes or from the microsomes is an excellent substrate for PKH1.

Griffin (OMRF): I do not see right away what is incompatible with having the histone phosphorylated right after translation. We have a very strong antibody. Some very good immunologists tell me its titer is about 1-2 million against this protein, and we have a cryptic form of the protein, I think, phosphorylated in some region or other and it has lost a lot of activity. It has also lost 8 dilutions versus this antiserum. So not seeing something with an antiserum is no proof that it is not there, I'm afraid, any more than not seeing activity is proof that the protein is not there.

Stricker: I agree with you. The results concerning the absence of fluorescence in the nuclei do not mean that the protein is not there. They can suggest that the PKH1 antigenic determinants are not available in the nuclei.

Chambers (U. of Colorado): I have been working with this enzyme too, as you know, and I think the point is that H1 phosphorylation does actually occur in the nucleus. We have made preparations from the cytoplasm and from the nucleus and we find that in fact probably 95% of the enzyme is located inside the nucleus. Do you feel that the antibody is not penetrating the nucleus?

Cohen: What cell type?

Chambers: Randomly growing Novikoff Hepatoma, which is in all stages of the cell cycle.

Stricker: My guess is that the antibody gets into the nucleus but the antigen is not available for interaction with the antibody. Can you further develop your technique in order to support the conclusion that only 5% of your enzyme is in the cytoplasm?

Chambers: We followed your procedure for enzyme purification.

Stricker: Have you retried it? You cannot test the enzyme after using the microsomal fraction. You need at least a partial purification to test the enzyme. As the quantity of the enzyme is much lower in the cytoplasm than in the nucleus in your cell culture conditions, the microsomal enzyme might become inactivated during the purification procedure.

Chambers: Well, I have been trying to purify the enzyme but have not achieved it from the nucleus.

Stricker: I would guess that you cannot find this activity on a crude microsomal fraction due to the presence of high levels of phosphatases, ATPases, inhibitory metal ions etc. present on microsomes.

Graziani: Is the enzyme that you have a phosphoprotein? Is it possible to phosphorylate it with cAMP dependent protein kinase?

Stricker: No, we have tried it and we would be very glad if it were because it could be a regulatory protein. We have tried to phosphorylate PKH1 with the cAMP dependent kinase of bovine heart and a cAMP-dependant kinase from plasmacytoma. In these cases PKH1 never could be phosphorylated; however, other cyclic nucleotide-dependent kinase(s) might phosphorylate it.

Graves (Iowa S.U.): You purify this enzyme from the microsomal membrane. Is it specifically associated with the microsomes as it comes down or does it just happen to come down with that fraction, as a lot of cytosolic enzymes do in crude microsomal preparations?

Stricker: I mentioned that the described enzyme is cytoplasmic. We never found any activity in the soluble cytoplasmic fraction, so my guess is that the kinase is membrane-associated and not cytosolic enzyme.

Cohen: Just one technical point: you say that there is much less phosphoserine than phosphothreonine in H1, but the problem with all of these analyses is that phosphothreonine is much more stable to acid hydrolysis. I'm suggesting that we might be underestimating the amount of phosphoserine in the H1.

Stricker: I agree.

H1 HISTONE PHOSPHORYLATION: AN ANALYSIS OF DIFFERENT PHOSPHORYLATION SITES AND THEIR RELATIONSHIP TO CHROMOSOME REPLICATION[1]

Philip Hohmann

Department of Experimental Biology
Roswell Park Memorial Institute
Buffalo, New York

INTRODUCTION

The phosphorylation of H1 histones has been observed in a variety of cells under two principal physiologically significant stimuli. In many non-replicating cells, such as cells of the adult rat liver, H1 histones are phosphorylated after stimulation by hormones which raise the intracellular level of cyclic AMP (Langan, 1971). The second major event which stimulates H1 histone phosphorylation is cell replication, and the phosphorylation is most vigorous during the S-phase (Balhorn *et al.*, 1972) and M phase (Bradbury *et al.*, 1974; Gurley *et al.*, 1974) of the cell cycle. These two general cases suggest that the phosphorylation of H1 histones changes the conformation-dependent interactions these proteins have with DNA, permitting chromatin activation, replication or condensation. Since it is apparent that the structures of active, replicating and condensed chromatin are different, it is important to ask if the phosphorylation reactions which H1 histones undergo at different stages in the life cycle of cells are also different. The present discussion will present evidence that the sites phosphorylated in H1 histones in replicating cells may differ at different stages in the cell cycle (Hohmann *et al.*, 1976), and that the site phosphorylated

[1]*The studies performed in the author's laboratory were aided by grant NP-222 D from the American Cancer Society.*

ISBN 0-12-387560-9

as a consequence of cyclic AMP stimulation is different from these former sites as well (Langan and Hohmann, 1975). Thus, H1 histones are proteins with a number of possible phosphorylation sites, and the ultimate structural role that H1 histones play in stabilizing transitions in chromatin organization may depend on the combination of sites modified. We will present evidence that there is no clock-like coupling between the phosphorylation of a particular region or amino acid in H1 histones and a fixed point in the cell cycle (Wilkinson *et al.*, 1982) as well as other evidence that suggests that the phosphorylation of H1 histones during cell replication serves to relax structures apparently needed for long term gene expression, rather than drive the formation of structures needed for correct chromosome condensation.

H1 HISTONE CHEMISTRY

The sequences of H1 histones have diverged much more than the sequences of the nucleosomal histones (Cole, 1977). A given cell line (Hohmann, 1980) or tissue (Kinkade, 1969) is capable of expressing 3-5 H1 histone subtypes in significant amounts. These can be readily fractionated by ion-exchange chromatography (Hohmann, 1980). The ratio of the various H1 subtypes is not constant, but varies in a tissue specific way (Kinkade, 1969). Changes in the ratio of H1 subtypes can be observed in developing rat testis cells (Kistler and Geroch, 1975), in developing sea urchins (Arceci *et al.*, 1976; Newrock *et al.*, 1977) and during a single round of hormone-sensitive cell division in mouse mammary cells *in vitro* (Hohmann and Cole, 1969). Since H1 histones are believed to promote supercoiling of the DNA nucleosome fiber (Thoma *et al.*, 1979), a particular ratio of H1 subtypes may establish a limited diversity in the possible packing of nucleosomes. This, in turn, may be related to the establishment of long-term states of genome organization important for differential gene activity (Hohmann, 1978). It can then be readily appreciated that the phosphorylation of H1 histones could profoundly influence the potential compaction of the DNA nucleosome fiber by either an active or passive mechanism.

SITES OF PHOSPHORYLATION IN H1 HISTONES

A number of phosphorylation sites in H1 histones are known. Figure 1 shows a composite H1 molecule which displays

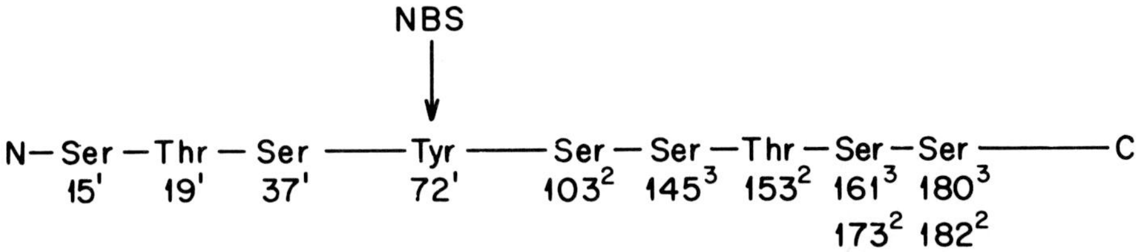

FIGURE 1. Phosphorylation sites in H1 histones. The positions of eight known serine and threonine phosphorylation sites are given in this representation of a composite H1 molecule. The tyrosine at position 72 is not known to be phosphorylated, but cleavage at this residue by N-bromosuccinimide (NBS) defines the NH_2- and COOH-terminal regions operationally. The superscript numbers denote the source of the H1 subtypes. 1. Rabbit thymus RTL 4; 2. Rabbit thymus RTL 3; 3. Trout H1 (see Cole, 1977 and Macleod et al., *1977, for further details).*

6 serine and 2 threonine phosphorylation sites. Not all H1 histones will have all of these sites. The source of the sequence data is given in the Figure. While the actual residue number of a site depends on the source of the H1 histone, all the major sites of phosphorylation are located in the basic NH_2 and COOH terminal regions of the H1 molecule, not in the central, hydrophobic region. The central region contains the sole tyrosine (tyr 74) in H1 histones, and this can be attacked by N-bromosuccinimide (NBS) to yield the NH_2 and COOH terminal fragments (Bustin and Cole, 1969). Using this bisection technique, it is possible to assign phosphorylation reactions to a particular basic region of the H1 histone (Hohmann *et al.*, 1976).

The phosphorylation site at ser 37 is the cyclic AMP sensitive site termed "site A" by Langan (1971). This site occurs in a peptide which is similar in sequence to peptides phosphorylated by cyclic AMP dependent protein kinase *in vitro* (Daile *et al.*, 1975). Not all histone subtypes have site A (Langan *et al.*, 1971), adding to the potential specificity of cyclic AMP dependent phosphorylation.

The phosphorylation site at ser 103 has been termed "site B" by Langan (1971). Site B is a substrate for a widespread cyclic AMP independent kinase *in vitro*, but site B has not been found to be phosphorylated to a significant degree *in vivo*.

The six remaining sites are phosphorylated during cell replication in a variety of cell types. These sites are found in peptides which are minor variations of the sequence Lys-Ser-(PO_4)-Pro-Lys described for trout H1 histone by Macleod *et al*. (1977). The positions of these sites favor the relative centers of the basic regions of the H1 molecules.

CYCLIC AMP AND H1 HISTONE PHOSPHORYLATION

The phosphorylation of H1 histones by cyclic AMP stimulated protein kinase has been reviewed in detail (Langan, 1971). It is important, however, to appreciate the technical approach which demonstrated that phosphorylation of H1 histones at site A was not a trivial consequence of minor fluctuations in cell replication. The general approach involves the construction of phosphopeptide maps of regions of H1 histones, and the protocol used is shown in Figure 2. An appli-

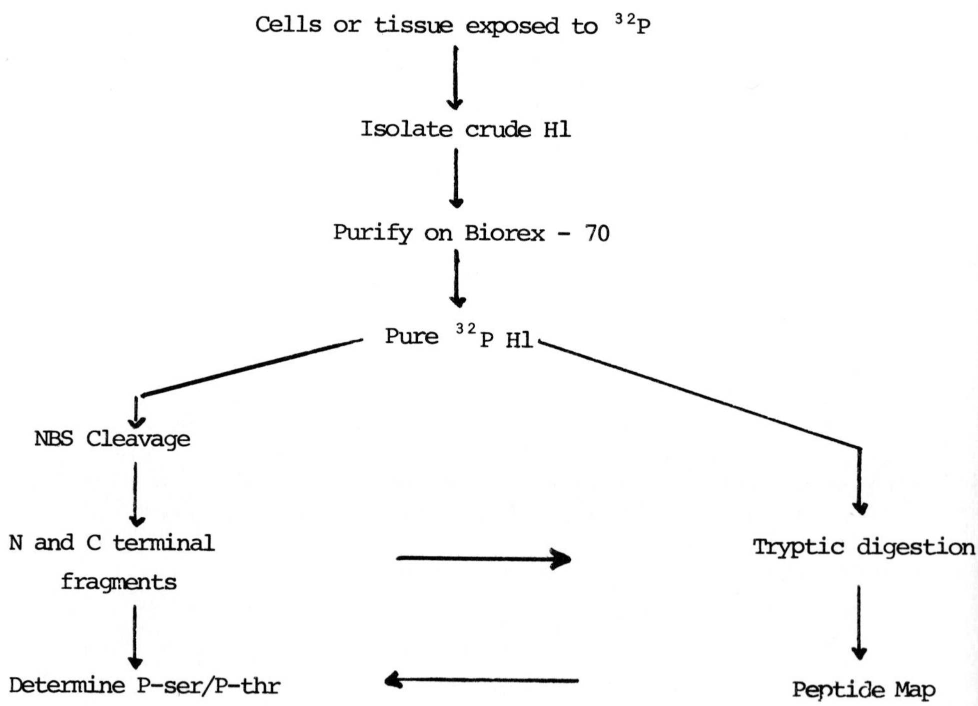

Figure 2

cation of these techniques to rat H-35 cells in culture, which are sensitive to cyclic nucleotides, clearly distinguished between replication dependent and cyclic nucleotide dependent phosphorylation events.

H1 histones were labeled with (^{32}P) in replicating cultures of H-35 cells in the presence or absence of dibutyryl cyclic AMP. The purified H1 histones were digested with trypsin and a phosphopeptide fraction was isolated on the basis of co-electrophoresis on paper at pH 7.9 with a marker peptide for site A. Subsequent electrophoresis of this peptide fraction on paper at pH 1.9 revealed numerous peptides in addition to a peptide with the mobility of the marker for site A (Figure 3). The peptide which co-electrophoresed with the marker responded strongly to the presence of dibutyryl cyclic AMP while the remaining phosphopeptides did not. These latter peptides were, however, strongly phosphorylated in actively growing cells and only marginally phosphorylated in stationary cells (c.f. Langan and Hohmann, 1975). On the basis of different peptide structures and the differing sensitivities of the sites to cell growth and cyclic nucleotide stimulation, it is clear that H1 histones can be phosphorylated at different sites for different physiologically significant reasons.

THE PHOSPHORYLATION OF H1 HISTONES DURING CELL REPLICATION

The phosphorylation of H1 histones is positively correlated with the rate of cell replication. Studies on the relationship between H1 histone phosphorylation and cell repli-

---

FIGURE 2. Flow sheet for tryptic phosphopeptide mapping studies. To avoid artifacts from ^{32}P labeled contaminants, H1 histones must be purified. Chromatography on biorex 70 resin is recommended, but preparative gel electrophoresis is also adequate. The various techniques, such as NBS cleavage before tryptic digestion, can be used to generate data on the regions of the H1 histone phosphorylated in combination with straightforward tryptic peptide mapping. The electrophoretic fractionations on paper can be done sequentially to yield finer resolution of peptides. For example, mixtures of peptides resolvable at pH 7.9 can be eluted and refractionated at pH 1.9 (see Hohmann et al., *1976, for further details).*

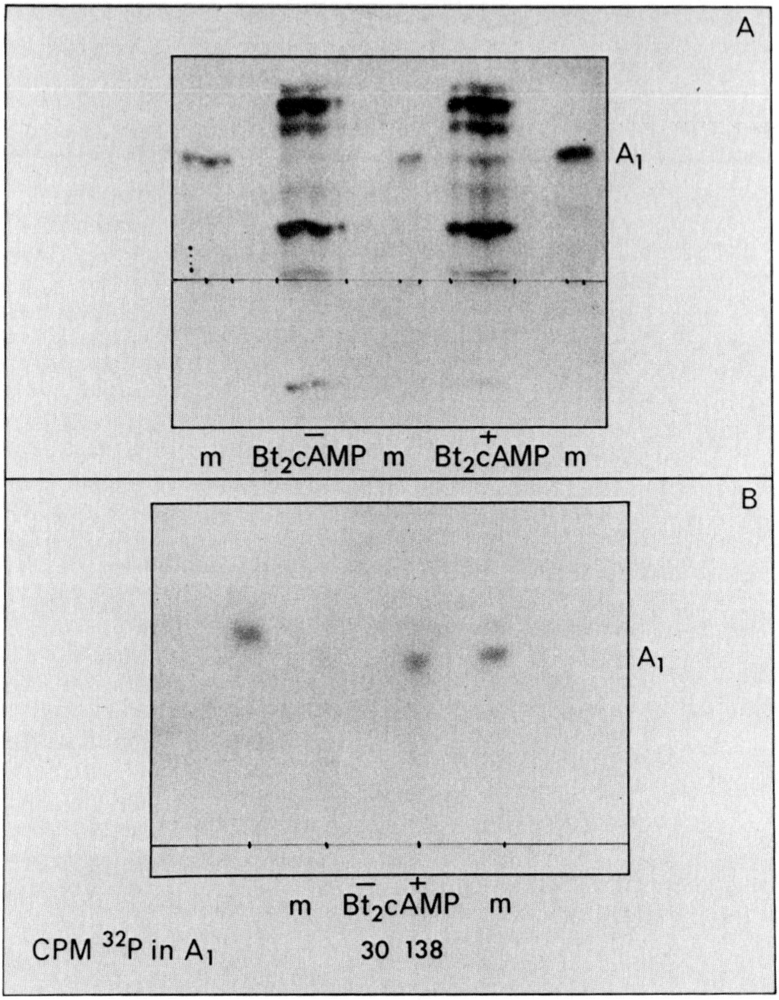

FIGURE 3. Effect of dibutyryl cyclic AMP on the phosphorylation of rat H1 histones. Exponentially growing Reuber H-35 cells were exposed to ^{32}P, and the purified H1 histones were digested with trypsin. The tryptic digests were initially fractionated by electrophoresis on paper at pH 7.9, and a fraction which co-electrophoresed with a marker peptide for site A (ser-37, Figure 1) was further fractionated by electrophoresis on paper at pH 1.9 (Plate A). In Plate A, m denotes the electrophoresis of the marker peptide, and its mobility is given as A1. The phosphorylation of a peptide at position A1 is stimulated by dibutyryl cyclic AMP (Bt2cAMP). The remaining peptides are unaffected by Bt2cAMP, but respond strongly to the rate of cell replication. Purification of A1

cation have been carried out mostly in tissue culture cells since established cell lines grow rapidly and are easily synchronized. The available evidence from such systems suggests that neither the quality nor the relative quantity of specific H1 histone subtypes is critically important for the stable replication and condensation of chromosomes characteristic of the species in question. There are, however, numerous species and cell specific aspects to H1 histone phosphorylation during cell replication. This suggests that there is more than one acceptable sequence of phosphorylation reactions during the cell cycle. In particular, transformed cells may differ from untransformed cells in the major sites phosphorylated during interphase.

In 1975, the observation was made in Chinese hamster CHO cells that threonine was strongly phosphorylated in H1 histones only during mitosis (Hohmann *et al.*, 1975). The application of tryptic phosphopeptide mapping techniques to H1 histones from synchronized CHO cells demonstrated that different regions and amino acids were phosphorylated in CHO cell H1 histone as a function of the cell cycle (Hohmann *et al.*, 1976). The results of these studies are shown in Figure 4.

About 2 hours before S phase, CHO H1 histone is phosphorylated on a serine residue in the COOH-terminal region of the molecule. This event is H1G1, and it involves previously synthesized, or "old", H1 molecules. About 30% of the H1 molecules contain 1 phosphate group; the remainder are unphosphorylated. The phosphorylation of serine in the COOH-terminal region continues in S phase (H1S), presumably involving both old and new H1 molecules, but about 10% of the H1 molecules pass through multiple phosphorylated states. The phosphopeptide maps of G1 and S phase CHO cell H1 histone are identical. This could be due to the insensitivity of the methods for the detection of low levels of transient phosphorylation, or to the phosphorylation of serine residues in identical tryptic peptides from different parts of the COOH-terminal region (Macleod *et al.*, 1977). During mitosis all CHO H1 histone is subject to a superphosphorylation event (H1M) which involves both serine and threonine in both the NH_2- and COOH-terminal regions of the H1 molecule. All CHO H1 molecules have at least 4 phosphate groups in mitosis (Hohmann *et al.*, 1976).

Figure 3 continued

by thin layer chromatography (Plate B) demonstrates that the peptide is pure and retains its responsiveness to Bt2cAMP. (See Langan and Hohmann, 1975, for further details).

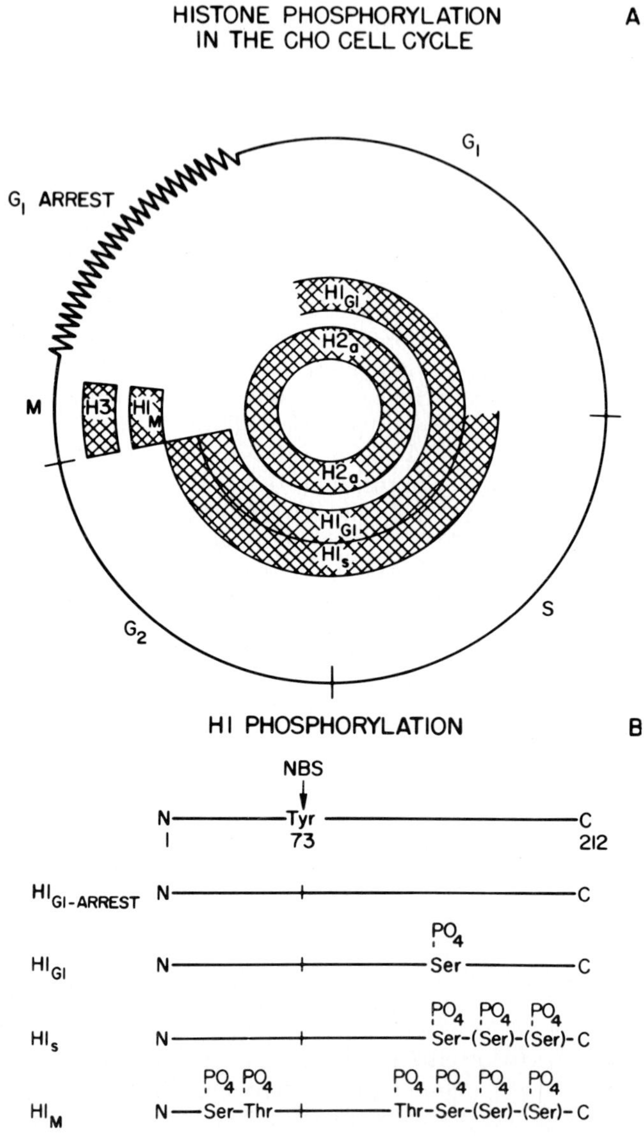

FIGURE 4. Histone phosphorylation in the CHO cell cycle. Plate A is a diagrammatic representation of the CHO cell cycle. The hatched areas represent stages in the cell cycle when histones are phosphorylated. The H1 histone is phosphorylated in G1, S and M. In Plate B the H1 histone is drawn in a manner consistent with that of Figure 1. The positions of the specific serine and threonine residues are not known for CHO H1 histones, and the point of NBS cleavage

The studies with CHO cells suggest that there are functional connections between the phosphorylation of the COOH-terminal region of the H1 molecules and DNA synthesis and the phosphorylation of the NH_2-terminal region and chromosome condensation. Studies by Sherod *et al.* (1970) also suggest that the COOH-terminal region of H1 histone is the major acceptor of phosphate in rat hepatoma HTC cells and mouse Ehrlich ascites cells in interphase. There are, however, contradictory studies, including studies by Lake (1973) in Chinese hamster V79 cells, studies by Ajiro *et al.* (1981a) in HeLa Cells, and studies from my own laboratory (Hohmann, 1979) in rat, mouse and Chinese hamster cells which suggest that the phosphorylation of the NH_2-terminal region of all or some H1 histone subtypes is not strongly related to mitosis. A comparison of Chinese hamster V79, CHW and CHO cells led to the conclusion that the phosphorylation of the NH_2-terminal region of H1 histones occurred in interphase and that the extent depended on the cell line in question (Wilkerson *et al.*, 1982).

The phosphopeptide maps of V79, CHW and CHO cell H1 histone subtypes were compared (Wilkinson *et al.*, 1982). As previously observed (Hohmann *et al.*, 1976), CHO H1 histone was not phosphorylated in the NH_2-terminal region during interphase while all V79 and CHW H1 subtypes incorporated about 30% of the ^{32}P into the NH_2-terminal region during interphase. The phosphopeptide maps of all the H1 subtypes were very similar, but enough differences were observed to account for the phosphorylation of the NH_2-terminal regions of the V79 and CHW H1 subtypes.

The notion that the phosphorylation of H1 histones during cell replication depends to some degree on the cell type studied has been extended to normal mouse mammary explants in organ culture. Such tissue explants contain H1 histone subtypes which are qualitatively identical with those of transformed LM/TK-cells. However, about 60% of the ^{32}P is incorporated into the NH_2-terminal region of the mouse mammary H1 histones while this same fraction is incorporated into the COOH-terminal region of the mouse LM/TK-cell H1 histones

Figure 4 continued

(tyr-73) is given relative to calf thymus CTL-1 (Cole, 1977). Note that, in CHO cells, interphase phosphorylation involves only serine in the COOH-terminal region of the H1 molecule while mitotic phosphorylation involves both serine and threonine in both the NH_2- and COOH-terminal regions of the molecule. (Reproduced from Gurley et al., *1978, with permission from Academic Press).*

(Hohmann, unpublished observations). Such a result suggests that normal and transformed cells may differ fundamentally in the preferred sites of H1 histone phosphorylation in interphase.

The studies discussed above show that there is no obligatory connection between the phosphorylation of the NH_2-terminal region of H1 histones and chromosome condensation or the phosphorylation of the COOH-terminal region and DNA synthesis. One can then ask the more fundamental question whether there is any obligatory connection between the species-specific nature of H1 histones and the ability to replicate and condense the genome at all. For example, the Chinese hamster genome is replicated and condensed equally well in V79 or CHO cells, yet the composition of H1 histone subtypes varies, as does the nature of their phosphorylation in interphase (Wilkinson *et al.*, 1982). We have therefore used a somatic cell hybrid of mouse and Chinese hamster to demonstrate that the Chinese hamster genome can be stably replicated and condensed using mouse H1 histones and their phosphorylated forms (Hohmann *et al.*, 1982).

The mouse-Chinese hamster hybrid LMC-1 (68 chromosomes) was isolated from the fusion of mouse (LM/TK-46 chromosomes) and Chinese hamster (CHW-1102, 22 chromosomes) cells. The chromosome composition was stable for over 1 year in continuous culture, and an analysis of isozymes demonstrated that both mouse and Chinese hamster functions were expressed. Characterization of H1 and H2b histones demonstrated, however, that these proteins were solely of mouse origin. Tryptic phosphopeptide maps of the H1 histones in hybrid cells revealed that they were identical with those of the mouse parent. Marker peptides previously assigned to Chinese hamster H1 histones (Hohmann, 1979) were absent. A diagram of the phosphopeptide maps is shown in Figure 5. We conclude that the stable replication and condensation of the Chinese hamster genome can occur in the presence of mouse H1 histones.

Other studies in mouse-human somatic cell hybrids (Ajiro *et al.*, 1978; Hohmann *et al.*, 1980) suggest that mouse and human chromosomes can be replicated and condensed in the presence of H1 histone from only one species, but chromosome loss occurs in these latter hybrids. We speculate that chromosome loss is not due to the presence of inappropriate H1 histones. More recent studies (Ajiro *et al.*, 1982) show that mouse-rat somatic cell hybrids make H1 histones from both parents. It is unclear whether they coexist on the same chromosome, but this would not be unexpected.

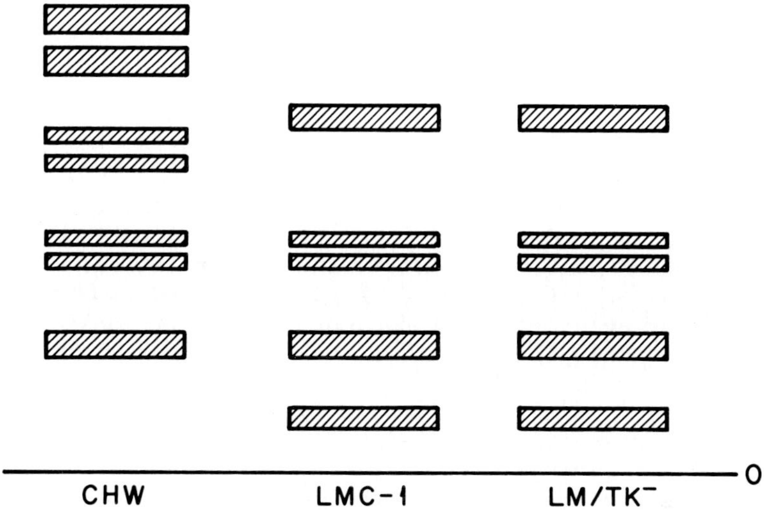

FIGURE 5. Diagrammatic representation of phosphopeptide maps of mouse, Chinese hamster and mouse-Chinese hamster hybrid cell H1 histones. The H1 histones were labeled with ^{32}P in exponentially growing cultures, and the purified H1 molecules were digested with trypsin. The tryptic phosphopeptides were fractionated by electrophoresis on paper at pH 1.9. O represents the origin, anode at the bottom. CHW: Chinese hamster parent; LM/TK-: mouse parent; LMC-1: stable hybrid of CHW x LM/TK-.

SUMMARY AND FUTURE PROSPECTS

It is commonly believed that the phosphorylation of H1 histones causes conformational changes in the protein. There can be no doubt that this occurs *in vivo* (Adler *et al.*, 1972; Glover *et al.*, 1981; Billings *et al.*, 1979). The major problem has been to determine if conformational changes extrapolated to living cells can cause the formation of specific structures, such as condensed chromosomes. The alternative viewpoint would be to say that the conformational changes were permissive of a specific change. If one chooses the permissive explanation, which I believe is strongly suggested by the data, then one is further challenged to accept the observations that the timing of certain phosphorylation reactions may not be rigorously connected with the ultimate change in conformation required. This follows from the observation of NH_2-terminal phosphorylation of H1 histones occurring throughout the cell cycle, but most strongly in mitosis.

Taken as a whole, the data presented here suggest that H1 histone phosphorylation is a passive reaction, necessary for the relaxation of interphase structures involved in long-term genome organization. The phosphorylation-dependent conformational changes may then allow short-term replication structures, such as condensed chromosomes, to be formed by as yet unknown mechanisms. In short, the phosphorylation of H1 histones may be a mechanism to get H1 histones "out of the way" without actually removing them from their position in the genome. What then is the function of the H1 variants so carefully deposited at different stages of development (Hohmann and Cole, 1969; Arceci *et al.*, 1976; Newrock *et al.*, 1977; Kistler and Geroch; 1975)? Ajiro *et al.* (1981b) suggest that, if H1 subtypes are placed non-randomly, they could aid the cell in controlling the structure of different regions of the genome. I have stated (Hohmann, 1978) that I believe, that in their unphosphorylated forms, H1 histone subtypes adapt the monililform (beads on a string) structure of DNA to a limited number of higher order packing states needed for the genesis of structures ultimately responsible for differential gene activity. The phosphorylation of H1 histones might relax these structures in order to allow cells of a given species to replicate and condense the genome in the same general way in all cells. Upon dephosphorylation of H1 histones, the overall genome structure could revert to the form it had before replication. These speculations suggest that H1 histones serve diverse roles in chromatin structure, both passive and active. Further experimentation is needed to elucidate the relationship between the content of H1 histone variants and the potential for replication in cells expressing different genetic functions.

ACKNOWLEDGEMENTS

I would like to thank Ms. Teresa A. Harrison for typing the manuscript. The studies in Figure 3 were performed in the laboratory of Dr. T.A. Langan, and his support is gratefully acknowledged.

REFERENCES

Adler, A.J., Langan, T.A., and Fasman, G.D. (1972). *Arch.*

Biochem. Biophls. 153, 769.
Ajiro, K., Borun, T.W., and Cohen, L.H. (1981a). *Biochemistry 20,* 1445.
Ajiro, K., Borun, T.W., Shulman, S.D., McFadden, G.M., and Cohen, L.H. (1981b). *Biochem. 20,* 1454.
Ajiro, K., Kano-Tanaka, K., Higashida, H., and Tanaka, T. (1982). *Somatic Cell Genet. 8,* 377.
Ajiro, K., Zweidler, A., Borun, T.W., and Croce, C.M. (1978). *Proc. Natl. Acad. Sci. USA 75,* 5599.
Arceci, R.J., Senger, D.R., and Gross, P.R. (1976). *Cell 9,* 171.
Balhorn, R., Bordwell, J., Sellers, L., Granner, D., and Chalkley, R. (1972). *Biochem. Biophys. Res. Commun. 46,* 1326.
Billings, P.C., Orf, J.W., Palmer, D.K., Talmage, D.A., Pan, C.G., and Blumenfeld, M. (1979). *Nucleic Acids Res. 6,* 2151.
Bradbury, E.M., Inglis, R.J., and Matthews, H.R. (1974). *Nature 247,* 257.
Bustin, M., and Cole, R.D. (1969). *J. Biol. Chem. 244,* 5291.
Cole, R.D. (1977). *In* "The Molecular Biology of the Mammalian Genetic Apparatus" (P. T'so, ed.), p. 93. Elsevier, Amsterdam.
Daile, P., Carnegie, P.R., and Young, J.D. (1975). *Nature, 257,* 416.
Glover, C.V.C., Vavra, K.J., Guttman, S.D., and Gorovsky, M.A. (1981). *Cell 23,* 73.
Gurley, L.R., Walters, R.A., and Tobey, R.A. (1974). *J. Cell Biol. 60,* 356.
Gurley, L.R., Tobey, R.A., Walters, R.A., Hilderbrand, C.E., Hohmann, P.G., D'Anna, J.A., Barham, S.S., and Deaven, L.L. (1978). *In* "Cell Cycle Regulation" (J.R. Jeter, I.L. Cameron, G.M. Padilla and A.M. Zimmerman, eds), p. 37. Academic Press, New York.
Hohmann, P. (1978). *Subcell. Biochem. 5,* 87.
Hohmann, P. (1979). *J. Biol. Chem. 254,* 9022.
Hohmann, P. (1980). *Arch. Biochem. Biophys. 205,* 198.
Hohmann, P., and Cole, R.D. (1969). *Nature 223,* 1064.
Hohmann, P., He, D., and Shows, T.B. (1983). *Exp. Cell Res. 143,* 207.
Hohmann, P., Hohmann, L.K., and Shows, T.B. (1980). *Somatic Cell Genet. 6,* 653.
Hohmann, P., Tobey, R.A. and Gurley, L.R. (1975). *Biochem. Biophys. Res. Commun. 63,* 126.
Hohmann, P., Tobey, R.A., and Gurley, L.R. (1976). *J. Biol. Chem. 251,* 3685.
Kinkade, J.M. JR. (1969). *J. Biol. Chem. 244,* 3375.

Kistler, W.S., and Geroch, M.E. (1975). *Biochem. Biophys. Res. Commun. 63,* 378.

Lake, R.S. (1973). *J. Cell Biol. 58,* 317.

Langan, T.A. (1971). *Ann. N.Y. Acad. Sci. 185,* 166.

Langan, T.A., Rall, S.C., and Cole, R.D. (1971). *J. Biol. Chem. 246,* 1942.

Langan, T.A., and Hohmann, P. (1975). *In* "Chromosomal Proteins and their Role in the Regulation of Gene Expression: Proceedings" (G.S. Stein and L.J. Kleinsmith, ed.), p. 113. Academic Press, New York

Macleod, A.R., Wong, N.C.W., and Dixon, G.H. (1977). *Eur. J. Biochem. 78,* 281.

Newrock, K.M., Alfageme, C.R., Nardi, R.V., and Cohen, L.H. (1977). *Cold Spring Harbor Symp. Quant. Biol. 42,* 421.

Sherod, D., Johnson, G., and Chalkley, R. (1970). *Biochemistry 9,* 4611.

Thoma, R., Koller, T., and Klug, A. (1979). *J. Cell Biol. 83,* 403.

Wilkinson, D.J., Shinde, B.G., and Hohmann, P. (1982). *J. Biol. Chem. 257,* 1247.

DISCUSSION

Langan (U. Colorado): In connection with your suggestion about the passive role of phosphorylation in the interaction of H1 with chromatin, we have recently begun a study of the properties of chromatins reconstituted with highly phosphorylated H1 histone subcomponents, which we prepare enzymatically, which allows us to produce large quantities of fully characterized material. In the reconstitution experiments we have done so far, we find exactly what you have suggested. First of all, highly phosphorylated H1 is not more effective than normal unphosphorylated H1 in promoting a condensation of chromatin and, in fact, if the preliminary results are correct, it even leads to a chromatin structure which is less condensed than chromatin reconstituted with nonphosphorylated H1. That would be quite in keeping with your suggestion.

Hohmann: I would also like to point out, in reconstitution, however, that if one tries to make native chromatin with a certain nucleosome repeat length, you generally fail in reconstitution. Reconstitution does not give you certain structural things, but I am sure it gives you others. I'm pleased that it doesn't change the condensation.

Langan: The chromatin reconstitution can now be carried out with the preservation of internucleosomal repeat lengths and, in addition, the nativeness of these reconstituted chromatins with respect to the positioning of H1 in the structure can be demonstrated by showing that the 166 base pair fragment obtained on nuclease digestion of the core particle is restored when the H1 is reconstituted, so that the protective action of H1 indicates that, as best we can tell, the active structure can be reconstituted.

Stricker: Are the phosphorylated sites the same during S phase and in mitosis? Did you analyze the sequence?

Hohmann: In CHO cells, the S-phase sites are retained and new sites occur in mitosis. In V79 cells, the S-phase sites and mitotic sites are very, very similar. In V79 cells, it is merely the quantity of phosphate which varies between S-phase and mitosis. There is some "mitotic" phosphorylation in S-phase, but almost 20% of the molecules remain unphosphorylated at all sites. I believe the turnover of ^{32}P gives you certain peptides in all phases of the cell cycle. However, the stability of the ^{32}P is such that, when you get mitosis, those 4-6 phosphates are very stable and you get a quantitative increase in peptides characteristic of mitosis. The peptide maps representing the actual sites of phosphorylation in mitotic CHO cells and interphase H-35 rat cells are virtually identical. There are minor variations due to species differences in peptide mobility, but to sum up, the interphase sites which are phosphorylated in the NH_2-terminal region appear to be the same as the mitotic sites, but reduced in the quantity of ^{32}P incorporated.

Allfrey: The question arises as to whether the differential phosphorylation patterns are due to differential activities of the kinases or to differential activities of the respective phosphatases. In various cell types, you could envision a system where different phosphatases were active in a site specific way to remove phosphates, and that this would give you an apparent alteration in phosphorylation pattern.

Hohmann: This is what we are forced to pursue, because it is apparent that the identical histones from two cells may appear identical for all practical terms, yet the phosphorylation is not. An alternative is that the sites could be buried and just unavailable due to conformation effects. That we cannot strictly rule out.

EFFECTS OF PHOSPHORYLATION ON THE RECYCLING OF EUKARYOTIC INITIATION FACTOR 2 (eIF-2)

Rosemary Jagus[1]

Laboratory of Molecular Hematology
Section on Protein Biosynthesis
NHLBI, NIH
Bethesda, Maryland

Eukaryotic initiation factor 2 (eIF-2) is a protein which plays an obligatory role in the initiation of protein synthesis (reviewed in Ochoa and deHaro, 1979; Jagus *et al.*, 1981). It functions to promote the binding of the initiator tRNA, Met-$tRNA_f$ to the small ribosomal subunit (Safer *et al.*, 1975). The factor, eIF-2, becomes reversibly phosphorylated in many circumstances in which the initiation of translation is blocked (reviewed in Hunt, 1979, 1980). However, phosphorylation/dephosphorylation of eIF-2 is not associated with activation/inactivation, but with perturbation of the recycling mechanism which promotes guanine nucleotide exchange on the factor (Jagus and Safer, 1981; Siekierka *et al.*, 1982a,b; Safer *et al.*, 1982a; reviewed in Safer and Jagus, 1981; Safer *et al.*, 1982b). Immunochemical analyses of the level and distribution of eIF-2 under varying protein synthetic conditions suggest that recycling takes place on the preinitiation complex formed between the small ribosomal subunit, eIF-2, GTP, Met-$tRNA_f$ and mRNA.

The factor, eIF-2, has a molecular weight of 120,000 and is composed of three nonidentical subunits designated α, β and γ (Lloyd *et al.*, 1980). The γ-subunit (M_r = 55,000, pI = 8.9) (Lloyd *et al.*, 1980) is the subunit which interacts with the initiator tRNA, Met-$tRNA_f$, to promote binding of the initiator tRNA to the small ribosomal subunit (Jagus *et al.*,

[1]*Present address: Laboratory of Cellular and Developmental Biology, Bldg. 6, Room B1-07, NIADDK, NIH, Bethesda, MD*

ISBN 0-12-387560-9

1982). The α-subunit (M_r = 32,000, pI = 5.1 (Lloyd *et al*., 1980) binds GTP (Barrieux and Rosenfield, 1977), which facilitates interaction of the γ-subunit with Met-$tRNA_f$. Thus, eIF-2 may be regarded as an allosteric protein. GDP is also bound, and inhibits the interaction of eIF-2 and Met-$tRNA_f$ (Safer *et al*., 1975). The α-subunit is the target of the heme-regulated protein kinase which catalyses the phosphorylation of one site per α-subunit and leads to the inhibition of initiation (Farrell *et al*., 1977; Levin *et al*., al., 1976; Gross and Mendelewski, 1977; Farrell *et al*., 1978), reviewed in (Hunt, 1980). The function of the β-subunit (M_r = 35,000, pI = 5.4) (Lloyd *et al*., 1980) is unclear, but it seems to play a role in the recycling of eIF-2 and may also react with guanine nucleotides (Jagus *et al*., 1982).

The interactions of eIF-2 in the formation of the first preinitiation complex (43S preinitiation complex) are well understood (Ochoa and deHaro, 1979; Jagus *et al*., 1981). However, the mechanism(s) whereby eIF-2 is released from the preincubation complex and recycled to take part in subsequent rounds of initiation is not well known. However, eIF-2 release occurs prior to the binding of the 60S subunit (Peterson *et al*., 1979) and depends upon hydrolysis of the bound GTP which is catalyzed by eIF-5. The existence of a promoter of guanine nucleotide exchange has long been anticipated (Ochoa and deHaro, 1979; Hunt, 1980), since the Kd for GDP is 3.1 x 10^{-8} M compared with a Kd for GTP of 2.5 x 10^{-6} M (Walton, 1975, 1976) and because GDP is a potent inhibitor of the interaction with initiator tRNA. This factor has now been identified and found to be a high molecular weight, polypeptide, which has been variously designated anti-HR1 (anti-heme regulated inhibitor) (Amesz *et al*., 1979; Goumans, 1981), eIF-2-SP (eIF-2-stimulatory protein) (Siekierka *et al*., 1982a,b; Siekierka *et al*., 1981), SRF (protein regulatory factor) (Gupta, 1981), eRF (eukaryotic regulatory factor) (Goumans *et al*., 1982; Voorma *et al*., 1982) and RF (regulatory factor) (Safer *et al*., 1982a). In view of its functional intimacy with eIF-2 the revised disignation, *eIF-2B*, has been proposed (Safer *et al*., 1982b; Konieczny and Safer, 1983) which accords with previously established nomenclature (Anderson *et al*., 1977).

eIF-2B was first isolated and characterized by Voorma's group in Utrecht (Amesz *et al*., 1979; Goumans, 1981; Goumans *et al*., 1982) and found to be the active component of the less well characterized activity ESP (Siekierka *et al*., 1982; Gupta, 1981; deHaro and Ochoa, 1979) and probably SRF (Gupta, 1981; Das *et al*., 1982). It is a multisubunit protein of molecular weight 280,000 (Safer *et al*., 1982a; Siekierka *et* al., 1982b; Goumans, 1981; Siekierka *et al.*, 1981; Goumans *et al.*, 1982; Voorma and Amesz, 1982; Konieczny and Safer, 1983). The exact subunit structure has not been demonstrated unambiguously, but is seems to consist of equimolar amounts of five dissimilar peptides of molecular weights, 26K, 39K, 54K, 67K and 82K (Safer *et al.*, 1982; Goumans, 1981; Siekierka *et al.*, 1981). The involvement of the 26K and 67K subunits in the structure is still uncertain. The factor, eIF-2B, is isolated as a stoichiometric complex with eIF-2 (Safer *et al.*, 1982a; Goumans, 1981; Konieczny and Safer, 1983). This factor catalyzes the exchange of bound GDP in inactive eIF-2: GDP, for GTP (Siekierka *et al.*, 1982a,b; Safer *et al.*, 1982a; Amesz *et al.*, 1979). The identification of eIF-2B and the elucidation of its role in the recycling of eIF-2 has contributed to the understanding of the role of eIF-2 phosphorylation in the regulation of protein synthesis.

The regulation of eIF-2 activity has been extensively studied in a cell-free protein synthesizing system from rabbit reticulocytes which retains the protein synthetic characteristics and high activity of the intact cell (reviewed in Ochoa and deHaro, 1979). The primary product of the system is globin and continued globin synthesis requires the addition of hemin to the cell-free system (reviewed in Ochoa and deHaro, 1979). In the absence of hemin, protein synthesis continues at the same rate as in its presence for 4-6' and then declines to a rate which is less that 5% of the initial rate (Zucker and Schulman, 1968; Adamson *et al.*, 1968). The decline in globin synthesis reflects a decrease in the rate of initiation (Zucker and Schulman, 1968; Adamson *et al.*, 1968). These biphasic kinetics reflect the gradual inactivation of eIF-2 (reviewed in Safer and Jagus, 1981). Table 1 shows the salient characteristics of the inhibition of protein synthesis in the absence of hemin which led to the conclusion that phosphorylation of eIF-2 involves an alteration of its ability to be reutilized.

During the lag period before protein synthesis becomes inhibited, the heme-regulated kinase is activated and eIF-2 is rapidly phosphorylated (Farrell *et al.*, 1978). However, a very active phosphatase activity is present (Safer and Jagus, 1979) and the steady state amount of eIF-2 phosphorylated reaches only 25-30% of the available pool (Jagus *et al.*, 1982; Safer *et al.*, 1980b; Levin *et al.*, 1979; Leroux and London, 1982). The $t_{1/2}$ of phosphorylated eIF-2 is 20-30 seconds (Safer and Jagus, 1979). The steady state level of eIF-2 phosphorylation can be manipulated experimentally

TABLE 1. *Characteristics of the Inhibition of Protein Synthesis Suggesting that Phosphorylation Inhibits the Recycling of eIF-2*

Characteristics implicating the impairment of eIF-2 activity	Characteristics implicating a defect in recycling of eIF-2
1. eIF-2 becomes rapidly phosphorylated.	1. Steady state level of eIF-2 phosphorylation is only 25–30% when system is inhibited by 95%.
2. Addition of eIF-2 to inhibited lysate overcomes the inhibition.	2. Steady state phosphorylation of eIF-2 does not correlate with final rate of protein synthesis, but with time taken for inhibition to occur.
3. Addition of purified eIF-2 kinase inhibits hemin-supplemented lysate.	3. eIF-2 added to inhibited lysate is only used stoichiometrically.
4. Steady state level of Met-$tRNA_f$ in preincubation complexes decreases.	4. eIF-2B restores activity.
	5. High levels of GTP (1–2 mM) restore activity.

(Safer *et al.*, 1980a). The level of phosphorylation does not simply determine the level of active eIF-2. Instead, phosphorylation determines the lag time, which is the time required for the reduced rate of protein synthesis to be established (Safer *et al.*, 1980a). This has been interpreted to reflect the progressive inactivation of eIF-2 resulting from its conversion, while transiently phosphorylated, to a form which has a limited capacity to recycle (Ochoa and deHaro, 1979; Jagus *et al.*, 1982; Safer *et al.*, 1980a). Recycling may be equated with guanine nucleotide exchange and consistent with this is the finding that high levels of GTP (1-2 mM immediately restore protein synthesis (Safer *et al.*, 1982a).

Restoration of protein synthetic activity to inhibited lysate by exogenous eIF-2 requires high levels of the factor. Approximately ten-fold higher quantities than the endogenous concentration are needed to significantly affect protein synthesis and this restoration is only temporary (Safer *et al.*, 1982a; Safer *et al.*, 1977). When eIF-2 stimulated globin synthesis is compared with the amount of factor added to the inhibited lysate, it is apparent that only stoichiometric utilization of eIF-2 occurs (Safer *et al.*, 1982a; Safer *et al.*, 1977). This is in contrast with the situation in hemin-supplemented lysate in which endogenous eIF-2 is reutilized rapidly and efficiently.

The factor, eIF-2B, was first isolated as an activity which could overcome the inhibition of protein synthesis caused by hemin deficiency (Amesz *et al.*, 1979; Goumans, 1981). The factor is not a phosphatase (Amesz *et al.*, 1979; Gupta, 1981). Its addition to hemin-deficient lysate does not result in a decrease in the steady-state phosphorylation of eIF-2 (Amesz *et al.*, 1979; Goumans, 1981). Table 2 illustrates the characteristics of eIF-2B which suggest it plays a part in eIF-2 reutilization. In contrast to the stoichiometric utilization of eIF-2 when added to inhibited lysate, eIF-2 added as the complex eIF-2:eIF-2B is utilized catalytically (Safer *et al.*, 1982a). The addition of eIF-2B alone also restores protein synthesis. This presumably occurs by facilitating the reutilization of endogenous eIF-2 (Safer *et al.*, 1982a; Amesz *et al.*, 1979; Goumans, 1981). Restoration of protein synthetic activity is achieved by 10-20 fold lesser quantities of eIF-2B than of eIF-2 (Safer *et al.*, 1982a) and the utilization of added eIF-2B is catalytic. In addition to facilitating eIF-2 reutilization in hemin-deficient lysate, eIF-2B also promotes catalytic utilization of eIF-2 in the *in vitro* assay of methionyl puromycin synthesis

TABLE 2.

Characteristics of eIF-2B	Characteristics implicating eIF-2B in catalytic utilization of eIF-2
1. M_r = 220-280,000	1. Restores protein synthetic activity to inhibited lysate by converting stoichiometric utilization of eIF-2 to catalytic utilization (Safer et al., 1982a)
2. Composed of five subunits: 26,000; 39,000; 54,000; 67,000; 82,000	2. Converts stoichiometric utilization of eIF-2 to catalytic utilization in the in vitro assay, methionyl puromycin synthesis (Goumans, 1981; Goumans et al., 1982; Voorma and Amesz 1982)
3. Isolated as a 1:1 complex with eIF-2	3. Promotes guanine nucleotide exchange of eIF-2 (Siekierka et al., 1982a,b; Safer et al., 1982a)
4. Also designated anti-HRI (Amesz et al., 1979; Goumans, 1981), eRF (Goumans et al., 1983; Voorma and Amesz, 1982), RF (Safer et al., 1982a), ESP (deHaro and Ochoa, 1979), eIF-2SP (Siekierka et al., 1982a,b; Siekierka et al., 1981)	

(Goumans, 1981; Goumans *et al.*, 1982; Voorma and Amesz, 1982).

Involvement of eIF-2B in catalytic utilization of eIF-2 suggests a role for the factor in guanine nucleotide exchange. The eIF-2:eIF-2B complex has been shown to have a lower affinity for GDP than eIF-2 alone (Siekierka *et al.*, 1982a,b; Safer *et al.*, 1982a; Siekierka and Ochoa, 1982). Also, the exchange of eIF-2 bound GDP for GTP has been shown to be facilitated by eIF-2B *in vitro* (Siekierka *et al.*, 1982a,b; Safer *et al.*, 1982a; Siekierka and Ochoa, 1982).

A role for phosphorylation in preventing eIF-2:eIF-2B interaction is suggested, since phosphorylation of eIF-2 decreases its ability to be recycled and since eIF-2B facilitates recycling of eIF-2. Table III lists the evidence which indicates that the inhibition of protein synthesis caused by phosphorylation of the α-subunit of eIF-2 results from the blockade of eIF-2B promoted guanine nucleotide exchange of eIF-2. Phosphorylation inhibits all known effects of eIF-2B on eIF-2 activity *in vitro*. It inhibits the eIF-2B promoted catalytic utilization of eIF-2 in methionyl puromycin synthesis (Goumans, 1982; Goumans *et al.*, 1982; Voorma and Amesz, 1982) and the eIF-2B promoted guanine nucleotide exchange *in vitro* (Siekierka *et al.*, 1982a,b; Safer *et al.*, 1982a). In addition, phosphorylation inhibits the formation of the eIF-2B complex *in vitro* (Safer *et al.*, 1982a; Goumans, 1981; Goumans *et al.*, 1982, Voorma and Amesz, 1982; Konieczny and Safer, 1983), as well as the exchange of α-subunits between eIF-2:eIF-2B and free eIF-2 *in vitro* (Safer *et al.*, 1982a; Goumans, 1981; Voorma and Amesz, 1982).

In consideration of these results, the consensus in the field is that phosphorylation inhibits eIF-2B mediated guanine nucleotide exchange on eIF-2, which prevents the catalytic utilization of eIF-2 (Safer *et al.*, 1982b). The manner in which phosphorylation of the α-subunit achieves this result remains to be clarified. One possibility is that phosphorylation prevents the association or interaction of eIF-2B with eIF-2. This direct inhibition of association is favored by Voorma *et al.* (Voorma and Amesz, 1982) and Siekierka *et al.* (Siekierka *et al.*, 1982b; Safer *et al.*, 1982b; Voorma and Amesz, 1982). A problem with this proposal is that any hypothesis must consider the low steady state level of eIF-2 phosphorylation (20-30%) and the kinetics of inhibition which are observed. A direct inhibition model does not easily accommodate the 70-80% of eIF-2 which is unphosphorylated,

TABLE 3.

Evidence that the inhibition of protein synthesis caused by phosphorylation of the α-subunit of eIF-2 results from the obstruction of eIF-2B promoted guanine nucleotide exchange on eIF-2

1. *Phosphorylation of the α-subunit inhibits eIF-2B promoted guanine nucleotide exchange* in vitro *(Siekierka* et al., *1982a,b; Safer* et al., *1982a)*

2. *Phosphorylation inhibits eIF-2, eIF-2B complex formation* in vitro *(Safer* et al., *1982a; Goumans, 1982; Goumans* et al., *1982; Konieczny and Safer, 1983; Voorma and Amesz, 1982)*

3. *Phosphorylation inhibits the exchange of α-subunits between eIF-2, eIF-2B and eIF-2* in vitro *(Safer* et al., *1982a; Goumans, 1981; Voorma and Amesz, 1982)*

4. *Phosphorylation inhibits eIF-2B promoted catalytic utilization of eIF-2 in methionyl puromycin synthesis (Goumans, 1981; Goumans* et al., *1982; Voorma and Amesz, 1982)*

and which is unable to interact with eIF-2B to maintain protein synthesis. Since eIF-2B has an increased affinity for phosphorylated eIF-2 at the physiological guanylate energy charge, another possibility is that when phosphorylated, eIF-2 sequesters a small pool of eIF-2B into an inactive complex (Safer *et al.*, 1982a). However, if the estimated levels of eIF-2B at 10-20% of the eIF-2 pool size are correct, this latter model is also hard to reconcile with the kinetics of inhibition. A sequestration model can only operate if the pool sizes of eIF-2 and/or eIF-2B are larger than suggested by current estimates.

Information on eIF-2 and eIF-2B pool sizes, both relative to each other and to the number of ribosomes, would assist in the generation of a more plausible model. Also of assistance would be an appreciation of the interaction of eIF-2 with eIF-2B during normal and defective initiation cycles. These questions have been approached using antibodies to eIF-2 both to determine eIF-2 levels and to monitor its distribution in active and inhibited lysates. Antisera to eIF-2 were raised in sheep, using eIF-2 of 90% purity. Most antisera contained antibodies to the α-and β-subunits of eIF-2 as well as to the 82K subunit of eIF-2B which was present as a minor contaminant in the eIF-2 preparation used to immunize the sheep. Quantitation is achieved by reacting anti-eIF-2 to lysate proteins transferred to nitrocellulose from SDS-polyacrylamide gels. Figure 1 compares the Coomassie blue staining pattern of eIF-2 and lysate proteins, separated by SDS-polyacrylamide

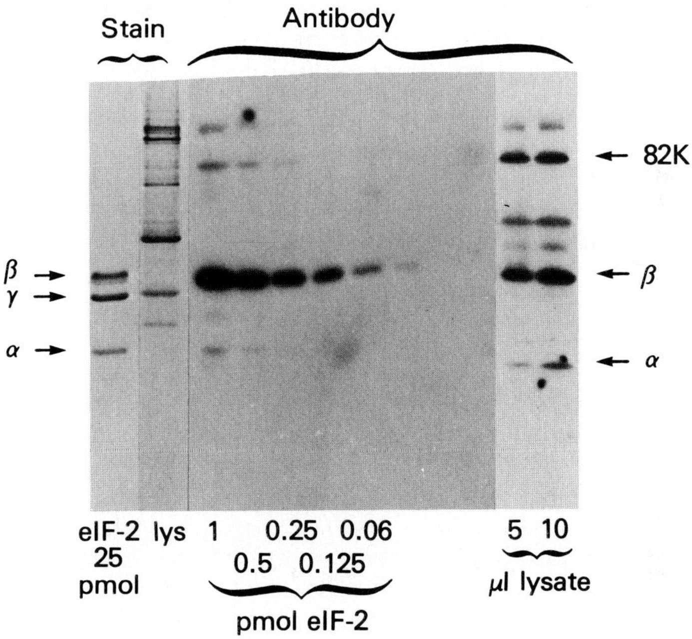

FIGURE 1. Comparison of Coomassie blue staining pattern and immunochemical visualisation of eIF-2 and lysate proteins. eIF-2 and a pH 5 precipitate of reticulocyte lysate were electrophoresed on SDS-polyacrylamide gels (12.5% acrylamide, 0.05% bis) and either stained or transferred to nitrocellulose in 250 mM glycine, 20 mM Tris, pH 8.8, 20% MeOH, 0.01% SDS, at 60V for 75 minutes. The nitrocellulose blots were preincubated in 3% BSA in tris-buffered saline (50 mM Tris, pH 7.4, 200 mM NaCl) followed by incubation with sheep anti-eIF-2, goat anti-sheep IgG (Cappel) and (^{125}I) protein A (NEN, ca. 400 mCi/mmol). Coomassie blue staining patterns (left) are compared with autoradiograms of the nitrocellulose blots (left).

gel electrophoresis, with immunochemical visualization of eIF-2 α-, β- and eIF-2B (82K) subunits. The staining pattern of 25 pmol of eIF-2 is compared with 0.06-1 pmol of eIF-2 after autoradiography of nitrocellulose blots incubated with sheep anti-eIF-2, goat anti-sheep (IgG) and finally (^{125}I) protein A. 0.06 pmol of eIF-2 and eIF-2B (82K) are easily detected. The staining pattern of lysate proteins does not reveal any of the subunits of eIF-2 or eIF-2B. Using antibodies, it is possible to detect both eIF-2 α-, β- and eIF-2B (82K) subunits in 1 μl of lysate. Quantitation of the amounts, using densitometric analyses of the autoradiograms, gives a concentration for eIF-2 of 55-65 pmol/ml, or 0.27 - 0.32 mol eIF-2/mol 80S ribosome equivalent. This is within the range of earlier estimates reviewed in Safer and Jagus, (1981). Comparison of the relative amounts of eIF-2 and eIF-2B (82K) present in the eIF-2:eIF-2B complex with those in the lysate (data not shown) suggest that at least the 82K subunit of eIF-2B is present in equimolar amounts to the β-subunit and presumably with other eIF-2 peptides. This means either that the ratio of eIF-2:eIF-2B is 1:1, or that the 82K peptide is present in larger amounts than the other eIF-2B subunits and exists in forms other than eIF-2B or the eIF-2:eIF-2B complex.

In order to investigate eIF-2 distribution, lysate was fractionated on sucrose density gradients under conditions chosen to give maximum resolution of 43S and 48S preinitiation complexes from free eIF-2 (7S) and the eIF-2:eIF-2B complex (17S). 80S monoribosomes and polysomes are sedimented by this procedure. Lysate proteins from the sucrose gradient fractions were separated by SDS-polyacrylamide gel electrophoresis, transferred to nitrocellulose and incubated with sheep anti-eIF-2, goat anti-sheep (IgG) and (^{125}I) protein A. Conditions were optimized using lysate incubated with initiation inhibitor, edeine. Edeine prevents 60S subunit joining and leads to the accumulation of a 48S intermediate containing equimolar amounts of small ribosomal subunit, eIF-2 and mRNA (Safer *et al.*, 1978). Figure 2 depicts an autoradiogram of a typical edeine-induced eIF-2 distribution pattern. Most of the eIF-2 (3 - 3.5 pmol/100 μl lysate) is found in the 48S preinitiation complex, along with most of the eIF-2B (82K). The comigration of eIF-2B (82K) with the small ribosomal subunit is surprising, since data from the fractionated system suggest that although eIF-2B forms stable complexes with eIF-2, it is never transferred or found with the 43S preinitiation complex *in vitro* (Goumans, 1981; Goumans *et al.*, 1982; Voorma and Amesz, 1982). Furthermore,

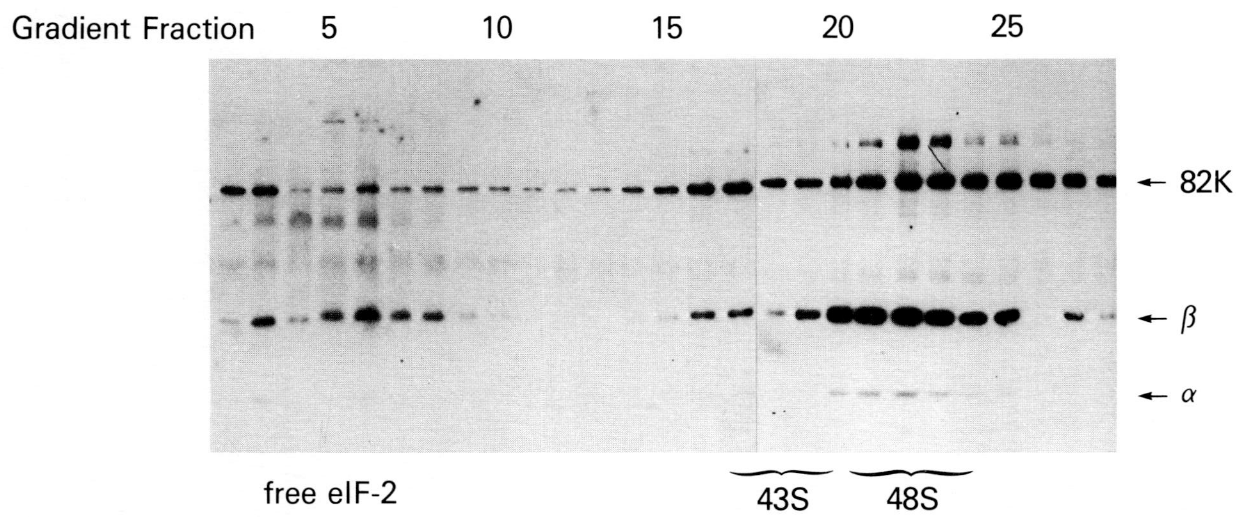

FIGURE 2. Effects of edeine on distribution of eIF-2 and eIF-2B (82K). Reticulocyte lysate was incubated in the presence of edeine for 10 minutes and fractionated on 15-35% sucrose gradients as described (42) at 25,000 rpm for 15 h. After fractionation, proteins in the upper 8 fractions were precipitated at pH 5 to remove hemoglobin. Proteins in the remaining fractions were precipitated with two vols acetone. The precipitated proteins were processed as described in the legend to Figure 1. An autoradiogram is shown.

binding of ^{14}C-labeled eIF-2B was not found in edeine-accumulated 48S preinitiation complexes in reticulocyte lysate (Konieczny and Safer, 1983). In addition to that found in the 48S preinitiation complex, smaller levels of eIF-2 and eIF-2B (82K) are present in the small pool of 43S preinitiation complexes. eIF-2 and eIF-2B (82K) not bound to ribosomal subunits are also seen near the top of the gradient.

Figure 3 compares the distribution of eIF-2 and eIF-2B (82K) in lysate incubated in the presence or absence of hemin. In the presence of hemin, 0.6 - 0.7 pmol of eIF-2 are found in the 43S preinitiation complex, along with an equivalent amount of eIF-2B (82K). Factor which is not bound to ribosomes occurs in two incompletely resolved pools near the top of the gradient. These probably represent free eIF-2 and the eIF-2:eIF-2B complex. By contrast, in the absence of hemin, 1.5 - 1.7 pmol of eIF-2 are found in the preinitiation complex, but this is not associated with equivalent amounts of eIF-2B (82K). Instead, the bulk of eIF-2B is found in the region between free eIF-2 and 43S preinitiation complexes, which may reflect a reduced affinity for eIF-2. This is pertinent to a defect in the preinitiation complexes which accumulate in the absence of hemin (Ochoa and deHaro, 1979; Jagus *et al.*, 1982; Gross, 1979).

As protein synthesis becomes inhibited in the absence of hemin, the levels of Met-$tRNA_f$ bound to the small ribosomal subunit decreases (Ochoa and deHaro, 1979; Jagus *et al.*, 1982; Gross, 1979), but mRNA accumulates (Ochoa and deHaro, 1979; Jagus *et al.*, 1982; Gross, 1979). This is a very unexpected finding, since mRNA can bind only to the small ribosomal subunit after its interaction with eIF-2 GTP and Met-$tRNA_f$ (reviewed by Ochoa and deHaro, 1979). This discrepancy between Met-$tRNA_f$ and mRNA bound indicates an instability of preinitiation complexes which are formed in the absence of hemin (Ochoa and deHaro, 1979; Jagus *et al.*, 1982; Gross, 1979). For instance, it is known that Met-$tRNA_f$ is deacylated (Balkow *et al.*, 1973) and that $tRNA_f^{Met}$ is gradually, but not completely, lost from the complex (Kaplansky, 1982).

These data, coupled with the accumulation of eIF-2 bound to the small ribosomal subunit suggest that in the absence of hemin, the activity cycle of eIF-2 is arrested at the stage appropriate for release and recycling. It is intriguing that at least one peptide of the factor involved in the recycling mechanism seems unable to interact with eIF-2 present in the defective complexes. This supports the model which proposes that phosphorylation of eIF-2 prevents its interaction with eIF-2B.

The presence of the 82K subunit of eIF-2B on the 43S and 48S preinitiation complexes in lysate which are active in protein synthesis raises the issue of whether the regeneration of active eIF-2 takes place only on the preinitiation complex. This notion is difficult to accept since hydrolysis of eIF-2 bound GTP also occurs on the 48S preinitiation complex, presumably to change the conformation of eIF-2 so that it no longer has a high affinity for ribosomes. However, consistent with the idea that eIF-2 accumulating in defective preinitiation complexes cannot interact with eIF-2B and recycle, preliminary data show that the addition of eIF-2B or GTP to inhibited lysate overcomes the accumulation of defective complexes, although the addition of eIF-2 has no effect (data not shown).

If the regeneration of active eIF-2 does indeed take place on preinitiation complexes, an older model of cumulative inhibition of eIF-2 by phosphorylation might be revised (Ochoa and deHaro, 1979; Jagus *et al.*, 1982). This model was consistent with the kinetics of inhibition and the low steady state phosphorylation, but did not consider the interaction of eIF-2 with eIF-2B. Briefly, the earlier model proposed that, in the absence of hemin, as eIF-2 proceeded through its activity cycle, that proportion which was transiently phosphorylated became converted irreversibly to a form which could not be recycled. With each successive initiation cycle, the remaining eIF-2 entering in a phosphorylated state exits in an inactive form. This leads to the accumulation of inactive eIF-2 and a steady reduction of protein synthesis. If the regeneration of active eIF-2 can only take place on preinitiation complexes and if eIF-2B cannot be associated with eIF-2 bound to the defective complexes formed with phosphorylated eIF-2, then eIF-2 will be discharged from such complexes in an active form.

It should be emphasized that many of these data on the distribution of eIF-2 and eIF-2B in lysate are inconsistent with information derived from fractionated reconstituted systems. For instance, regeneration of active eIF-2 can be promoted by eIF-2B *in vitro* without the involvement of other components of the preinitiation complex (Siekierka *et al.*, 1982a,b; Safer *et al.*, 1982a). However, although partial reactions of the eIF-2 activity cycle can be recreated in

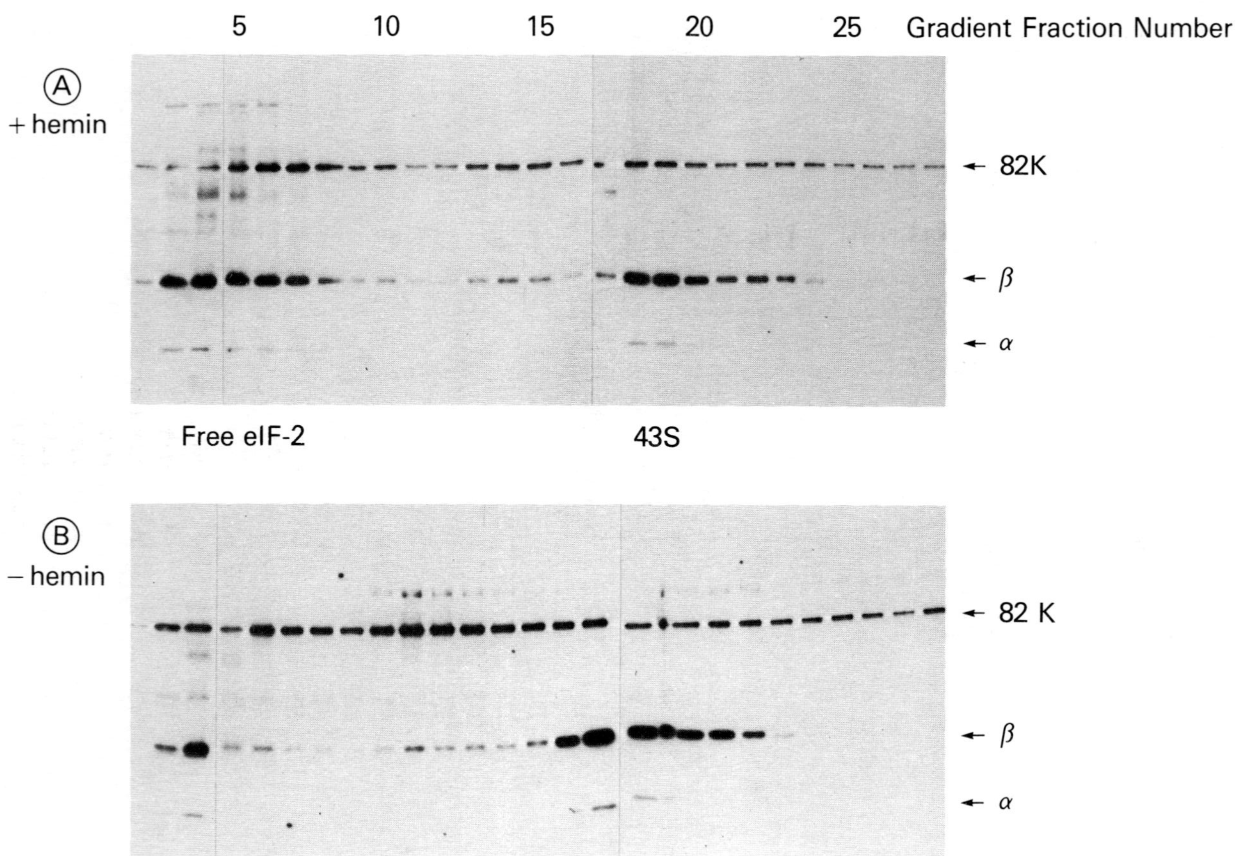

5
10
15
20
25
Gradient Fraction Number
A
+ hemin
← 82K
← β
← α
Free eIF-2
43S
B
− hemin
← 82 K
← β
← α

solution, it is difficult to know at what point in the cycle such reactions normally take place.

In summary, investigations using antibodies have provided some novel insights into the interactions of eIF-2 and eIF-2B and extended our understanding of the role of phosphorylation in the inhibition of protein synthesis. At least one subunit of eIF-2B, the 82K peptide, is present in equimolar amounts with eIF-2. Both eIF-2 and eIF-2B occur in stoichiometric amounts in the 43S preinitiation complex in lysate which is active in protein synthesis. However, in the absence of hemin, eIF-2 accumulates in defective complexes, whereas eIF-2(82K) does not. Thus, the phosphorylation of eIF-2 appears to prevent its association with eIF-2B in the preinitiation complex causing the accumulation of inactive eIF-2 and the inhibition of protein synthesis.

Acknowledgements

The author is very grateful to Drs. Julian Gordon, Peter Nielson and Hans Trachsel for instruction in the use of immunochemical techniques. The electrophoresis and transfer conditions were established by these investigators at the FMI, Basel, Switzerland. The author also wishes to thank Dr. Brian Safer for encouragement in the further development of these techniques and helpful discussions. Thanks are also extended to Mr. Stewart and colleagues at the NIH farm, Beltsville, MD, for production of the sheep antiserum to eIF-2, and to Dr. D.M.F. Cooper for careful reading of the manuscript and to Ms. Bonnie Richards for expert secretarial assistance.

FIGURE 3. Effects of hemin deficiency on the distribution of eIF-2 and eIF-2B(82K). Reticulocyte lysate was incubated in the presence or absence of hemin for 15 minutes and fractionated on sucrose density gradients as described in the legend to Figure 2. Transfer of proteins to nitrocellulose and further processing was performed as described under Figure 1.

REFERENCES

Adamson, S.D., Herbert, E., and Godchaux, W. (1968). *Arch. Biochem. Biophys. 125*, 671.

Amesz, H., Goumans, H., Haubrich-Morree, T., Voorman, H.O., and Benne, R. (1979). *Eur. J. Biochem. 98*, 513.

Anderson, W.F., Bosch, L., Cohn, W.E., Lodish, H., Merrick, W.C., Weissbach, H., Wittmann, H.G., and Wool, I.G. (1977). *FEBS Lett. 76*, 1.

Balkow, K., Mizuno, S., and Rabinovitz, M. (1973). *Biochem. Biophys. Res. Commun. 54*, 315.

Barrieux, A., and Rosenfeld, M.G. (1977). *Biochemistry 16*, 514.

Das, A., Bagchi, M., Roy, R., Ghosh-Dastidar, P., and Gupta, N.K. (1982). *Biochem. Biophys. Res. Commun. 104*, 89.

deHaro, C., and Ochoa, S. (1979). *Proc. Natl. Acad. Sci. USA 76*, 2163.

Farrell, P.J., Balkow, K., Hunt, T., Jackson, R.J., and Trachsel, H. (1977). *Cell 11*, 187.

Farrell, P.J., Hunt, T., and Jackson, R. (1978). *Eur. J. Biochem. 89*, 517.

Goumans, H. (1981). Ph.D. Thesis, University of Utrecht, The Netherlands.

Goumans, H., Amesz, H., Voorma, H.O., and Benne, R. (1983). *Current Topics in Cellular Regulation 22*, in press.

Gross, M. (1979). *J. Biol. Chem. 254*, 2370.

Gross, M., and Mendelewski, J. (1977). *Biochem. Biophys. Res. Commun. 74*, 559.

Gupta, N.K. (1981). *In* "Protein Synthesis in Eukaryotes", R. Bercoff, ed., Plenum Press, New York.

Hunt, T. (1979). *Miami Winter Symp. 16*, 321.

Hunt, T. (1980). *In* "Recently Discovered Systems of Enzyme Regulation by Reversible Phosphorylation", P. Cohen, ed., Mol. Aspects of Cell Regulation Vol. 1, p. 175. Elsevier/North Holland, Amsterdam.

Jagus, R,. and Safer, B. (1981). *J. Biol. Chem. 256*, 1324.

Jagus, R., Anderson, W.F., and Safer, B. (1981). *Prog. Nucleic Acid Res. Mol. Biol. 25*, 127.

Jagus, R., Crouch, D., Konieczny, A., and Safer, B. (1982). *Current Topics in Cellular Regulation 21*, 35.

Kaplansky, D.A., Kwan, A., and Gross, M. (1982). *J. Biol. Chem. 257*, 5722.

Konieczny, A., and Safer, B. (1983). *J. Biol. Chem.*, in press.

Leroux, A., and London, I.M. (1982). *Proc. Natl. Acad. Sci. 79*, 2147.

Levin, D.H., Ranu, R.S., Ernst, V., and London, I.M. (1976). *Proc. Natl. Acad. Sci. USA 73,* 3112.

Levin, D., Ernst, V., Leroux, A., Petryshyn, R., Fagard, R., and London, I.M. (1979). *Proc. Natl. Acad. Sci. 73,* 128.

Lloyd, M.A., Osborne, J.R., Safer, B., Powell, G.M., and Merrick, W.C. (1980). *J. Biol. Chem. 255,* 1189.

Ochoa, S., and deHaro, C. (1979). *Annu. Rev. Biochem. 48,* 549.

Peterson, D.T., Safer, B., and Merrick, W.C. (1979). *J. Biol. Chem. 254,* 7730.

Safer, B., and Jagus, R. (1979). *Proc. Natl. Acad. Sci. 76,* 1094.

Safer, B., and Jagus, R. (1981). *Biochimie 63,* 709.

Safer, B., Adams, S.L., Anderson, W.F., and Merrick, W.C. (1975). *J. Biol. Chem. 250,* 9076.

Safer, B., Peterson, D., and Merrick, W.C. (1978). *In* "Translation of Natural and Synthetic Polynucleotides, A.B. Legogki, ed., 24. Univ. of Agriculture, Poznan, Poland.

Safer, B., Kemper, W., and Jagus, R. (1978). *J. Biol. Chem. 253,* 3384.

Safer, B., Jagus, R., and Crouch, D. (1980a). *J. Biol. Chem. 255,* 6913.

Safer, B., Jagus, R., Crouch, D., and Kemper, W. (1980b). *In* "Protein Phosphorylation and Bioregulation", G. Thomas, E.J. Podesta and J. Gordon, eds., 142. EMBO Workshop, S. Karger, Basel.

Safer, B., Jagus, R., Konieczny, A., and Crouch, D. (1982a). *In* "Interaction of Translational and Transcriptional Controls", M. Grunberg-Manago and B. Safer, eds., 311. Elsvier Biomedical, New York.

Safer, B., Grunberg-Manago, M., Badman, D., Bergmann, F., Freeman, C., Galasso, G., Jagus, R., and Williams, B. (1982b). *FEBS Lett. 147,* 1.

Siekierka, J.J., and Ochoa, S. (1982). *In* "Symposium on Hemoglobin Biosynthesis", E. Goldwasser, ed. Elsevier/ North Holland, in press.

Siekierka, J., Mitsui, K.I., and Ochoa, S. (1981). *Proc. Natl. Acad. Sci. USA 78,* 220.

Siekierka, J.J., Mauser, L., and Ochoa, S. (1982a). *Proc. Natl. Acad. Sci. USA 79,* 2537.

Siekierka, J.J., Mauser, L., and Ochoa, S. (1982b). *Proc. Natl. Acad. Sci. 79,* 327.

Trachsel, H., Erni, B., Schreier, M.H., and Staehelin, T. (1977). *J. Mol. Biol. 116,* 755.

Voorma, H.O., and Amesz, H. (1982). *In* "Interaction of Translational and Transcriptional Controls", M. Grunberg-Manago and B. Safer, eds., 297. Elsevier Biomedical, New

York.
Walton, G.M., and Gill, G.N. (1975). *Biochim. Biophys. Acta 390*, 231.
Walton, G.M., and Gill, G.N. (1976). *Biochim. Biophys. Acta 447*, 11.
Zucker, W.V., and Schulman, H.M. (1968). *Proc. Natl. Acad. Sci. USA 59*, 582.

DISCUSSION

Olson: Let me ask you a very naive question. How does heme inhibit the protein kinase?

Jagus: It is really not known, mainly because it is very difficult to purify the kinase in a reversibly active form. Any purification procedure seems to cause permanent activation making it impossible to use the purified kinase for the study of the activation process. There is some data in the literature suggesting that heme does bind directly to the kinase to keep it inactive, but it is not really very convincing evidence since the concentration of hemin used was very high. I'm not really sure that the effect of hemin represents a physiological mechanism. Reticulocytosis is induced by phenylhydrazine which induces hydrogen peroxide generation in the red cells causing membrane lipid peroxidation and lysis. It is likely therefore that the reticulocytes used to produce the cell-free protein synthesis system have suffered some oxidative damage and that hemin overcomes this by some kind of redose function.

Cohen: Is it true that if eIF-2 is added to hemin-deficient lysate, protein synthesis is restored, although if phosphorylated eIF-2 is added there is not restoration?

Jagus: No, it is not possible to add phosphorylated eIF-2.

Cohen: The problem is that eIF-2 is dephosphorylated very rapidly?

Jagus: Yes.

Cohen: So I would suggest adding thiophosphorylated eIF-2 to see whether that would restore protein synthesis. Thiophosphorylated eIF-2 wouldn't be dephosphorylated, so if you added high levels and found that it was used stoichiometrically, this would support your observation?

Jagus: Yes.

Johnson: I'm wondering, does the heme in your system do the same as it does in our carboxylation system; that is, run the system out of oxygen almost immediately and therefore nothing goes?

Jagus: It's possible and an attractive idea, but with the very high concentration of hemoglobin in the lysate, I don't think it's quite so simple.

Kahn: There are a couple of parallels I noticed between your protein and adenylate cyclase. I wonder if you have done anything, for example, with ADP-ribosylation or 8-azido GTP binding, to try and see which of the subunits bind guanine nucleotides?

Jagus: A little has been done with 8-azido GTP binding but I have not been very happy with the results. In fact it binds to both the alpha subunit and the beta. It does look as if the beta subunit is involved in the recycling mechanism. For instance, I can ADP-ribosylate the β-subunit of eIF-2 which converts it to a form which is utilized stoichiometrically. So of these three subunits, one (α-) seems to be involved in GTP-binding and regulation by phosphorylation. One (γ-) is the catalytic subunit and the β-subunit seems probably to be the one which interacts with eIF-2B.

Kahn: Which one of those is alpha or beta? Alpha is phosphorylated?

Jagus: Yes.

Kahn: Beta interacts with eIF-2B?

Jagus: Yes, with eIF-2B, I think.

Kahn: And gamma does what?

Jagus: It is the catalytic subunit which binds Met-$tRNA_f$.

Roach (Indiana U.): Does phosphorylation of the β-subunit of beta do anything?

Jagus: The beta subunit is phosphorylated; it has two moles of phosphate in the subunit. As far as anyone knows, the phosphorylation state does not change in lysate. I have measured the phosphate content of β-subunit purified from hemin-supplemented and hemin-deficient lysate and there appears to be no difference.

Cohen: Under what conditions did you take the phosphate out?

Jagus: I haven't done that.

Cohen: You haven't removed phosphate from the β-subunit *in vitro* and then put it back into lysate?

Jagus: No.

Cohen: Do you know what happens if you take it out? Does it affect the phosphorylation of the alpha subunit?

Jagus: Not *in vitro*, but I don't know about *in vivo*. I have some intriguing hints that the phosphorylation of the β-subunit may be a more complicated story than we previously suspected. When I first tried to quantitate the levels of the eIF-2 in the preinitiation complexes, I used phosphorylated eIF-2, phosphorylated in the beta subunit, because this is not dephosphorylated when added to the lysate. Using this method of quantitation, the answer I got was totally the opposite to the answer with the antibodies; the amount of the eIF-2 in the detectable preintiation complexes seemed to go down and so I think there may be some normal phosphorylation-dephosphorylation event on the beta subunit in the preinitiation complex, which may well have something to do with the recycling mechanism.

Cohen: Are any of the eIF-2B subunits phosphorylated?

Jagus: Yes, the 67K subunit is.

Cohen: By what?

Jagus: Partially purified activities from lysate. This doesn't seem to affect activity of the protein *in vitro*. The problem about this is that in some investigations, Harry Voorma's for instance, while the 67K is really an eIF-2B subunit. He finds that the reversal activity doesn't co-elute with that particular peptide.

Cohen: If someone tells me that the 65K subunit of a protein is phosphorylated, my reaction is that impurities affect this complex.

Jagus: Ah, the 67K is subunit and the rest is eIF-2.

Kahn: The blotting and antibody experiments you showed us toward the end; were those all done with polyclonal antibody?

Jagus: Yes. The problem with monoclonals is that they recognize only six or seven amino acids. Many initiation factors bind GTP or ATP, and they all have very immunogenic hydrophobic regions. This is probably why monoclonal antibodies cross-react with so many other proteins. It's really disappointing. So I do, in fact, get much cleaner looking pictures with polyclonals.

Ingebritsen (Colorado): Why does the polyclonal antibody recognize the 82K subunit of eIF-2B?

Jagus: It reflects the fact that the eIF-2 preparation, which I injected into the sheep, had some 82K contamination; perhaps something that is extremely immunogenic.

AFFINITY PROBES FOR PROTEIN ACETYLATION AND PHOSPHORYLATION

Vincent G. Allfrey
Richard Sterner
Irene Y.C. Sun

The Rockefeller University
New York, New York

The postsynthetic modifications of chromosomal proteins are dynamic and complex. They include multiple forms of phosphorylation, acetylation, methylation, ADP-ribosylation, and covalent attachments to DNA. In addition, there are glycosylations and peptide chain branching reactions of high specificity. In analyzing the enzymatic mechanisms, substrate specificities, and functions of these reactions it would be very useful to have probes that permit the selective recovery of the most recently modified polypeptide chains and derivative fragments containing the modifying group. Two such recovery probes will be described; one for protein acetylation reactions, and the other for protein phosphorylations. Both probes were developed for studies of the *in vivo* modifications of nuclear proteins, but they are also applicable to proteins in other parts of the cell.

ACETYLATION OF LYSINE ε-AMINO GROUPS OF DNA-BINDING PROTEINS

Interest in the acetylation of the histones and high mobility group (HMG) proteins stems from the recognition that

[1]*This research was supported in part by BRSG S07 RR07065 awarded by the Biomedical Research Support Grant Program, Division of Research Resources, National Institutes of Health, by NIH Grant GM 17383, and by Grant NP-288 from the American Cancer Society.*

ISBN 0-12-387560-9

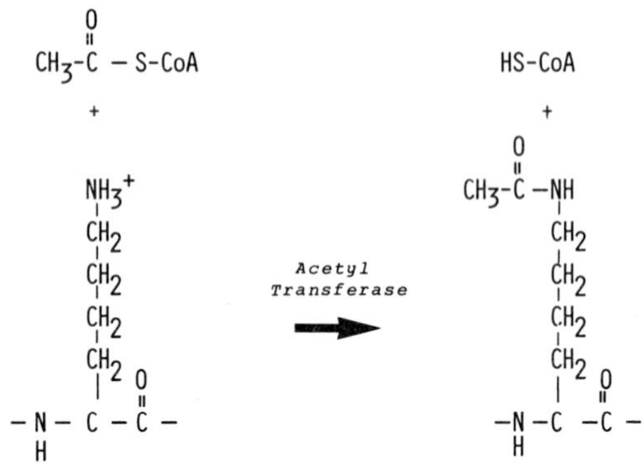

FIGURE 1. Schematic representation of the postsynthetic modification of protein lysine residues by acetylation. Note that acetyl group transfer from acetyl-Coenzyme A to the ε-amino group of lysine results in the neutralization of its positive charge.

this type of postsynthetic modification could play an important role in the regulation of chromatin structure by modulating the interactions between the basic domains of the proteins and the negatively-charged phosphate groups of DNA. The subject has been reviewed recently (Allfrey, 1980) and only a few highlights will be summarized here. The reaction involves the transfer of an acetyl group from acetyl-Coenzyme A to an ε-amino group of a lysine residue within the polypeptide chain, as shown in Figure 1. Perhaps the most significant aspect of this is the charge neutralization associated with the reaction, since it would be expected to weaken the electrostatic interactions between the postively-charged lysine residues and the negatively-charged DNA strand. This role of acetylation is made more likely by the fact that the sites of acetylation occur in clusters in the basic amino-terminal regions of all the histone classes comprising the nucleosome "core". Histone H2A has an acetyl group at position 5; histone H2B has acetyl groups at positions 5, 10, 13 and 18; histone H3 is acetylated at positions 9, 14, 18 and 23, and ε-N-acetyllysine may occur at positions 5, 8, 12 and 16 of histone H4, as shown in Figure 2. Thus the eight histones of the nucleosome "core" could contain as many as 26 acetyl groups on the appropriate lysine residues. A distinction has to be made between these "internal" acetylations affecting lysine ε-amino groups within the polypeptide and the acetylation of the amino-terminal serine residues in histones H2A and H4. The latter reaction takes place in the

HISTONE H4

```
   P              Ac*          Ac*             Ac*             Ac*              Me
*Ac-Ser-Gly-Arg-Gly-Lys-Gly-Gly-Lys-Gly-Leu-Gly-Lys-Gly-Gly-Ala-Lys-Arg-His-Arg-Lys-
    1              5            8    10    12              16                    20

-Val-Leu-Arg-Asp-Asn-Ile-Gln-Gly-Ile-Thr-Lys-Pro-Ala-Ile-Arg-Arg-Leu-Ala-Arg-Arg-Gly-
                                            30                                   40

-Gly-Val-Lys-Arg-Ile-Ser-Gly-Leu-Ile-Tyr-Glu-Glu-Thr-Arg-Gly-Val-Leu-Lys-Val-Phe-Leu-
                                     50                                  60

-Glu-Asn-Val-Ile-Arg-Asp-Ala-Val-Thr-Tyr-Thr-Glu-His-Ala-Lys-Arg-Lys-Thr-Val-Thr-Ala-
                             70                                  80

-Met-Asp-Val-Val-Tyr-Ala-Leu-Lys-Arg-Gln-Gly-Arg-Thr-Leu-Tyr-Gly-Phe-Gly-Gly-COOH
                         90                                  100       102
```

FIGURE 2. Amino acid sequence of calf thymus histone H4 showing its modifications by acetylation, phosphorylation, and methylation. There are multiple sites of acetylation in the amino-terminal region of the molecule, but this is a dynamic reversible process and individual H4 molecules may contain 0 to 4 ε-N-acetyllysine residues.

cytoplasm at the time of histone synthesis (Ruiz-Carrillo *et al.*, 1975) and the acetyl-serine linkage remains intact for the life of the protein.

In most cells, the internal acetyl groups are subject to turnover, and the dynamics of histone acetylation and deacetylation are rapid and complex. There are a number of enzymes which carry out the acetyl group transfer reactions, some specific for different proteins in the nucleus, such as histone H4 (Wiegand and Brutlag, 1981) and some with broader specificities (Belikoff *et al.*, 1980). Multiple deacetylase activities have also been described. Both acetylation and deacetylation take place in the cell nucleus and autoradiographic experiments clearly indicate that the incorporation of radioactive acetate occurs within the chromosomes of *Drosophila melanogaster* (Allfrey *et al.*, 1968).

Electrophoretic techniques are employed to monitor the transitions in histone acetylation levels. The usual way is to separate the histones in acid-urea-polyacrylamide gels (Panyim and Chalkley, 1969). The method not only separates the major histone classes, but also resolves molecules of a given class according to their charge. The neutralization of lysine residues due to acetylation of the positively-charged ε-amino groups reduces the electrophoretic mobility in proportion to the number of acetyl groups present. Histone H4, for example, shows a rapidly migrating non-acetylated form followed by four bands corresponding to the mono-, di-, tri-, and tetra-acetylated forms, respectively. After labeling experiments with radioactive acetate, the leading band is non-radioactive while the slowly-moving bands are isotopically labeled. So it is a very simple matter to monitor the level of acetylation of a histone and to measure the uptake of acetate into the various histone subfractions. This is true for all of the nucleosomal "core" histones, but it is particularly easy to do for histone H4, which travels out ahead of the others.

An interesting phenomenon is that when cells are placed in media containing millimolar concentrations of sodium butyrate, the activity of the histone deacetylases is inhibited. The dynamic balance between acetylation and deacetylation is shifted to favor accumulation of the multiacetylated forms of both the histones and the HMG proteins. Depending upon butyrate concentration and the time of exposure, most of the histones, H3 and H4 in particular, appear as multiacetylated species (Riggs *et al.*, 1977). When butyrate is removed, there is a rapid reversal of the effect, and the histones revert to less acetylated forms within minutes (Vidali *et al.*, 1978b; Covault *et al.*, 1982). This is a matter of some concern when you are trying to analyze histone acetylation. The deacetylases operate very rapidly and it is essential to add butyrate immediately in the preparation of nuclei, chromatin, or chromosomal proteins. Otherwise, kinetic studies of acetylation become essentially meaningless, and one may fail to detect the modification of lysine residues with high rates of acetyl group turnover (Sterner *et al.*, 1981).

The hyperacetylation of histones in butyrate-treated cells is associated with changes in chromatin structure and transcriptional activity (Allfrey, 1980). However, the question of causation remains open, because butyrate also affects the phosphorylation of histones and other nuclear proteins (Boffa *et al.*, 1981). Nevertheless, there are many close correlations between increased levels of histone acetylation and gene activation for RNA synthesis that strongly support the view that acetylation provides a reversible enzymatic mechanism for the control of DNA accessibility to RNA-synthesizing enzymes.

Just a few examples of this relationship will be cited here. Among the earliest systems to be investigated was the activation of transcription in lymphocytes stimulated by mitogens such as phytohemagglutinin (PHA) (Pogo *et al.*, 1966). The kinetics of RNA synthesis in PHA-stimulated cells, as measured by pulse-labelling experiments with radioactive uridine, were compared with the corresponding incorporations of acetate into the arginine-rich histones, H3 and H4. The

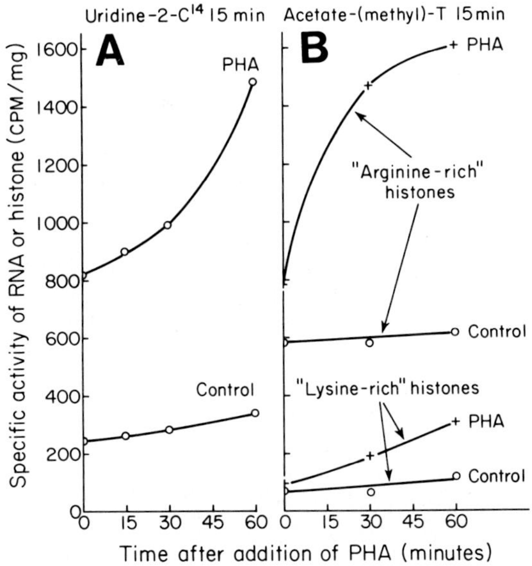

FIGURE 3. Comparative kinetics of uridine incorporation into RNA (A) and acetylation of histones (B) in control and phytohemagglutinin-stimulated lymphocytes. Note that the increase in acetylation of the arginine-rich histones (H3 and H4) precedes the increase in RNA synthetic activity in the mitogen-stimulated cells.

results (Figure 3) clearly indicate that histone acetylation precedes the increase in RNA synthesis, as would be expected if this modification acts to release constraints on DNA template function. The increase in DNA accessibility can be demonstrated in other ways, for example, by an increased binding of Acridine Orange (Killander and Rigler, 1969) or Actinomycin D (Darzynkiewicz *et al.*, 1969). Similar temporal correlations between increased acetylation and gene activation have been noted in many other systems, most notably in tissues responding to hormonal stimulation (Allfrey, 1980). The response is usually seen within minutes. For instance, there is over a ten-fold increase in the acetylation of uterine histones only 10 minutes after the injection of estradiol (Pasqualini *et al.*, 1981).

Another way to look at the phenomenon is to ask whether acetylation takes place preferentially in the transcriptionally active regions of chromatin. Experiments on *Planococcus* show convincingly that the uptake of radioactive acetate into the transcriptionally active maternal chromosomes greatly exceeds that in the transcriptionally suppressed paternal chromosomes (Berlowitz and Pallotta, 1972) (see Figure 4). Similarly, comparisons of the histones from the RNA-synthesizing macronuclei of *Tetrahymena* with those of the transcriptionally inert micronuclei clearly localize the acetylated histones in the active macronuclei while the micronuclear histones are not acetylated (Gorovsky *et al.*, 1973). Indeed, the micronuclei appear to lack the necessary enzymes for acetyl group transfer (Vavra *et al.*, 1982).

The results are consistent with the view that acetylation is part of the mechanism for relaxing the grip of the nucleosomal histones on the DNA strand. The increased accessibility of the DNA in active chromatin is evident in the susceptibility of transcribed genes to degradation by DNAse I (Weintraub and Groudine, 1976; Garel and Axel, 1976). Limited DNAse I digestions not only degrade the transcriptionally committed regions of chromatin but also release preferentially the multiacetylated forms of histones H3 and H4, thus confirming their localization on the transcriptionally active DNA sequences (Vidali *et al.* 1978a; Sealy and Chalkley, (1978).

The degradation of active chromatin by DNAse I also results in a selective release of other chromosomal proteins (Vidali *et al.*, 1977). Two proteins of the high mobility group, HMG-14 and HMG-17, appear to be particularly concerned with the maintenance of the DNAse-sensitive state (Weisbrod and Weintraub, 1979; Weisbrod *et al.*, 1980; Weisbrod, 1982). Both of these HMG proteins have multiple sites of acetylation and, as in the histones, the modification is restricted to lysine residues in the basic amino-terminal regions (Sterner *et al.*, 1981). Similar acetylations occur in the larger HMG proteins, HMG-1 and HMG-2, and are also restricted to their presumptive DNA-binding domains (Sterner *et al.*, 1978; 1979). Thus the horizons of acetylation have expanded to include not just the histones but other DNA binding proteins, some of which are known to be associated with transcriptionally active regions.

An important question is whether the acetylation of the HMG proteins is positively correlated with increases in transcription. We have just completed some experiments (in collaboration with Dr. Jorge Pasqualini of the C.N.R.S. Hormone Research Unit, Paris) on the stimulation of fetal and neonatal guinea pig uteri by estrogens. A sharp increase in the acetylation of uterine HMG's was evident within 10 minutes after injection of the hormone, but there was no corresponding stimulation of acetate uptake into the HMG's of nontarget tissues. We conclude, tentatively, that the acetylation of HMG proteins also serves to release constraints on DNA accessibility for transcription.

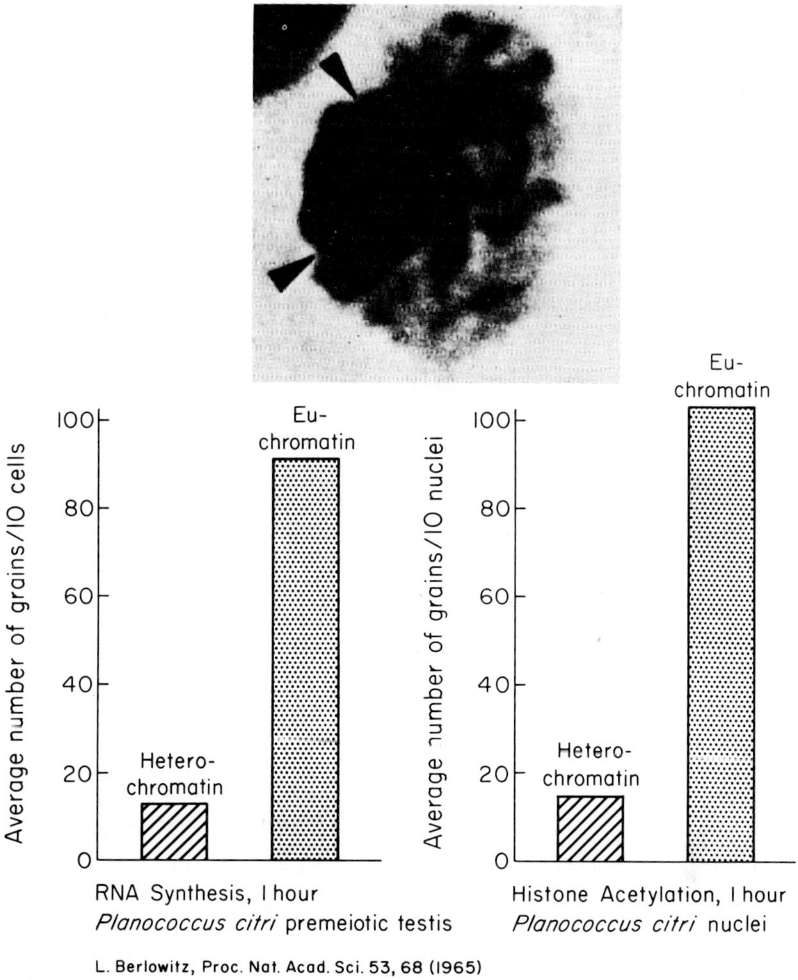

FIGURE 4. Comparisons of histone acetylation and RNA synthesis in the hetero- and eu-chromatin of Planococcus citrii. Quantitative autoradiography of the nuclei after pulse-labelling in vivo *with radioactive uridine or radioactive acetate shows that both RNA synthesis and histone acetylation are predominantly localized in the active (eu-) chromatin (which comprises the maternal chromosome set).*

In dealing with a subject like this, where there are clearly multiple sites of acetylation within a single polypeptide chain, as well as other types of postsynthetic modifications (such as phosphorylation, ADP-ribosylation, methylation and glycosylation), it would be a great advantage to be able to recall the molecules that were just acetylated. Richard Sterner and I have developed a new approach which employs the acetate analogue, 2-mercaptoacetate, as an affinity probe. The cell converts the analogue to mercaptoacetyl-Coenzyme A, which then donates the thio-derivatized acetyl group to the lysine residues of histones and HMG proteins (Figure 5). The advantage of this approach is that the protein molecules which have just been modified can be recovered selectively by affinity chromatography on organomercurial-Sepharose columns. Because of their newly acquired sulfhydryl groups, the mercaptoacetylated histones and HMG molecules are retained by the Hg-column while the unmodified protein molecules pass through. The thio-derivatized proteins are then eluted with dithiothreitol and analyzed in various ways to determine which part of the molecule contains the mercaptoacetylated lysine residue (Sterner and Allfrey, 1982).

Examples of this approach are illustrated in Figures 6 and 7, which compare the electrophoretic patterns of the histones and HMG proteins prepared from duck erythrocytes following incubations with either radioactive acetate or non-radioactive mercaptoacetate. Hg-affinity chromatography of the radioactive histones from the acetate-labelled cells shows that only histone H3 is retained by the column (Figure

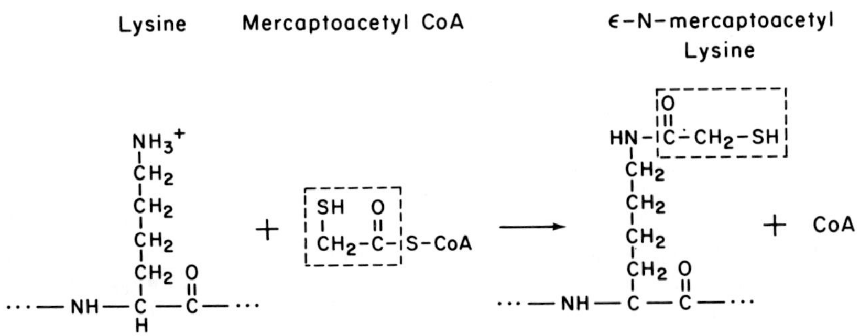

FIGURE 5. Schematic representation of the mercapto-acetylation of protein lysine residues, showing the reaction analogous to that depicted in Figure 1. The moiety enzymatically transferred from Coenzyme A to lysine-ε-amino groups is enclosed by the dashed lines. The sulfhydryl group permits the isolation of the modified protein by Hg-affinity chromatography.

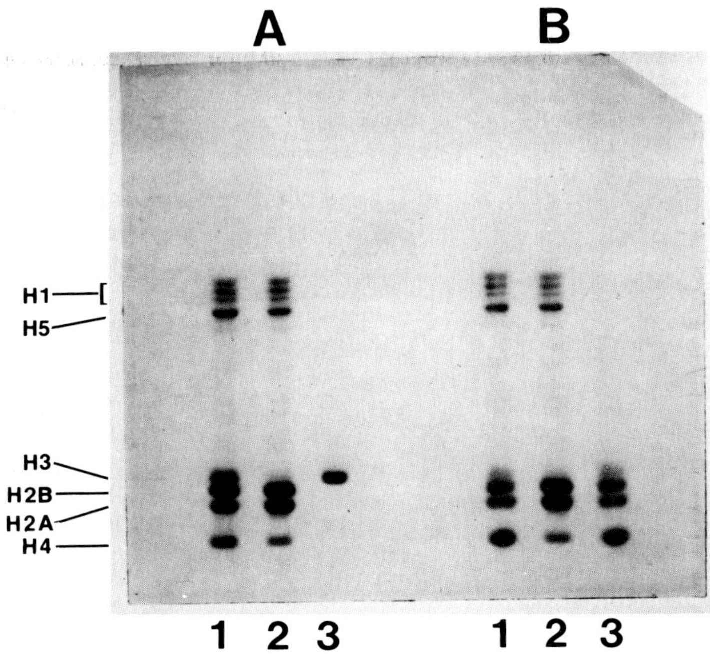

FIGURE 6. Electrophoretic analysis of avian erythrocyte histone fractions separated by Hg-affinity chromatography. Normally acetylated and mercapto-acetylated histones were separately chromatographed on organomercurial-Sepharose columns. The proteins seen in (A) are normally acetylated histones. 1, histone standard (sample of the total histone extract prior to Hg-affinity chromatography). 2, protein peak eluting in the void volume. 3, protein peak initially retained by the column and subsequently eluted with 10 mM 2-mercaptoethanol. Proteins seen in (B) are from mercapto-acetate-treated cells. 1, histone mixture (depleted of H3) before Hg-affinity chromatography. 2, protein peak eluting in the void volume. 3, proteins initially retained on the column and subsequently eluted in 10 mM 2-mercaptoethanol. Note that histones H2A, H2B and H4 are retained by the column only if they are mercaptoacetylated.

6A, lane 3), as expected, since H3 contains cysteine in its polypeptide chain and is the only histone which has sulfhydryl amino acids. Contrast this result with the separations of mercaptoacetylated histones in Figure 6B. A mixture of these histones (omitting H3) was subjected to Hg-affinity chromatography in the same way, but the retained histones now include H2A, H2B and H4 (Figure 6B, lane 3). Not all of these histones are retained by the column, only those containing the mercaptoacetyl groups. The number of -SH groups can be titrated by reaction with tritiated iodoacetate at pH 8.6 (Sterner and Allfrey, 1982), and it was calculated that about 1 in 800 H4 molecules were mercaptoacetylated under these conditions.

It has also been established that the mercaptoacetyl groups of histone H4 occur only in the chymotryptic fragment (residues 1-22) containing the known sites of enzymatic acetylation of the protein (Sterner and Allfrey, 1982).

The mercury-affinity chromatographic behavior of mercaptoacetylated HMG proteins is compared with that of normally acetylated HMG's in Figure 7. In the experiment using radioactive acetate only HMG-1 and HMG-E were retained by the column (Figure 7A, lane 3) and these proteins were adsorbed quantitatively (as expected, since both contain cysteine residues in their polypeptide chains). In experiments using mercaptoacetate, all four of the HMG proteins were retained by the mercury column (Figure 7B, lane 3), indicating that the transfer of mercaptoacetyl groups to HMG-14 and HMG-17 (which do not contain cysteine) had occurred. Based on the amount of protein adsorbed onto the column relative to the amount originally applied, it could be calculated that about 1 in 1000 HMG molecules had been mercaptoacetylated under these conditions.

The ability to recover selectively the most recently modified HMG molecules opens the way to the solution of some fundamental problems. For example, we know that HMG-14 and HMG-17 are also phosphorylated, but we do not know if the same molecules can be phosphorylated and acetylated at the same time. Using the mercaptoacetate in a double-labeling protocol with radioactive phosphate, one can recover the mercaptoacetylated molecules and determine if they are also phosphorylated. Such experiments are now in progress. Similar approaches can be used to study the simultaneity of acetylation and other modifications, such as ADP-ribosylation. Other applications of the mercaptoacetylation procedure include the identification and recovery of other acetyltransferase substrates such as polyamines and, possibly, proteins to which they become attached.

Further applications of the mercaptoacetate probe are now being tested. They include the development of affinity chromatographic techniques for the separation of protein molecules differing in their degree of thio-derivatization. It may also be possible to extend these procedures to permit the separation of chromatin subunits according to their contents

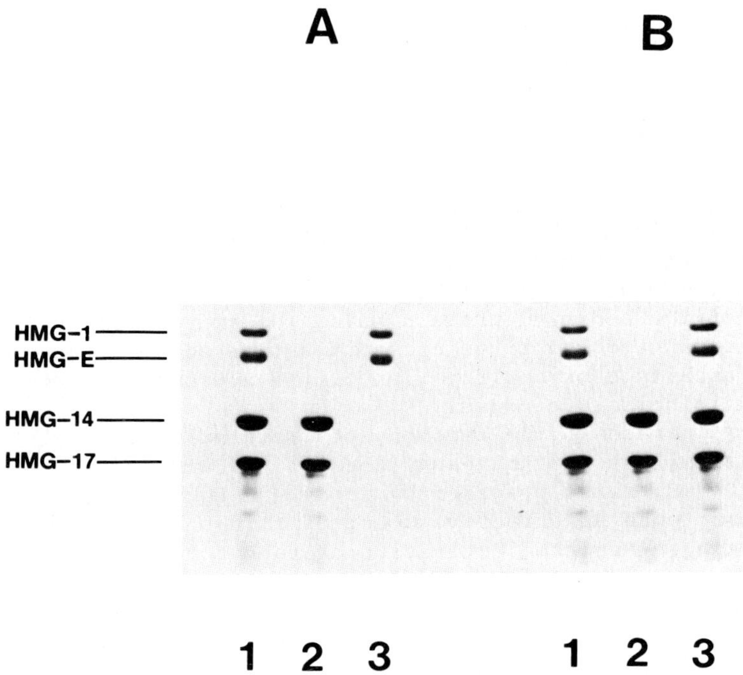

FIGURE 7. Electrophoretic analysis of avian erythrocyte HMG proteins separated by Hg-affinity chromatography. Normally acetylated and mercaptoacetylated HMG fractions were separately chromatographed on organomercurial-Sepharose columns. The proteins depicted in (A) are normally acetylated HMGs. 1, HMG standard (sample of the normally acetylated HMG extract prior to Hg-affinity chromatography). 2, protein peak eluting in the void volume. 3, protein peak initially retained on the affinity column and subsequently eluted in 10 mM 2-mercaptoethanol. Proteins depicted in (B) are from mercaptoacetate-treated cells. 1, sample of the total mercaptoacetylated HMGs prior to Hg-affinity chromatography. 2, protein eluting in the void volume. 3, protein initially retained on the column and subsequently eluted in 10 mM 2-mercaptoethanol. Note that HMG-14 and HMG-17 are retained by the column only if they are mercaptoacetylated.

of mercaptoacetylated histones and HMG proteins. If so, a direct test can be made of the association between hyperacetylation of chromosomal proteins and the transcriptional activity of the associated DNA sequences.

AFFINITY PROBES FOR PROTEIN PHOSPHORYLATION

The postsynthetic modification of nuclear proteins by phosphorylation involves both histone and nonhistone components of chromatin. The functional significance of these reactions is largely unknown, but clearly complex. Even with well-defined substrates, such as histone H1, the functional role of phosphorylation cannot be deduced simply by measuring overall levels of phosphate incorporation, because there are multiple sites of modification in different domains of the polypeptide chain. What is needed are procedures for the recovery of the modified protein molecules and derivative peptides containing the sites of modification so that one can relate each type of modification to the appropriate stimulus and physiological response.

Our approach to the recovery of proteins phosphorylated *in vitro* employs the thio-analogues of ATP and GTP (Figure 8). ATP-γ-S, first prepared by Goody and Eckstein (1971), was shown to be an effective donor of thio-phosphoryl groups in the kinase-mediated activation of rabbit muscle phosphorylase (Gratecos and Fischer, 1974). It was also effective in the thiophosphorylation of myosin light chains (Morgan *et al.*, 1976). We decided to combine kinase-mediated thicphosphorylations with procedures for the recovery of the thiophosphorylated proteins by Hg-affinity chromatography (Sun *et al.*, 1980).

The first reaction studied employed calf thymus histone H1 as a substrate for a highly purified cyclic AMP-dependent protein kinase from calf thymus lymphocytes. The enzyme was activated with cAMP and incubated with histone H1 and either [γ-^{32}P]ATP or [γ-^{35}S]ATP as the group donor. The respective

ATP-(γ-thio), [^{35}S]- **GTP-(γ-thio), [^{35}S]-**

FIGURE 8. Structures of the thio-derivatized nucleoside triphosphates ATP-γ-S and GTP-γ-S. These serve as thiophosphoryl group donors in the appropriate kinase reactions.

reaction products were then analyzed by affinity chromatography on organomercurial-Sepharose columns (Ruiz-Carrillo and Allfrey, 1973). Figure 9 compares the elution profiles of the ^{32}P-phosphate labeled and ^{35}S-thiophosphate labeled histones. Since calf thymus H1 has no cysteinyl residues it does not bind to the mercury column unless modified by the transfer of thiophosphate groups from ATP-γ-S. Under the conditions employed, virtually all of the histone was thioderivatized and retained by the column (Figure 9A), while the corresponding ^{32}P-labelled histone passed through the column in the void volume (Figure 9B).

Similar studies of the enzymatic transfer of thiophosphoryl groups from ATP-γ-S to other histones and HMG proteins confirmed the general utility of this procedure for selective recovery of the thio-derivatized protein molecules. In the case of proteins containing cysteinyl residues, such as histone H3, the -SH groups can be effectively blocked by carboxymethylation under conditions in which the thiophosphate does not react with iodoacetate (Sun *et al.*, 1980). This makes it possible to recover the recently thiophosphorylated protein molecules despite the presence of reduced cysteinyl residues in the protein sequence.

The use of ATP-γ-S as a thiophosphoryl group donor is also feasible in isolated cell nuclei (Sun *et al.*, 1980) or in permeabilized cells, but because of the restricted permeation of nucleotides through the plasma membrane, it is not an effective precursor for the study of protein phosphorylation in living cells. What is needed is a sulfur-derivatized precursor that can readily enter cells in culture or in the intact animal. We have employed a radioactive analogue of sodium orthophosphate, Na [^{35}S]thiophosphate, and tested for its uptake and utilization by intact cells (Sun and Allfrey, 1982a). HeLa S-3 cells were shown to incorporate this precursor into the γ-positions of ATP and GTP and into other nucleotides. The *in vivo* formation of ATP-γ-S results in the subsequent transfer of thiophosphoryl groups to cellular proteins, including histones H1, H2A, H3 and H4. The newly thiophosphorylated molecules were recovered by Hg-affinity chromatography as described previously.

This approach has been exploited to analyze the *in vivo* phosphorylations of histone H1. In log phase cultures the phosphorylation of H1 takes place at multiple growth-associated sites in the polypeptide chain (Langan, 1978; Hohmann *et al.*, 1976; Hohmann, 1978). The advantage of the thiophosphate labeling procedure, apart from the fact that one recovers only the modified proteins, is that it is possible to identify the sites of modification by Hg-affinity chromatography of the peptides generated by chemical or proteolytic

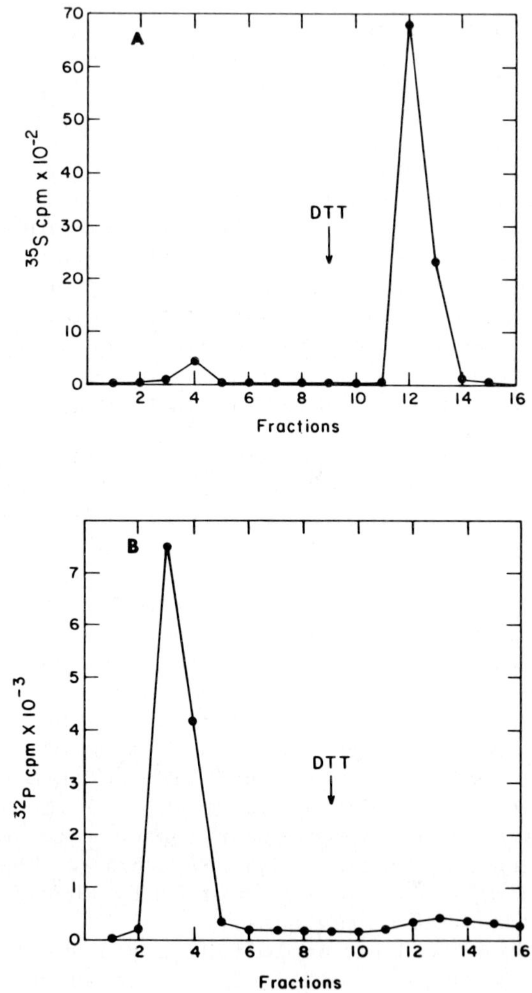

FIGURE 9

degradation of the modified protein (Sun and Allfrey, 1982a). The validity of this approach was tested by comparing the thiophosphorylated peptides of histone H1 with the corresponding ^{32}P-labelled peptides, as shown in Figure 10. It is evident that the sites of phosphorylation are in the same positions in the polypeptide chain.

The thiophosphorylated H1 was cleaved with N-bromosuccinimide to generate the NH_2-terminal and COOH-terminal halves which were then degraded by tryptic digestion and analyzed (Hohmann *et al.*, 1976). Thiophosphate groups were present in both halves of the molecule, and the modified amino acids were identified as thiophosphoserine and thiophosphothreonine (Sun and Allfrey, 1982a).

The use of thiophosphate as an affinity probe also provided an unequivocal answer to the question of whether newly synthesized histone H1 molecules are phosphorylated. When [^{3}H]threonine and nonradioactive thiophosphate were used simultaneously as precursors, the thiophosphorylated H1 molecules retained on the mercury column also showed the presence of the [^{3}H]threonine label.

One additional advantage of thiophosphate labeling should be mentioned, namely, that it permits the separation of recently modified phosphoproteins (or peptides) from similar proteins (or peptides) containing previously incorporated phosphate groups. Since only the sites of most recent modification are recovered, this should simplify studies of site-specific phosphorylations in response to cyclic nucleotides, divalent cations, hormones, growth factors, and other agents affecting cell growth and progression through the cell cycle.

THIOPHOSPHATE AS A RECOVERY PROBE FOR DNA SYNTHESIS

Thiophosphate, once introduced into the cell, goes in many directions. Once it enters ATP it is rapidly trans-

---

FIGURE 9. Affinity chromatographic separation of thiophosphorylated histone H1 molecules. (A), separation on organomercurial-Sepharose of enzymatically thiophosphorylated histone H1 following incubations with ATP-γ-[^{35}S] and purified cAMP-dependent protein kinase. The radioactive thiophosphorylated histone is retained on the column and subsequently eluted in 10 mM dithiothreitol. (B), chromatography of histone H1 enzymatically labeled with ATP-γ-[^{32}P]. In this case the radioactive histone is not retained on the column and emerges in the run-off peak.

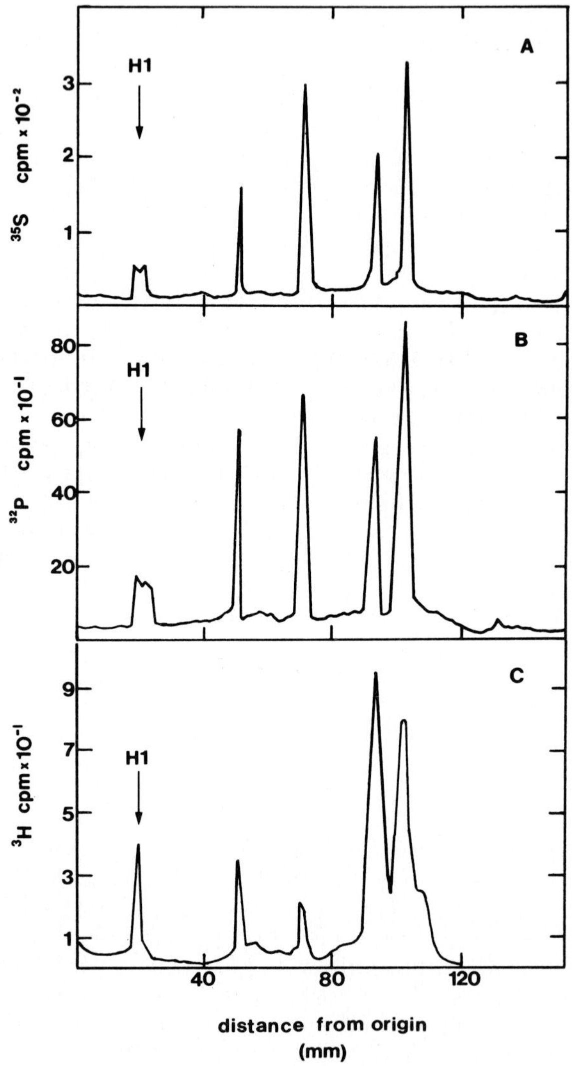
A
H1
^{35}S cpm x 10^{-2}
3
2
1
B
H1
^{32}P cpm x 10^{-1}
80
60
40
20
C
H1
^{3}H cpm x 10^{-1}
9
7
5
3
1
40
80
120
distance from origin
(mm)

FIGURE 10

ferred to other nucleotides and deoxynucleotides. We have recently studied whether thiophosphate can be used to label and recover DNA. CV-1 cells were infected with SV40 virus and incubated in the presence of sodium [^{35}S]thiophosphate. The viral DNA was prepared and the newly synthesized SV40 DNA molecules containing phosphorothioate groups were selectively recovered by affinity chromatography on Hg-Sepharose. The thio-derivatized DNA molecules were synthesized with fidelity, as judged by their patterns of cleavage with the restriction nucleases *Kpn* I and *Hin*d III (Sun and Allfrey, 1982b). Hydrolysis of the ^{35}S-labeled viral DNA to its component 5'-deoxy-nucleotides showed that 86% of the ^{35}S was present in thymidylic acid. It follows that most of the ^{35}S incorporation into DNA represents the phosphorylation of thymidine in the thymidine kinase reaction, in which ATP-γ-S would serve as the thiophosphoryl group donor.

Although the recovery of thio-derivatized DNA is not quantitative, about a third of the newly-labelled molecules were recovered by Hg-affinity chromatography when only 3 to 12 phosphorothioate groups were incorporated per viral genome (5,226 base-pairs).

We anticipate that this technique will find many applications in the study of the mechanism of DNA synthesis (e.g. it will allow the direct recovery of Okazaki fragments) and in analyzing temporal aspects of the replication of different DNA sequences during the cell cycle.

FIGURE 10. Electrophoretic separations of tryptic peptides derived from histone H1 following labeling of HeLa cells with ^{35}S-thiophosphate (A), ^{32}P-orthophosphate (B), or ^{3}H-threonine plus nonradioactive thiophosphate (C). Histone H1 was electrophoretically purified from each culture. The thiophosphorylated forms (in A and C) were recovered by Hg-affinity chromatography. Each histone preparation was digested with trypsin and the fragments were separated electrophoretically. Note the parallel distribution of ^{35}S-thiophosphate and ^{32}P-orthophosphate labeled peptides in A and B, respectively. A similar size distribution of labeled peptides is seen in the newly synthesized and thiophosphorylated H1 molecules (C).

ABBREVIATIONS USED

HMG - High mobility group
ATP-γ-S - Adenosine 5'-O-(3-thiotriphosphate)
GTP-γ-S - Guanosine 5'-O-(3-thiotriphosphate)
CV-1 cells - African green monkey CV-1 cell line
SV40 virus - Simian virus 40

REFERENCES

Allfrey, V.G. Goldstein, L. and Prescott, D.M., eds. (1980). *In* "Cell Biology: A Comprehensive Treatise" Vol. 3, p. 347. Academic Press, New York.

Allfrey, V.G., Pogo, B.G.T., Littau, V.C., Gershey, E.L., and Mirsky, A.E. (1968). *Science 159*, 314.

Belikoff, E., Wong, L.-J., and Alberts, B.M. (1980). *J. Biol. Chem. 255*, 11448.

Berlowitz, L., and Pallotta, D. (1972). *Exp. Cell Res. 71*, 45.

Boffa, L.C., Gruss, R.J., and Allfrey, V.G. (1981). *J. Biol. Chem. 256*, 9612.

Covault, J., Perry, M., and Chalkley, R. (1982). *J. Biol. Chem. 257*, 13433.

Darzynkiewicz, Z., Bolund, L., and Ringertz, N.R. (1969). *Exp. Cell Res. 55*, 120.

Garel, A., and Axel, R. (1976). *Proc. Natl. Acad. Sci. USA 73*, 3966.

Goody, R.S., and Eckstein, F. (1971). *J. Am. Chem. Soc. 93*, 6252.

Gorovsky, M.A., Pleger, G.L., Keevert, J.B., and Johmann, C. A. (1973). *J. Cell Biol. 57*, 773.

Gratecos, D., and Fischer, E.H. (1974). *Biochem. Biophys. Res. Commun. 58*, 960.

Hohmann, P. (1978). *In* "Subcellular Biochemistry" E.B. Roodyn, ed. Vol. 5, p. 87. Academic Press, New York.

Hohmann, P., Tobey, R.A., and Gurley, L.R. (1976). *J. Biol. Chem. 251*, 3685.

Killander, D., and Rigler, R. (1969). *Exp. Cell Res. 54*, 163.

Langan, T.A. (1978). *Methods Cell Biol. 18*, 127.

Morgan, M., Perry, S.V., and Ottaway, J. (1976). *Biochem. J. 157*, 687.

Panyim, S., and Chalkley, R. (1969). *Arch. Biochem. Biophys. 130*, 337.

Pasqualini, J.R., Cosquer-Claveul, C., Vidali, G., and

Allfrey, V.G. (1981). *Biol. Reprod. 25,* 1035.
Pogo, B.G.T., Allfrey, V.G., and Mirsky, A.E. (1966). *Proc. Natl. Acad. Sci. USA 55,* 805.
Riggs, M.G., Whittaker, R.G., Neumann, J.R., and Ingram, V.M. (1977). *Nature London 268,* 462.
Ruiz-Carrillo, A., and Allfrey, V.G. (1973). *Arch. Biochem. Biophys. 154,* 185.
Ruiz-Carrillo, A., Wangh, L.J., and Allfrey, V.G. (1975). *Science 190,* 117.
Sealy, L., and Chalkley, R. (1978). *Nucleic Acids Res. 5,* 1863.
Sterner, R., and Allfrey, V.G. (1982). *J. Biol. Chem. 257,* 13872.
Sterner, R., Vidali, G., Heinrikson, R.L., and Allfrey, V.G. (1978). *J. Biol. Chem. 253,* 7601.
Sterner, R., Vidali, G., and Allfrey, V.G. (1979). *J. Biol. Chem. 254,* 11577.
Sterner, R., Vidali, G., and Allfrey V.G. (1981). *J. Biol. Chem. 256,* 8892.
Sun, I.Y.-C., and Allfrey, V.G. (1982a). *J. Biol. Chem. 257,* 1347.
Sun, I.Y.-C., and Allfrey, V.G. (1982b). *Proc. Natl. Acad. Sci. USA 79,* 4589.
Sun, I.Y.-C., Johnson, E.M., and Allfrey, V.G. (1980). *J. Biol. Chem. 255,* 742.
Vavra, K.J., Allis, C.D., and Gorovsky, M.A. (1982). *J. Biol. Chem. 257,* 2591.
Vidali, G., Boffa, L.C., and Allfrey, V.G. (1977). *Cell 12,* 409.
Vidali, G., Boffa, L.C., Bradbury, E.M., and Allfrey, V.G. (1978a). *Proc. Natl. Acad. Sci. USA 75,* 2239.
Vidali, G., Boffa, L.C., Mann, R.S., and Allfrey, V.G. (1978b). *Biochem. Biophys. Res. Commun. 82* 223.
Weintraub, H., and Groudine, M. (1976). *Science 193,* 848.
Weisbrod, S.T. (1982). *Nucl. Acids Res. 10,* 2017.
Weisbrod, S.T., and Weintraub, H. (1979). *Proc. Natl. Acad. Sci. USA 76,* 630.
Weisbrod, S.T., Groudine, M., and Weintraub, H. (1980). *Cell 19,* 289.
Wiegand, R.C., and Brutlag, D.L. (1981). *J. Biol. Chem. 256,* 4578.

DISCUSSION

Klee: You mentioned the fact in the early part of your talk that acetylation was very slow as opposed to deacetylation. Later on when you talked about acetylation of the histone in HMG, you mentioned the fact that it was very fast when the gene is turned on. Do you think there is an effect on the DNA level or on the enzyme which it acetylates?

Allfrey: I think these enzymes are subject to very subtle control. Both the acetylases and the deacetylases appear to be phosphoproteins and it may well be that their activity is controlled in turn by levels of phosphorylation. When I stress the deacetylation problem, I stress it because it is a problem; that unless one takes particular precautions to get butyrate in very quickly, the levels of acetylation of the histones can change within minutes. In terms of the kinetics of labeling, if you put in radioactive acetate as a probe, the kinetics of uptake are also rapid, but they differ in different regions of the chromatin.

Klee: Looking at the acetylating enzymes could be a potential tool to study the mechanism of turning on a gene.

Allfrey: I think that a qualification has to be made here. This is a mechanism for relaxing the grip on the DNA, not necessarily causative in transcription. I always draw the analogy of the file drawer - that this is the equivalent of opening the file drawer without selecting a file, then more subtle forms of information have to be fed into the system to turn on a particular gene. So what we are doing is making the DNA accessible, not necessarily using it.

Graves (Iowa State): Do you have any information that you can tell us about the use of thioacetylation and thiophosphorylation techniques for the study of the relationship of one modification to another in the same protein molecule?

Allfrey: We are doing those experiments right now and I do not have the answer.

Graves: You do expect to find some differences, don't you?

Allfrey: There will be differences. The question is to define those differences in some meaningful way. What we are doing is looking at SV40 minichromosomes and looking at the modifications of the chromosomal proteins, as a function of transcription in different regions of the chromosome.

Cohen: When you thiophosphorylate proteins, this makes them resistant to dephosphorylation.

Allfrey: That is correct.

Cohen: Now I want to ask you two questions. One is: thioacetylation, is that resistant to deacetylation?

Allfrey: We are not sure yet. I'll tell you why I'm hesitating and that is that the only probe we have used is a nonradioactive mercaptoacetate and to do that type of experiment, we need a radioactive mercaptoacetate to monitor the loss. We are making that, but we have not done it yet.

Cohen: Presumably those proteins that are thiophosphorylated *in vivo* will also be phosphatase-resistant. So you have here a method to tell whether changes in thiophosphorylation result from activation of the kinase or inhibition of phosphatase. Can you comment on that?

Allfrey: Well, I think that is a safe thesis in some cases, but I have no experimental evidence on that - that is, comparing a system with regard to thiophosphate uptake and thiophosphate loss.

Wold: I noticed that in one of the histones, one site was both a methylation and an acetylation site. Is that correct?

Allfrey: I am not sure of that. When I made the slide, the acetylation site at lysine-9 of histone H3 was certain, but the methylation of that residue was more doubtful. There is a methylation site in histone H3, but it is further up the chain, at lysine-26.

Wold: So there is no evidence that any of these acetylated lysines ever are methylated?

Allfrey: Quite the opposite. In both H3 and H4, the acetylation sites are in the amino-terminal region and the next lysine residue in that region is modified by methylation. I am wondering whether the function of that methylation is to prevent the acetyltransferase from proceeding further up the chain.

Hohmann: I was wondering if there is any evidence that you know of that the introduction of phosphate in the DNA backbone changes in any way the stability or interactions of the newly synthesized DNA with nucleosomal "particles"?

Allfrey: There is no evidence for that. What we have done in the SV40 system is to ask how many thiophosphates enter the SV40 DNA. About 30% of the DNA can be recovered when only 3 to 12 phosphorothioate linkages occur in the entire genome (5,226 base-pairs). We have tested to see whether the thioderivatized DNA would be cut appropriately by restriction nucleases *Kpn* I and *Hind* III. The cuts were in the right places. Thus with low levels of substitution one can say that DNA synthesized is apparently normal and will behave in the expected way. But with high levels of thio-substitution, I don't think you can make those assumptions. A recent paper by Vosberg and Eckstein, (1982) shows that if the phosphorothioate linkages happen right at the cutting site, the restriction nuclease activity is inhibited.

Griffin (OMRF): I have two quick related questions: Could one envision butyrate induction, as revolving around the inhibition of histone acetylation? To get some supporting data for that, is there any enzyme induced without a thioacetic acid?

Allfrey: There are no long-term studies of the effects of mercaptoacetate on cells. Most of them have employed short exposures and recovery of the proteins modified during the pulse. With regard to the effects of butyrate on acetylation, this situation is complicated. We first reported that butyrate inhibits the activity of histone deacetylase both *in vitro* and *in vivo*. The accumulation of the hyperacetylated forms of histones H3 and H4 in butyrate treated cells is commonly assumed to have something to do with gene activation under those conditions. But the problem is that butyrate also affects the phosphorylation of chromosomal proteins, e.g., cells placed in butyrate show inhibitions of phosphorylation of histones H1 and H2 and HMG proteins, while phosphorylations of other proteins may be stimulated. It is a very complicated situation involving more than acetylation.

Langan (U. Colorado): I wonder if you know from the use of these techniques how rapidly - say H4 - molecules enter the newly acetylated pool and whether there is a subfraction of those molecules which enters it at a rapid rate compared to the remainder and how large that might be?

Allfrey: No. So far all we know is that the newly modified molecules can be recovered with high efficiency but we don't know anything about their distribution in different chromatin regions. We are investigating that, but we have decided not to use whole chromatin. We are focusing on a particularly well-defined gene sequence. What we are looking at is the ribosomal genes in blood serum where we can isolate the minichromosome and also the SV40 minichromosome. There we can actually say which part of the gene is being transcribed and which is not and then see if we can correlate levels of acetylation.

Langan: In a normal cell in culture, how rapidly do H4 molecules enter the newly acetylated pool?

Allfrey: Within 15 minutes, easily.

Langan: And what fraction of the total is involved in that?

Allfrey: That is a number I am not able to estimate, especially in erythrocytes where most of the chromatin is inert. It would be a small fraction.

VITAMIN K-DEPENDENT CARBOXYLASE FROM RAT LIVER[1]

Jean-Marie Girardot
Jolanta M. Kanabus-Kaminska
B. Connor Johnson

Oklahoma Medical Research Foundation
Oklahoma City, Oklahoma

INTRODUCTION

Vitamin K has been shown to be an essential cofactor for the *in vivo* carboxylation of prothrombin (Girardot *et al.*, 1974). This reaction was found to be gamma carboxylation of specific glutamyl residues in proteins (Stenflo *et al.*, 1974; Nelsestuen *et al.*, 1974). The reaction can be studied *in vitro* after detergent solubilization of the rat liver microsomes (Mack *et al.*, 1976; Esmon *et al.*, 1976). The specific requirements are vitamin K_1 + NAD(P)H, or the reduced form of the vitamin (vitamin K_1 hydroquinone) (Friedman and Shia, 1976; Girardot *et al.*, 1976), CO_2 and oxygen. In addition to carboxylation of endogenous precursor proteins, various glutamic acid containing synthetic substrates will serve as carboxyl acceptors (Suttie *et al.*, 1976).

It has been demonstrated that the vitamin K hydroquinone dependent reaction is the removal of a hydrogen from the γ-carbon of specific glutamyl residues prior to carboxylation (Friedman *et al.*, 1979) with concomitant formation of vitamin K 2,3-oxide (Larson *et al.*, 1981). Early attempts to purify the carboxylase after Triton X-100 solubilization of the microsomes have not been very successful. However, the

[1]*This work was supported in part by NIH Grant HL-17619.*

ISBN 0-12-387560-9

new detergent CHAPS, has proven to be a better solubilization agent (Girardot and Johnson, 1982).

Some properties of the crude CHAPS solubilized carboxylase, and the partially purified enzyme are presented in this paper. The presence of iron in the enzyme has also been investigated.

MATERIALS AND METHODS

Materials

Sodium [^{14}C]carbonate (55 mCi/mmole) was purchased from Amersham. Vitamin K_1, 3-(N-morpholino)propanesulfonic acid (MOPS), soybean trypsin inhibitor, L-α-phosphatidylcholine from egg yolk, carboxymethyl-Sepharose, dithiothreitol (DTT) and 6-aminohexanoic acid-activated-Sepharose-4B(ACH4B) were obtained from Sigma. Ultrogel AcA_{34} and AcA_{202} were purchased from LKB. Phe-Leu-Glu-Glu-Leu was obtained from Vega Fox and 3-([3-cholamidopropyl]dimethylammonio)-1-propane sulfonate (CHAPS) from Serva or synthesized in our laboratories.

Methods

Preparation of CHAPS Solubilized Rat Liver Microsomes. Vitamin K deficiency was produced in male, Sprague-Dawley rats, and the microsomes prepared as previously described (Girardot and Johnson, 1982). The CHAPS solubilized microsomal suspension was prepared by dilution of one volume of microsomes (approximately 30 mg protein/ml) with one volume of 50 mM MOPS buffer, pH 7.4, containing 2% CHAPS, 2 M NaCl, 2.5 mM DTT, 100 μg/ml soybean trypsin inhibitor, stirred for 20 min in ice and centrifuged at 200,000 x g for 60 min. The resulting supernatant contained all the proteins necessary for carboxylation and is termed "soluble system" (S_{200}).

Determination of vitamin K_1-2,3-oxide by High Performance Liquid Chromatography. After incubation, the samples were diluted to 0.2 ml and extracted 3 times with 0.4 ml of an Isopropanol-hexane (2:1) mixture (Matschiner *et al.*, 1967). The hexane layer was collected, dried under nitrogen and the residue dissolved in 0.5 ml methanol. Aliquots of 50 μl were injected into a Dupont 850 HPLC equipped with a sphericel ODS C-18 column (0.25 x 20 cm) and eluted with methanol at a flow rate of 1 ml/min. The retention time was 6.5 min for vitamin

K_1-2,3-oxide and 9 min for vitamin K_1. Each incubation was made in duplicate and each injection in triplicate using a Dupont automatic sampler. Vitamin K hydroquinone incubated in the absence of enzyme yielded 1% or less vitamin K_1-2,3-oxide. This has been subtracted.

Carboxylation Assay. Routinely, carboxylation was carried out in 50 mM MOPS, pH 7.4, 0.5% CHAPS, 0.5 M NaCl (or as indicated), 2.5 µCi sodium [^{14}C]carbonate (5 µl) at the total added carbonate concentration indicated in the text, and Phe-Leu-Glu-Glu-Leu (1 mM to 10 mM in 5 µl). The final volume was 100 µl and the reaction was initiated by the addition of 50 µg vitamin K_1 hydroquinone. For kinetic studies, larger volumes were incubated and aliquots were withdrawn at the times indicated. After incubation at 17°C for 30 min, the reaction was stopped by the addition of 0.8 ml of 10% (w/v) trichloroacetic acid followed by 0.1 ml of 5% (w/v) of bovine serum albumin. After 15 min at 4°C the protein precipitate was centrifuged and the TCA supernatant, containing the carboxylated peptide, was bubbled with carbon dioxide for at least 20 min to eliminate any unreacted labelled CO_2. The protein prepicitate was solubilized with 1 ml of 2% (w/v) Na_2CO_3 and reprecipitated with 2 ml of 10% (w/v) trichloroacetic acid. This step was repeated twice, the final protein pellet was dissolved in 0.5 ml 2% (w/v) Na_2CO_3 and neutralized by the addition of 0.1 M HCl solution. Five milliliters of Scintisol were added to both the trichloroacetic acid supernatant containing exogenous peptide substrate and to the redissolved endogenous protein substrate and the solutions were counted for 10 min with an 85% efficiency in a liquid scintillation counter (Packard). The results quoted are means of duplicate or triplicate assays varying by less than 10% from which a control incubated in the absence of added vitamin K_1 hydroquinone (KH_2) has been subtracted. The minus KH_2 control was not greater than 2% of the plus KH_2 assay. Protein was determined using coomassie blue reagent from Bio-Rad.

Preparation of the animals for ^{55}Fe experiment. A group of 10 rats were fed an iron deficient diet for four weeks and injected with desferal for four days at a daily dose of 30 mg/kg body weight. Two days later the rats were injected with 100 µCi of ^{55}Fe over a three day period and killed 24 hours after the last injection of ^{55}Fe. The microsomes were prepared as reported above.

RESULTS AND DISCUSSION

Comparison of Carboxylase Activity in the Presence of CHAPS and Triton X-100.

Both Triton X-100 and CHAPS were tested for their abilities to solubilize the carboxylase activity of the microsomes from the livers of normal rats (N mic), normal rats supplemented with vitamin K_1 (N mic + K), nutritionally vitamin K deficient rats (-K mic) and deficient rats supplemented with vitamin K_1 (-K mic + K) and Warfarin treated rats. The results are presented in Table I and as expected for Triton X-100 solubilization, only -K liver microsomes and Warfarin liver microsomes showed a good level of vitamin K dependent activity for both endogenous protein and peptide carboxylation, while the microsomes from normal, normal + K_1 and -K mic + K showed little activity following Triton X-100 extraction. On the contrary CHAPS solubilized microsomes showed a 4-to 40-fold increase for Phe-Leu-Glu-Glu-Leu carboxylation and up to 28-fold increase for endogenous protein carboxylation. These results indicate that CHAPS is a better detergent than Triton X-100 whether the microsomes contain low or high levels of endogenous substrate protein.

Carboxylase Activity in the Liver Microsomes of Vitamin K Deficient Rats Supplemented In Vivo with Vitamin K_1.

The high level of carboxylase activity detected in the CHAPS solubilized liver microsomes of normal rats and vitamin K deficient rats supplemented with vitamin K_1 for 2 hours are in disagreement with the previous reports of Dubin *et al.*, (1980) using Triton X-100 for the solubilization of the liver microsomes. These results led us to reexamine the carboxylase activity of the liver of vitamin K deficient rats supplemented with vitamin K_1. The liver microsomes prepared from these animals were solubilized with Triton X-100 (Figure 1B) and CHAPS (Figure 1A) and analysed for carboxylase activity. The results presented in Figure 1B are in agreement with previous report (Dubin *et al.*, 1980) and show that after Triton X-100 solubilization of the enzyme complex peptide carboxylation and endogenous substrate carboxylation rapidly decrease after injection of vitamin K_1, and after 1 hour, the low carboxylase activity remains constant. However, after CHAPS solubilization a different situation is encountered. The carboxylase activity toward pentapeptide and endogenous sub-

TABLE I. Solubilization of Vitamin K-Dependent Rat Liver Microsomal Carboxylation Activity by CHAPS and Triton X-100

	Carboxylase activity[a]					
	Peptide (FLEEL)			Endogenous protein		
Microsomes	CHAPS[b]	Triton X-100[c]	Ratio	CHAPS[b]	Triton X-100[c]	Ratio
Normal	21,800	1,136	19:1	2,450	620	4:1
Normal + Vit. K_1[e]	22,250	1,830	12:1	1,400	134	10:1
Vit. K deficient	52,630	14,200	4:1	9,800[d]	7,550[d]	1:1
Vit. K deficient + Vit. K_1[e]	38,025	926	40:1	3,364	120	28:1
Warfarin treated	48,940	11,955	4:1	9,450[d]	8,412[d]	1:1

The rat liver microsomes were solubilized and centrifuged as indicated under Materials and Methods. The incubations were conducted at 17°C for 60 min in 0.1 ml 50 mM phosphate buffer, pH 7.4, containing 1 mM FLEEL, 2.5 μCi $Na_2{}^{14}CO_3$, and 0.5 mM vitamin K_1 hydroquinone. For each detergent these were determined to be the optimum conditions for extraction and for assay.

[a]The carboxylase activity is expressed as ^{14}C cpm incorporated in the peptide or endogenous protein per mg of solubilized microsomal protein, minus the control incubated in the absence of vitamin K_1H_2.

[b]CHAPS was 1% and NaCl 1 M during solubilization, 0.5% and 0.75 M during incubation.

[c]Triton X-100 was 2% and NaCl 0.8 M during solubilization, 1% and 0.4 M during incubation.

[d]Not statistically different between CHAPS and Triton X-100.

[e]Vitamin K_1, 4 mg/kg body weight was injected ip 2 h before killing.

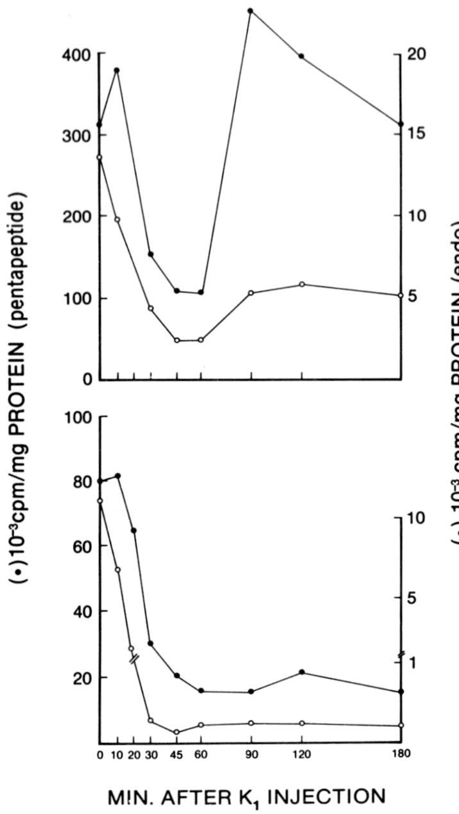

FIGURE 1. Effect of vitamin K_1 administration to vitamin K-deficient rats on the vitamin K-dependent carboxylation activity of their isolated and solubilized liver microsomes. Vitamin K-deficient rats were injected intraperitoneally with vitamin K_1 (4 mg/kg body weight), 10 to 180 min before decapitation. The values are typical of one experiment (3 rats in each group), however similar results were obtained for 3 other experiments. Panel A; the microsomes were solubilized with 1% CHAPS, 1 M NaCl. Panel B; the microsomes were solubilized with 2% Triton X-100, 0.8 M NaCl. The protein content of each batch of microsomes was adjusted to 30 mg/ml before solubilization. The carboxylation reaction was carried out in 0.1 ml final volume at 17°C for 60 min. The carboxylation assay was performed as described in Materials and Methods except 2mM pyridoxal-5'-phosphate was present and the reaction was initiated by vitamin K_1 hydroquinone produced by sodium dithionite reduction (Fieser 1940). FLEEL carboxylation (●—●); endogenous protein carboxylation (○—○).

TABLE II. *Purification Scheme of a Chaps Solubilized Vitamin K-Hydroquinone Dependent Carboxylase from Rat Liver Microsomes*

Fractions	Vol	mg protein total	Phe-Leu-Glu-Glu-Leu carboxylation CO_2 pmol/mg	fold puri*	Yield %	Protein[d] carboxylation CO_2 pmol/mg	Epoxidase activity/carboxylase ratio
Microsomes[a]	42	1120	43	1	100	47	1.65
Ultrogel AcA-34 chromatography[b]	128	162.5	640	15	220	100	3.83
CM -Sepharose chromatography[c]	18	3.6	15400	358	117	2900	4.05
Leu-Glu-Glu-Leu-Phe-ACH-48 chromatography[c]	3.5	1.3	17600	409	48	3850	ND+

*puri - purification
+ND - Not determined in this experiment

[a]The carboxylase activity of the microsomes has been determined at 0.5% CHAPS, 0.5M NaCl, conditions identical to those used after AcA-34 chromatography. The fractions were incubated at 17°C for 30 min in the presence of 1 mM Phe-Leu-Glu-Glu-Leu, 1 mM sodium [^{14}C]carbonate (specific activity approximately 30 cpm/pmol) as indicated in Materials and Methods. Although the activity of the microsomes varied from batch to batch, the specific activity of the purified enzyme substrate complex was similar for at least three different experiments, thus leading to variability in the degree of purification.

[b]The concentration of CHAPS was 0.5% and NaCl 0.5 M during carboxylation.

[c]The pooled fractions were assayed without the addition of NaCl. Readdition of CHAPS detergent, phospholipids or NaCl did not significantly increase the activity of the carboxylase.

[d]The carboxylation of protein substrate is not absolutely linear over the 30 min period used in the assay and the results in this column mainly reflect the total amount of endogenous precursor.

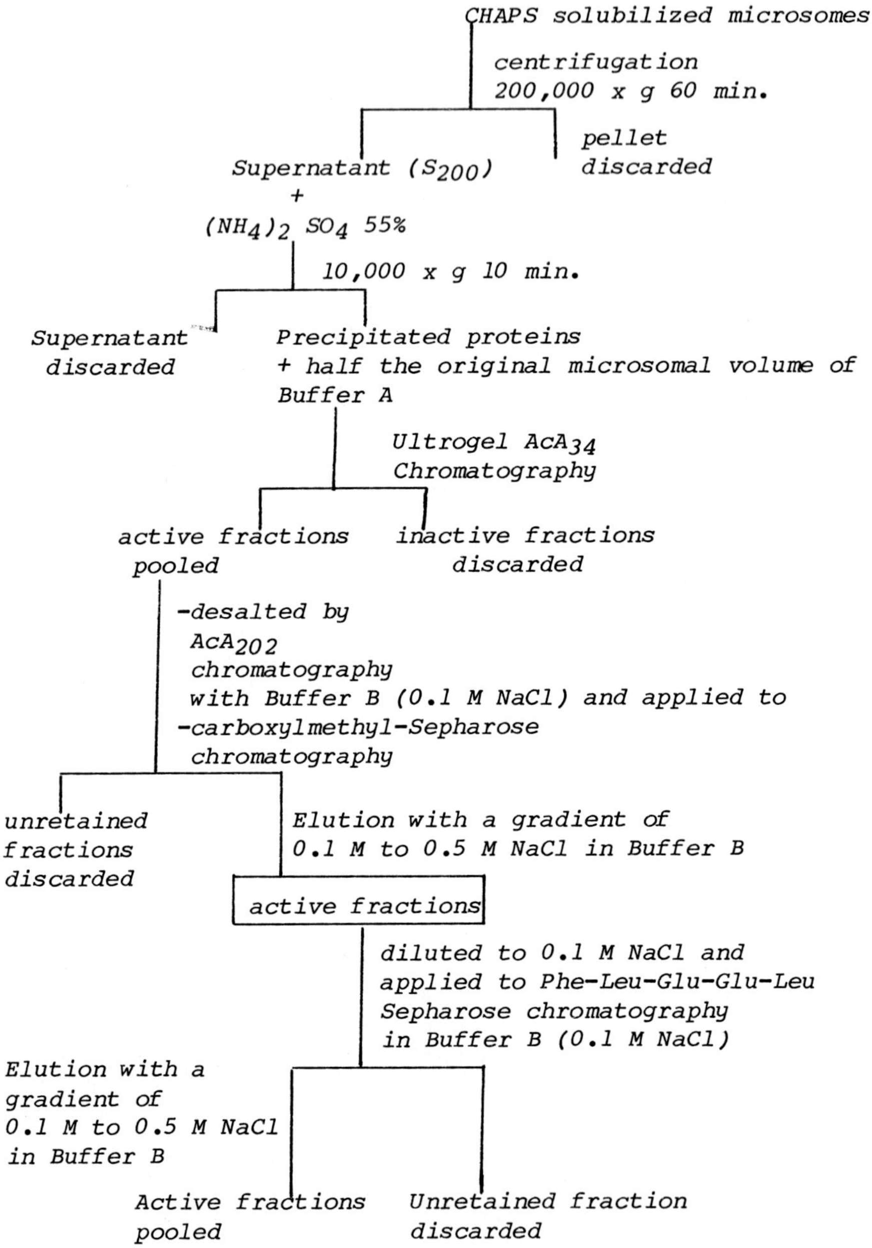
CHAPS solubilized microsomes
centrifugation
200,000 x g 60 min.
Supernatant (S_{200})
+
$(NH_4)_2$ SO_4 55%
pellet
discarded
10,000 x g 10 min.
Supernatant
discarded
Precipitated proteins
+ half the original microsomal volume of
Buffer A
Ultrogel AcA_{34}
Chromatography
active fractions
pooled
inactive fractions
discarded
-desalted by
AcA_{202}
chromatography
with Buffer B (0.1 M NaCl) and applied to
-carboxylmethyl-Sepharose
chromatography
unretained
fractions
discarded
Elution with a gradient of
0.1 M to 0.5 M NaCl in Buffer B
active fractions
diluted to 0.1 M NaCl and
applied to Phe-Leu-Glu-Glu-Leu
Sepharose chromatography
in Buffer B (0.1 M NaCl)
Elution with a
gradient of
0.1 M to 0.5 M NaCl
in Buffer B
Active fractions
pooled
Unretained fraction
discarded

Scheme 1

strate decreased rapidly during the first hour following vitamin K_1 injection. Carboxylation of pentapeptide then returns to the original high level, and while it then slowly decreases it remains very active. The carboxylation of endogenous substrate also increases one hour after vitamin K_1 injection, probably due to newly synthesized precursor protein substrates.

The low level of carboxylase activity seen one hour after vitamin K_1 treatment could be due to the depletion of endogenous precursor proteins in the liver or to inhibition of the carboxylase by endogenous carboxylated proteins. To test the latter hypothesis, vitamin K deficient liver microsomes were added to the microsomes obtained one hour after vitamin K_1 treatment, solubilized with CHAPS and tested for carboxylase activity. Stimulation but no inhibition of the vitamin K-dependent carboxylase activity was observed indicating the absence of an inhibitor in the vitamin K-supplemented rat liver microsomes (not shown). These results suggest that carboxylase is a constitutive enzyme whose activity is modulated by the level of endogenous protein substrate.

Purification of the Vitamin K-Dependent Carboxylase from Rat Liver Microsomes.

The purification of the carboxylase has been described (Girardot 1983) and is summarized in the following flow chart. (Scheme 1)

Scheme 1. Purification of carboxylase (Girardot, 1983). Buffer A: 50 mM MOPS, pH 7.4, 0.5 M NaCl, 2 mM DTT, 0.1% L-α-d-phosphatidylcholine, 20 µg/ml soybean trypsin inhibitor. Buffer B: identical to buffer A but no NaCl is included.

Characteristics of the Partially Purified Complex. A summary of the purification scheme is presented in Table II. The increased activity recovered after Ultrogel AcA-34 suggests that an inhibitor has been separated from the carboxylase complex. The enzyme(s) required for peptide carboxylation, protein carboxylation and vitamin K_1 2,3-oxide formation co-purified during the different steps. The ratio of epoxidase activity to carboxylase activity, when the carboxylation is measured at approximately 1 mM sodium carbonate, are identical to the ratio previously observed (Larson *et al.*, 1981).

Time Course of Carboxylation. The rate of carboxylation of peptide by partially purified enzyme is linear for 90 min at 20°C (Figure 2) while endogenous protein carboxylation is

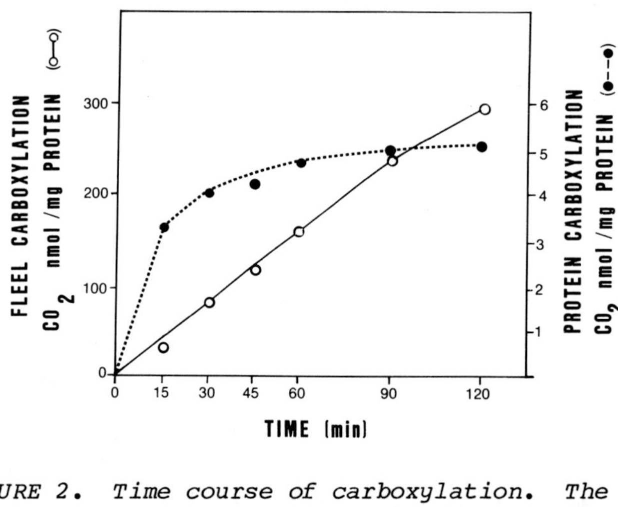

FIGURE 2. Time course of carboxylation. The partially purified enzyme was incubated at 20°C in the presence of 5 mM Phe-Leu-Glu-Glu-Leu, 2.5 mM Na_2CO_3 (specific activity of [^{14}C] 13 cpm/pmol), 0.135 M NaCl in buffer B. The final volume was 1 ml and contained 190 μg of protein. At the time indicated, aliquots of 0.1 ml were withdrawn and the incorporation of $^{14}CO_2$ was monitored as described in Materials and Methods. Each time point represents the average of duplicates from which a control without added vitamin K_1 hydroquinone has been substrated. o—o peptide carboxylation; ●---● protein carboxylation.

practically completed within 15 min incubation as in the solubilized microsomes, indicating that enzyme and precursor proteins are tightly bound.

Carboxylation as a Function of Enzyme Concentration. The peptide carboxylation is linear with enzyme concentrations from 3.6 to 360 µg of protein per ml of incubation mixture (Figure 3). The carboxylation of endogenous substrate is also linear in the same range of protein concentration (not shown).

Carboxylation as a Function of Peptide Concentration. The carboxylase responds to increased concentrations of peptide with an apparent Km calculated by double reciprocal plots of approximately 3 mM and the inhibition of endogenous substrate protein carboxylation by 10 mM peptide is 70% (Figure 4). The absence of cytochrome P_{450} and P_{420} was ascer-

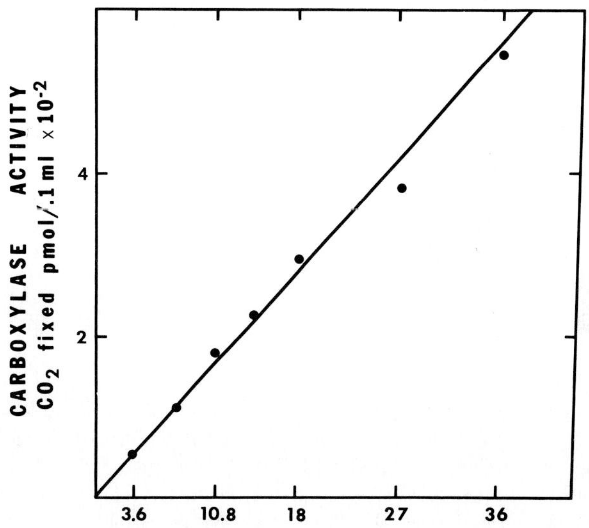

FIGURE 3. Carboxylation of peptide with increasing concentration of enzyme. The partially purified enzyme was incubated at various concentrations indicated in the Figure in the presence of 1 mM peptide, 1 mM $Na_2{}^{14}CO_3$ (specific activity 55 cpm/pmol) 0.135 M NaCl and 0.55 mM vitamin KH_2 in a final volume of 0.1 ml in buffer B. After 30 min incubation at 20°C the reaction was stopped and the samples treated as indicated in Materials and Methods.

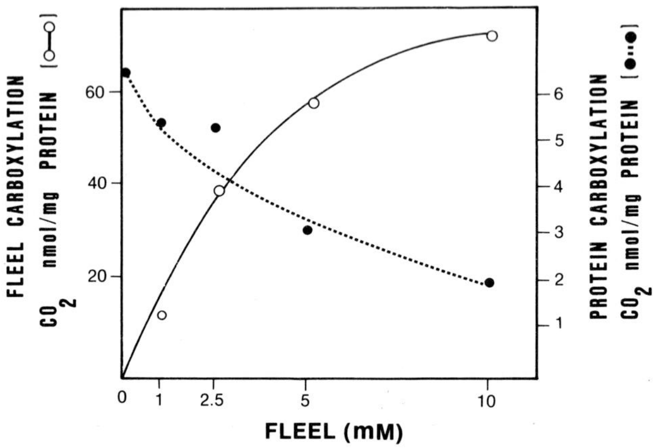

FIGURE 4. Carboxylation as a function of pentapeptide concentration. The purified carboxylase was incubated in 0.1 ml of Buffer B containing 19 μg of partially purified enzyme protein, 1 mM $Na_2{}^{14}CO_3$ (specific activity 33 cpm/pmol) at the peptide concentration indicated in the figure. The samples were incubated for 30 min at 20°C and the samples were processed as described in Materials and Methods.

tained by the difference in the spectrum of the samples after addition of Na-dithionite and CO bubbling. The NADPH cytochrome P_{450} reductase activity tested by reduction of cytochrome c is separated from the carboxylase after carboxymethyl-Sepharose chromatography (results not shown). These results definitively rule out a role of cytochrome P_{450} in vitamin K-dependent carboxylase.

Evidence for a High Molecular Weight Substrate Associated with the Carboxylase

During purification, both endogenous and exogenous carboxylation activities coelute indicating the presence of protein substrate(s) associated with the carboxylase. In

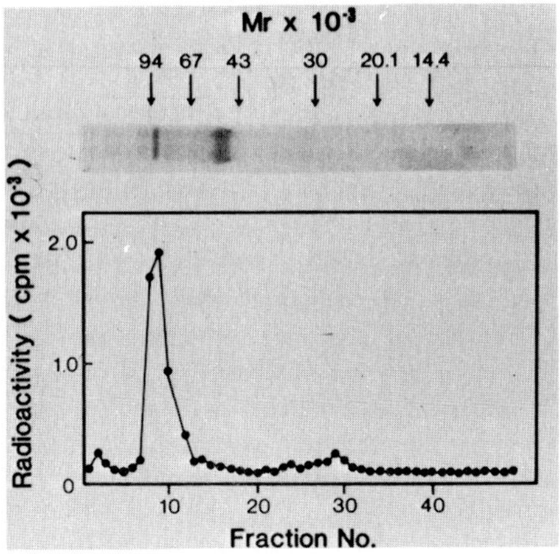

FIGURE 5. Determination of the size of the endogenous protein substrate associated with the carboxylase. One milliliter of partially purified enzyme containing 200 µg of protein was incubated at 17°C for 30 min in the presence of 25 µCi of $Na_2{}^{14}CO_3$ and 20 µl of vitamin K_1 hydroquinone (25 mg/ml in ethanol). The reaction mixture was then applied to an Amicon Centrifuge micropartition system and centrifuged 10 min at 2000 x g. The proteins retained by the membrane were solubilized in 2% SDS, 2% β-mercaptoethanol in 0.5 M Tris-glycine buffer pH 6.9 and applied in two slots to a 10% polyacrylamide slab gel containing 0.1% SDS. After electrophoresis one track was stained with Coomassie blue, the other was cut in 2.5 m/m strips. The strips were incubated overnight in tissue solubilizer and counted for radioactivity.

order to determine the molecular weight of the protein(s) substrate present in the purified carboxylase, carboxylation of the carboxylase complex was performed and the product of the reaction submitted to SDS slab gel electrophoresis. The result presented in Figure 5 clearly indicate the presence of one radioactive protein band of approximately 87 K. Similar results were obtained by Wallin and Suttie (Wallin and Suttie, 1982).

Effects of Stimulators and Inhibitors of the Purified Carboxylase Complex

The properties of the purified carboxylase complex are in general identical to those of solubilized microsomes (Table III). The purified complex responds qualitatively similarly to stimulators and inhibitors as does the vitamin K-dependent carboxylase of the solubilized microsomes. Pyridoxal-5'-phosphate stimulates pentapeptide carboxylation at 1 mM, above this level inhibition occurs. In contrast with the report of Larson and Suttie (Larson and Suttie, 1980) Mn^{2+} up to 10 mM does not stimulate carboxylation of peptide when added to solubilized microsomes (results not shown). However, 5 mM Mn^{2+} stimulates 1.7-fold the activity of the enzyme when added to partially purified carboxylase. The stimulatory effects of PLP and Mn^{2+} are not additive. Tetrachloro-4-pyridinol and hematin at a concentration of 100 μM completely block the carboxylation of peptide. The NAD(P)H dependent vitamin K_1-reductase is not a part of the carboxylase complex since vitamin K_1 and NAD(P)H do not support carboxylation by the purified enzyme complex.

Effects of Hematin and Vitamin K Hydroquinone on the Oxygen Concentration during Carboxylation

Hematin (Table III) and methemoglobin (Canfield *et al.*, 1980) are potent inhibitors of carboxylation. Therefore an attempt was made to determine whether the inhibition obtained was due to a direct effect on the carboxylase enzyme or a side reaction depleting the system of a necessarily component. The carboxylating system (Figure 6A) and buffer alone (Figure 6B) were incubated in an oxygraph cell. The level of oxygen was monitored after addition of hematin, followed by vitamin K_1 hydroquinone or vitamin K_1 hydroquinone first followed by hematin.

Addition of 1 μM hematin to buffer alone does not result in oxygen consumption; however, the addition of vitamin K_1 hydroquinone to buffer which was 1 μM in hematin depleted the system of oxygen in less than 20 sec (Figure 6B left panel). When vitamin K_1 hydroquinone is added first no change in oxygen level is seen, but addition to this system of hematin to make the solution 1 μM again produced a rapid decrease in oxygen level (Figure 6B right panel). Similar results are obtained when the buffer is replaced by solubilized vitamin K deficient microsomes. Although the decrease in oxygen level

TABLE III. Effect of Inhibitors and Stimulators on the Purified Carboxylase Complex

Additive	Concentration	Peptide Carboxylation	% Activity	Protein Carboxylation	% Activity
Vitamin K	0.55 mM	0	0	0	0
+ NADH	2.00 mM	0	0	0	0
Vitamin K_1H_2	0.55 mM	59	100	3.5	100
2,3,4,6-tetrachloro-pyridinol	1 M	48.7	82	3.3	94
	10 M	6.6	11	1.42	40
	100 M	0.9	1.5	0.44	12
Hematin	10 M	42	71	2.8	80
	50 M	12	20	1.7	48
	100 M	0.4	0.6	0.4	11
Pyridoxal-5'-phosphate	1 mM	100	169	3.1	88
	2.5 mM	71	120	2.85	82
	5 mM	54	91	3.0	85
Mn^{2+}	5 mM	101	171	3.87	110
	10 mM	98	166	3.95	113

The conditions for incubation are as described in Materials and Methods, CHAPS was 0.5%, NaCl 135 mM; FLEEL 5 mM and $Na_2{}^{14}CO_3$ 1 mM, with a specific activity of 19.3 cpm/pmol. Samples were incubated at 17°C for 30 min.

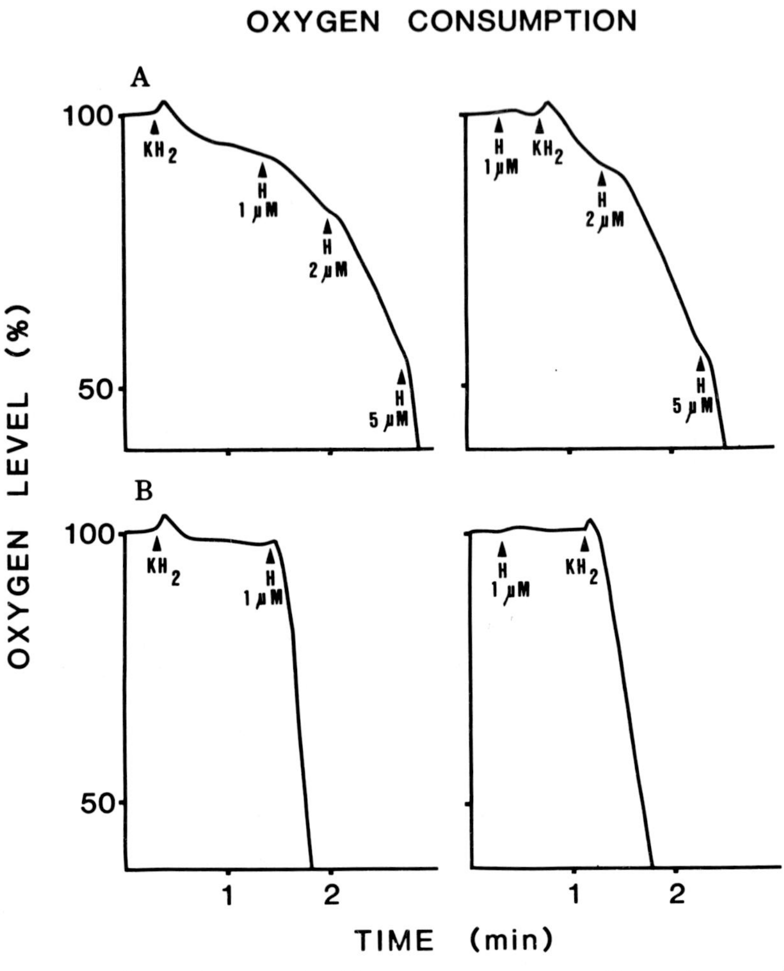

FIGURE 6. Changes in the oxygen level after the addition of vitamin K_1-hydroquinone and hematin. The oxygen concentration was recorded using a Gilson K-1C Oxygraph equipped with a Clark type electrode; polarizing voltage 0.6V; chart speed 0.2 mm/sec. 100% oxygen level is attributed to solution A or B saturated with air. The cell (2.5 ml) was filled with A or B containing 1 mM sodium carbonate. The remaining components were added with a syringe through a capillary opening. Vitamin K_1-hydroquinone solution in ethanol was added to a final concentration at 0.55 mM. Hematin (H) solution (1 mM) was added (to a final concentration of 1, 2 and 5 μM) before or after the addition of vitamin K_1 hydroquinone solution. A: Microsomes solubilized with 1% CHAPS, 1 M NaCl as described in Methods. B: The solubilized microsomes were replaced by buffer B.

does not proceed at the same rate, the system is depleted of oxygen in less than five minutes. These results suggest that hematin inhibits the carboxylation by depleting the system of oxygen. Since this reaction is dependent on the presence of vitamin K_1 hydroquinone, it is equally possible that the level of vitamin K_1 hydroquinone rapidly decreases thus depleting the system of the reduced vitamin. The following reactions can account for these effects.

Fe^{3+}-protoporphyrin is reduced to Fe^{2+}-protoporphyrin by vitamin K_1 hydroquinone to give vitamin K quinone. Then Fe^{2+}-protoporphyrin is spontaneously reoxidized by oxygen to Fe^{3+}-protoporphyrin. The reaction stops when all the vitamin K_1 hydroquinone is converted to the quinone or when oxygen concentration is too low to sustain reoxidation of Fe^{2+}-protopophyrin.

It was postulated that copper ion and copper complexes inhibit carboxylation by reacting with an oxygenated intermediate necessary for the reaction (Esnouf, 1979). Copper salts and copper aspirinate were tested for their ability to decrease the oxygen level during carboxylation and similar results to those obtained with hematin indicate that oxygen and probably vitamin K_1 hydroquinone are depleted. Since oxygen consumption increases up to 250-fold after addition of hematin, copper ion or copper-aspirinate in the absence (buffer volume) or presence of carboxylase, these results suggest that inhibition by hematin and copper is probably due to depletion of neccessary components (vitamin K_1 hydroquinone and oxygen) rather than direct scavenging of an enzymatically formed oxygenated intermediate. Copper inhibition is prevented and reversed by dithiothreitol which is compatible with this hypothesis (Johnson *et al.*, 1979). These data do not, however, rule out the possibility of the chemical formation of an oxygenated intermediate necessary for the carboxylation. Experiments are now in progress to elucidate this phenomenum.

Presence of Iron in the Carboxylase Enzyme

It has been postulated that vitamin K-dependent carboxylase is a iron containing protein (Hall *et al.*, 1981). However that carboxylase preparation was contaminated by cytochrome P_{450} and presumably by other iron containing proteins. Since our purification procedure enables us to separate cytochrome P_{450} and NADPH cytochrome P_{450} reductase (Girardot,

1983) from the carboxylase complex, the presence of iron in the carboxylase was investigated. Iron deficient rats were injected with ^{55}Fe and the microsomes were prepared as described in experimental procedures.

Ultrogel AcA_{34} Chromatography

The S_{200} fraction obtained after centrifugation (See Methods) was adjusted to 55% $(NH_4)_2 SO_4$ and the pH was maintained constant by the addition of solid MOPS. After 20 min of stirring in ice, the precipitated proteins were separated by centrifugation at 20,000 x g for 10 min., solubilized in buffer A and applied on to a Ultrogel AcA_{34} column chromatography. The elution profile is presented in Figure 7. The

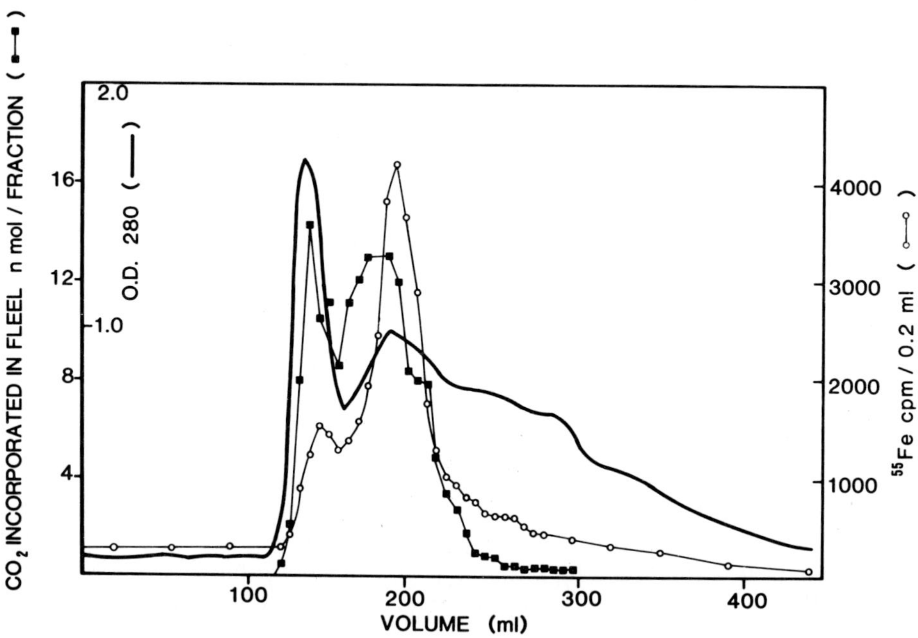

FIGURE 7. Ultrogel AcA-34 chromatography. Thirty ml of S_{200} of microsomes prepared from ^{55}Fe treated rats were adjusted to 55% $(NH_4)_2SO_4$ and the protein separated by centrifugation at 20,000 x g for 10 min. The precipitated proteins were solubilized in 4 ml buffer-A, and applied to an 2.6 x 80 cm column filled with Ultrogel AcA-34 and eluted with buffer-A at a constant flow rate of 14 ml per hour. Fractions of approximately 15 ml were collected. Absorbances at 280 nm (—). FLEEL carboxylation (■—■). ^{55}Fe (o—o).

carboxylase activity is eluted in two distinct peaks, one at the exclusion volume (135 ml) and representing a particulate form of the enzyme (Girardot 1983) and one peak of activity in the inclusion (180 ml) volume. Radioactive iron coelutes only with the first peak of activity and a second peak of ^{55}Fe appears later than the carboxylase (192 ml). The fractions containing the active carboxylase of peak II were pooled, concentrated using a nitrogen pressure cell equipped with a P.M. 10 membrane and diluted with Buffer B (no NaCl) to give a final concentration of 0.1 M NaCl and submitted to a carboxylmethyl-Sepharose chromatography.

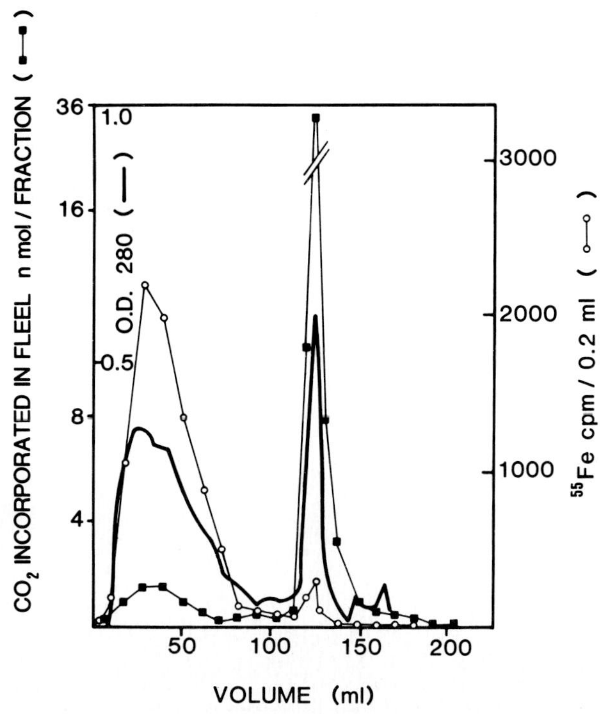

FIGURE 8. Carboxylmethyl-Sepharose chromatography. The fractions containing carboxylase activity (from 152 ml to 200 ml) were pooled, concentrated using a nitrogen pressure cell equipped with a PM 10 membrane, diluted with buffer B (no NaCl) to a NaCl concentration of 0.1 M, and applied into a 1.6 x 20 cm column filled with C M -Sepharose equilibrated with buffer B (0.1 M NaCl). The gel was washed until the 280 nm absorbance was constant. The carboxylase was then eluted with buffer B containing 0.25 M NaCl. Absorbance 280 mM (——); FLEEL carboxylation (■—■); ^{55}Fe (o—o).

The fractions obtained from Ultrogel AcA_{34} chromatography were applied to a CM-Sepharose column equilibrated with buffer B (0.1M NaCl) and the active carboxylase was eluted with 0.25 M NaCl in buffer B. The results are presented in Figure 8. Most of the ^{55}Fe is unretained by the gel and only 0.5% of the applied ^{55}Fe is eluted with the carboxylase. Further purification of the carboxylase using Fractogel TSK-55 (Figure 9) shows that ^{55}Fe does not specifically coelute with the carboxylase complex and might be associated with low amounts of contaminating proteins. A scheme of purification is presented in Table IV and it can be

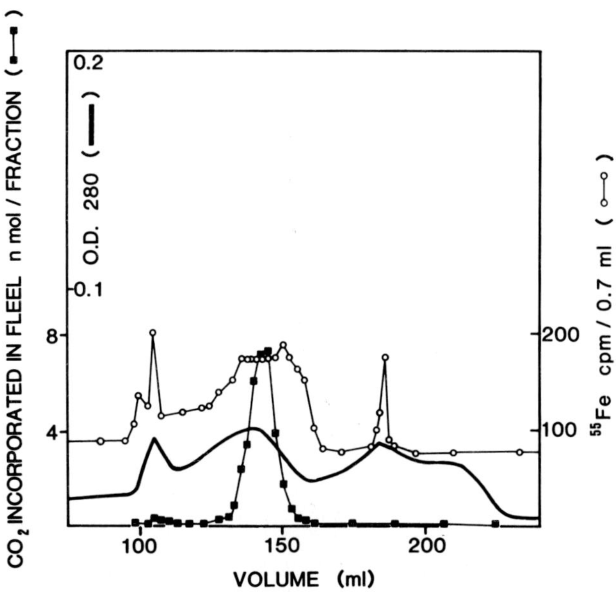

FIGURE 9. Fractogel TSK-55 chromatography. The fractions eluted from the CM-Sepharose chromatography were concentrated and 2.0 ml were applied to a 2.0 x 80 cm column filled with Fractogel TSK-55S. The proteins were eluted with buffer B (0.1 M NaCl) at a flow rate of 10 ml/hour and fractions of 3 ml were collected. Absorbance 280 nm (—); FLEEL carboxylation (■—■); ^{55}Fe (o—o).

TABLE IV. *Purification Scheme of Carboxylase from Liver Microsomes of ^{55}Fe Treated Rats.*

Fractions	mg Protein	FLEEL Carboxylation CO_2 pmol/mg protein	yields	purification	^{55}Fe cpm/mg protein	Total ^{55}Fe cpm
Microsomes[a]	1054	72	100%	1	3871	4.08×10^6
Ultrogel AcA-34 chromatography	49.5	1276	83	17	29130	1.44×10^6
CM -Sepharose[b] chromatography	7.06	3630	34	50	1122	7920
Fractogel TSK-55 chromatography	1.18	21672	33	301	1694	2000

[a]Conditions identical to those reported in Table II.
[b]The pooled fractions were assayed at 250 mM NaCl.

seen that although the enzyme activity increases 60-fold after Fractogel Chromatography of the carboxylmethyl-Sepharose eluate only a marginal increase in ^{55}Fe is detected. The partially purified carboxylase complex does not contain detectable amounts of cytochrome P_{450} or cytochrome P_{420}. Although it appears that the carboxylase does not contain iron, purification of the carboxylase to homogeneity is necessary in order to determine whether or not it is an iron containing enzyme.

ACKNOWLEDGEMENTS

We express our deep appreciation to Mrs. Julia Watson and Ms. Gale Patton for expert technical assistance and to Ms. Lisa Gropf for the preparation of this manuscript.

REFERENCES

Canfield, L.M., Sinslay, T.A., and Suttie, J.W. 1980. *Arch. Biochem. Biophys. 202,* 515.

Dubin, A., Suen, E.T., Delaney, R., Chiu, A., and Johnson, B.C. 1980. *J. Biol. Chem. 255,* 349.

Esmon, C.T., and Suttie 1976. *J. Biol. Chem. 251,* 6238.

Esnouf, N.P., Green, N.R., Hill, H.A.O., and Walter, S.J. 1979. *FEBS Letters 107,* 146.

Fieser, L.F. 1940. *J. Biol. Chem. 133,* 391.

Friedman, P.A., and Shia, N. 1976. *Biochem. Biophys. Res. Commun. 70,* 647.

Friedman, P.A., Shia, M.A., Gallop, P.A. and Griep, A.E. 1979. *Proc. Natl. Aea. Sci. U.S.A. 76,* 3126.

Girardot, J.M. 1983. *J. Biol. Chem.* 257, 15008.

Girardot, J.M., and Johnson B.C. 1982. *Anal. Biochem. 121,* 315.

Girardot, J.M., Delaney, R., and Johnson, B.C. 1974. *Biochem. Biophy. Res. Comm. 59,* 1197.

Girardot, J.M., Mack, D.O., Floyd, R.A., and Johnson, B.C. 1976. *Biochem. Biophy. Res. Commun. 70,* 655.

Hall, A.L., Bettger, W.S. and Olson, R.E. 1981. *Fed. Proc. 40,* 873 abst. 3600.

Johnson, B.C., Mack, D. O., Delaney, R., Wolfensberger, M.R., Esmon, C., Price, J.A., Suen, E., and Girardot J.M. 1979. *In* "Vitamin K Metabolism and Vitamin K-dependent Proteins," (J.W. Suttie, ed.), p. 455. University Park Press, Baltimore.

Larson, A.E., and Suttie, J. W. 1980. *FEBS Letters 118*, 91.
Larson, A.E., Friedman, P.A., and Suttie, J.W. 1981. *J. Biol. Chem. 256*, 11032.
Mack, D.O., Suen, E.T., Girardot, J.M., Miller, J.W., Delaney, R., and Johnson, B.C. 1976. *J. Biol. Chem. 251*, 3260.
Matschiner, J.T., Taggart, W.V., and Amelotti, J.N. 1967. *Biochem. 6*, 1243.
Nelsestuen, G.L., Zytkovics, T.H., and Howard, J.B. 1974. *J. Biol. Chem. 249*, 6347.
Stenflo, J., Fernlund, P., Egan, W., and Roepstorpp, P., 1974. *Proc. Nat. Acad. Sci., USA 71*, 2730.
Suttie, J.W., Hageman, J.N., Lehrman, S.R., and Rich, D.H., 1976. *J. Biol. Chem. 251*, 5827.
Wallin, R., and Suttie J.W. 1982. *Arch. Biochem. Biophys. 214*, 155.

DISCUSSION

Friedman: Could you tell us something about the stability of the preparation? Can you store it? Can you freeze it?

Girardot: We did not test all the preparations, but the enzyme is relatively stable. In one case we could keep the enzyme for about three weeks at 4° with no loss in activity in the absence of glycerol. We tried to freeze the enzyme in the presence and absence of glycerol and we observed a 50-60% decrease in activity, similar to 4°C with glycerol. The stability of the enzyme varies from preparation to preparation.

Friedman: Is it possible with the amount of activity that you do have, to actually run a gel, a non-denaturing gel perhaps, and cut the bands out and try to look for a small amount of carboxylase activity?

Girardot: Yes, well somebody's smiling in the audience.

Friedman: Because there are a lot of bands there besides the two big ones.

Girardot: I am not saying that it is pure. The fact that if it is a heme or sulfur iron enzyme the concentration there was about 130 µg/ml. If that was pure enzyme - pure heme or pure iron sulfur enzyme - then we should see the spectrum, and we don't. If 10% of the protein represents pure enzyme and 90% contaminants, then the band we see are not what we are looking for, although the precursor protein is one of the two big bands. With this low enzyme concentration, it is difficult to say, but we would like to do non-denaturing gel and then cut the bands and then try to assay for activity, because the purified enzyme is very active. Therefore, even if we lose 90% of the activity on the gel, we should be able to detect it and have something really more precise for the molecular weight. Target size analysis indicates a molecular weight of approximately 60,000.

Olson: Do I understand that in your purification scheme, you found that in fractions off the column, protein substrate and carboxylase are closely adherent throughout all these steps. In fact you never really get rid of the prothrombin precursor.

Girardot: That is correct, and the ratio of exogenous carboxylase activity to endogenous protein substrate carboxylation remains constant after the first step.

Olson: Is it also true that in one of your chromatographic separations you got two peaks of carboxylase activity, and if so, how do you explain that?

Girardot: After centrifugation at 200,000 g for 60 minutes, we obtained the clear supernatant and near the bottom of the tube some heavy colored fractions overlayered the unsolubilized microsome. This fraction contains activity and I think this is not completely solubilized microsomes. If you pool your tubes directly after centrifugation and then go on with the fractionation, then you obtained that first peak.

Olson: So you think this is a particulate form instead of being one of two isozymes?

Girardot: Yes, some could be because the solubilization is not good enough and secondly since the CHAPS and the salt are lower during chromatography than they were during solubilization, some reconstitution of microsomes could occur on the ultrogel AcA_{34}.

Olson: How much protein is there in the fraction that represents the 400-fold purification? How much of weight of protein?

Girardot: About 1 - 1.3 mg.

Wallin (U. Tromso): If you are successful with your peptide affinity column, how was the peptide hooked to Sepharose?

Girardot: There is an amino hexanoic acid are between the peptide and the Sepharose. It is Leu-Glu-Glu-Leu-Phe and then the arm and then Sepharose.

Wallin: What kind of arm are you using?

Girardot: It is a six carbon chain?

Gallin: Are you using the commercial Sepharose?

Girardot: Yes.

Knauer (OMRF): Can you remove the CHAPS, and if you do, what happens to the enzyme activity?

Girardot: We have not done the CHAPS dialysis to see what happens to the activity.

Knauer: The assay is always done in the presence of detergent?

Girardot: Yes. I'm anxious to do what you suggested, because with adenylate cyclase you can remove detergent and get activity although it is a hydrophobic molecule, so I would like to see if the enzyme will be soluble in the absence of detergent. That would simplify the procedure, although I'm not sure, because phospholipids need to be present for the activity of the enzyme.

Vermeer: When your enzyme and the endogenous substrate form a tight complex, it might be possible that the enzyme and the product are less tightly coupled. Did you ever try to incubate your enzyme in the presence of FLEEL and then a purification step in order to separate the enzyme and the carboxylated endogenous substrate?

Girardot: No, not yet. Well, as I said, I would like to have a homogenous preparation before I do anything, because, looking at the *in vivo* experiments, I think the enzyme needs a substate for activity. So unless you have a homogenous preparation, you cannot dissociate fractions and then try to do reconstitution. After obtaining a homogenous complex, you separate those two molecules, and then isolate them. Then you try to reconstitute the system. Unless I have a homogenous preparation, I will not attempt to do it.

STUDIES ON VITAMIN K-DEPENDENT CARBOXYLASE FROM THE COW[1]

C. Vermeer[2]
B.A.M. Soute
M. de Metz

Department of Biochemistry
State University of Limburg
The Netherlands

INTRODUCTION

Vitamin K-dependent carboxylase is a membrane-bound enzyme which is found in cells from various tissues (Vermeer *et al.*, 1982b). It is mainly localized in the endoplasmic reticulum from which, after disruption of the cells, it may be obtained by differential centrifugation (Vermeer *et al.*, 1982a). Carboxylase is then recovered in the microsomal fraction. The enzyme may be solubilized with the aid of detergents, but further purification has been shown to be difficult (Wallin and Suttie, 1982).

When experimental animals are treated with vitamin K-antagonists, endogenous substrates accumulate in the liver and in most cases non-carboxylated (or descarboxy-) proteins appear in the blood stream (Carlisle *et al.*, 1975). Descarboxyprothrombin has been purified from bovine plasma and used as a substrate for carboxylase. Although descarboxyprothrombin as such had a rather high K_M (0.4 mM), the quality of this substrate could be greatly improved by limited proteolysis (Soute *et al.*, 1981). This resulted in a pep-

[1]*This research was supported in part by Grant MD 82145 from the Thrombosestichting Nederland.*
[2]*To whom all correspondence should be addressed: Department of Biochemistry, State University of Limburg, P.O. Box 616, 6200 MD Maastricht, The Netherlands.*

ISBN 0-12-387560-9

tide consisting of the amino acid residues 13-29 in descarboxyprothrombin; six glu-residues, which can be carboxylated, are still left. This substrate had a K_M of about 0.001 mM.

Analysis of the hepatic endogenous substrates for carboxylase showed that they mainly consist of precursors of the clotting factors X (50%) and prothrombin (20%) and that they occur complexed to carboxylase (de Metz *et al.*, 1981a). By using highly purified Sepharose-bound antifactor X antibodies, we were able to extract the enzyme/substrate complexes from the solution. After washing away the non-bound microsomal proteins (more than 99.5% of the material), a greatly purified colorless Solid-phase carboxylase (Sp-carboxylase) was obtained. The Sepharose-bound enzyme is active as such, and is rather stable, because at 25°C and in the presence of an excess of exogenous substrate, the carboxylation reaction proceeds in a linear way for more than 8 hours.

Sp-carboxylase was used for studying the mechanism of the carboxylation (de Metz *et al.*, 1982a). The latter is thought to occur in two steps. In the first step, reduced vitamin K is oxidized into a hydroperoxide (Figure 1). This reaction does not necessarily take place on the enzyme, because it already occurs when vitamin K hydroquinone is solubilized in water. In a subsequent step the cleavage of the peroxide bond is coupled to the carboxylation of a glutamic acid residue (Figure 2). The function of carboxylase may be that it promotes the proper alignment of the reaction components and the subsequent heterolytic cleavage of the peroxide bond. The shift of the electron pair thus results in the formation of vitamin K epoxide, a gla-residue and water. Obviously, a similar mechanism may be proposed in which a homolytic cleavage of the peroxide initiates the carboxylation. In that case there is a one-electron shift and an additional mechanism has to be proposed for the formation of a CO_2-radical.

MATERIALS AND METHODS

Materials

Vitamin K and [^{3}H]-vitamin K (which was a kind gift) were obtained from Hoffmann-La Roche. Triton X-100, menadione, warfarin, 2-mercaptoethanol, dithiothreitol (DTT), phosphatidylinositol and sphingomyelin were from Sigma Chemical Co. Vitamin K and menadione were transformed into the hydroquinone form by adding 0.5 mg $NaBH_4$ to 1 ml of a solution containing 10 mg of the quinone and 0.005 ml 2-mercaptoethanol.

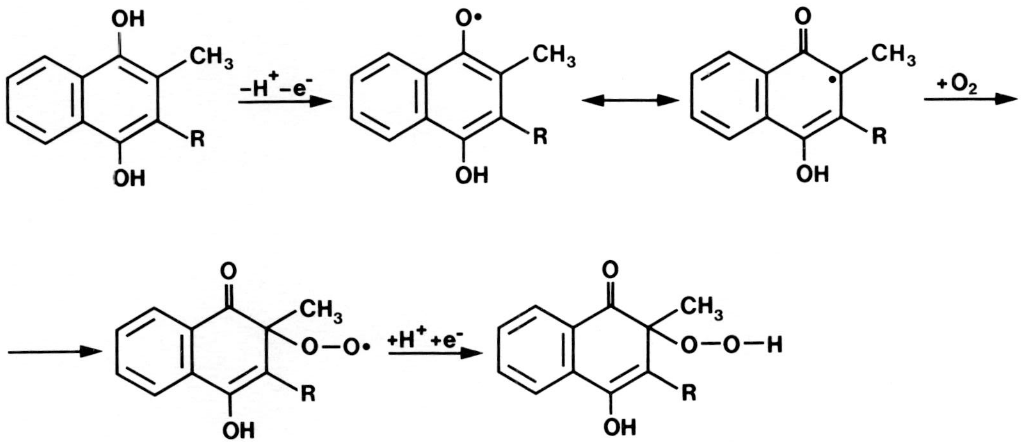

FIGURE 1. Proposed mechanism for the formation of vitamin K hydroperoxide.

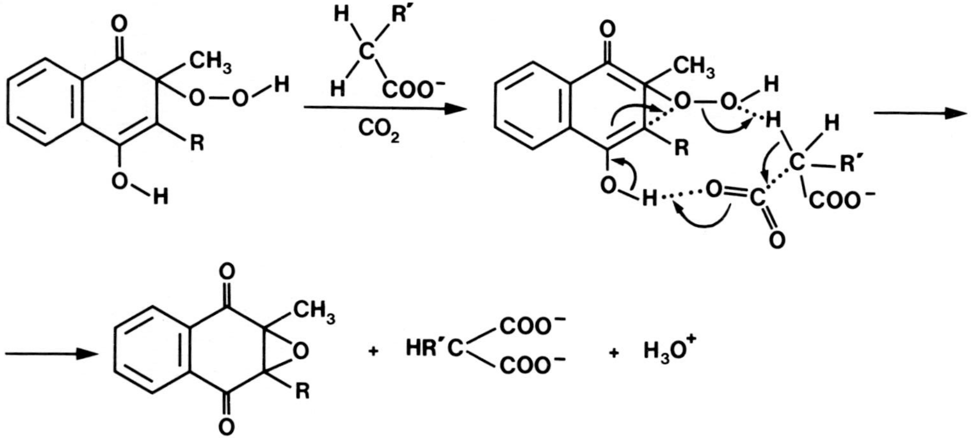

FIGURE 2. Proposed mechanism for the carboxylation of glutamic acid residues.

Vitamin K epoxide was prepared as described by Vermeer *et al.* (1982a) and 3-DTT-MK-0 (the thioester adduct of DTT and menadione) as by Mack *et al.* (1979). $NaH^{14}CO_3$ (40 Ci/mol) and Atomlight were obtained from New England Nuclear and the synthetic pentapeptide Phe-Leu-Glu-Glu-Leu (F L E E L) from Vega Fox. Fragment Su was prepared from purified descarboxyprothrombin (Soute *et al.*, 1981). Phospholipase A_2 (from *Naja naja*), phospholipase C (from *Bacillus cereus*) and phospholipase D (from Savoy cabbage) were prepared as described by Zwaal *et al.* (1975). Phosphatidylcholine was prepared from egg yolk (Singleton *et al.*, 1965) and used for the production of phosphatidic acid, phosphatidylserine, phosphatidylethanolamine and phosphatidylglycerol (Comfurius and Zwaal, 1977). Lyso-PC was prepared by hydrolysis of phosphatidylcholine with phospholipase A_2. Methyl ethyl ketone (2-butanone), methylpropyl ketone (2-pentanone) and diethyl ketone (3-pentanone) were from Aldrich and the synthetic phospholipids (di-C12:0 PC, di-C14:0 PC, di-C16:0 PC, di-C18:0 PC, di-C18:1 cis PC and di-C18:1 trans PC) were purchased from Calbiochem. CNBr-activated Sepharose 4B was obtained from Pharmacia. All other chemicals were from Merck.

Methods

Preparation of Carboxylase. Solubilized microsomes were prepared from the livers of either normal or warfarin-treated cows (Vermeer *et al.*, 1982a). Solid-phase carboxylase (Sp-carboxylase) was prepared by extracting solubilized microsomes from warfarin-treated cows with highly purified Sepharose-bound antifactor X antibodies (de Metz *et al.*, 1981a). Solubilized microsomes from the livers of various human donors were prepared in the same way as described for the bovine preparations. All donors had previously signed a donor codicil, and it was verified that they had not recently received anti-vitamin K drugs.

Carboxylase Assay. The vitamin K-dependent carboxylation reaction was measured in standard reaction mixtures (0.25 ml) containing either 0.1 ml Sp-carboxylase (0.015 mg of protein) or solubilized microsomes (1 mg of protein) in the presence of: 150 mM NaCl, 20 mM Tris-HCl, pH 7.5, 0.5 mM EDTA, 10 mM DTT, 10 mM F L E E L, 0.2 mM vitamin K hydroquinone (if added) and 0.01 mCi $NaH^{14}CO_3$. In some experiments vitamin K hydroquinone was replaced by similar concentrations of either vitamin K quinone, vitamin K epoxide, or 3 DTT-MK-0 hydroquinone. Unless indicated otherwise, all mixtures were incubated for 2 h at 25°C.

Phospholipids. The phospholipids of Sp-carboxylase were extracted and analyzed as described by (de Metz *et al.*, 1981b). Incubations with phospholipase A_2 and C were carried out in 10 mM $CaCl_2$, 150 mM NaCl and 20 mM Tris-HCl, pH 7.5. Incubations with phospholipase D were performed in 30 mM NaCl, 50 mM $CaCl_2$ and 50 mM sodium acetate, pH 6.3. All phospholipases were used in a concentration of 5 IU per ml, incubated with Sp-carboxylase (0.5 ml of slurry) for 1 h at 25°C and subsequently removed by washing with a buffer containing 150 mM $NaCl_2$, 2 mM EDTA, 20 mM Tris-HCl, pH 7.5, and 2 mg/ml ovalbumin. Phospholipids were added to Sp-carboxylase in the form of mixed micelles, in which the phospholipids and cholate were present in a 1:1 ratio (w/w). The micelles were prepared by evaporating the solvent from the various phospholipids and resuspending the residue in a buffer containing 150 mM NaCl, 20 mM Tris-HCl, pH 7.5 and cholate. The suspension was subsequently sonicated until a clear solution was obtained.

Analytical Techniques. Protein concentrations were determined according to Lowry *et al.* (1951). Polyacrylamide gel electrophoresis was performed in SDS as described by Laemmli (1970). After electrophoresis the gels were stained with Coomassie Brilliant Blue or sliced and counted. Vitamin K epoxide was quantitated as described earlier (Vermeer *et al.*, 1982a).

RESULTS

Comparison Between Bovine and Human Liver Carboxylase

Most studies concerning vitamin K-dependent carboxylase are performed in animal model system. In general it is assumed tacitly that these systems are comparable with human carboxylase and thus will give us some insight into the mechanism of vitamin K action in man. We have tested the reliability of this hypothesis by comparing carboxylase present in the solubilized microsomal fraction from bovine liver with that from human liver. Human livers were obtained from normal, healthy donors, who had died from brain trauma after traffic accidents. The donors had not recently been under medical treatment, nor had they received vitamin K-antagonists. Consequently, these livers were compared with livers from normal, non-anticoagulated cows. The optimal reaction conditions for both types of carboxylase were similar: pH 7.5, 25°C, 0.15 M KCl, 10 mM DTT and 10 mM $MnCl_2$. An addi-

TABLE I. Comparison Between Carboxylating Systems Obtained from Bovine and Human Liver[a]

Variable measured	*Bovine liver*	*Human liver*
K_m *(mM) of:*		
vitamin K	*0.013*	*0.016*
vitamin KH_2	*0.010*	*0.022*
F L E E L	*3.3*	*2.8*
Fragment Su	*0.001*	*0.0008*
Warfarin (nM) required for 50% inhibition at:		
0.025 mM vitamin K	*0.002*	*0.003*
0.025 mM vitamin KH_2	*0.4*	*1.5*
Amount of:		
carboxylase (pCi/min, mg)	*16*	*45*
endogenous substrate (pCi/mg)	*0*	*551*

[a]*All tests were performed as described in Materials and Methods. Vitamin K hydroquinone is designated as vitamin* KH_2. *The* K_M *of the various reaction components was calculated as described by Soute* et al. *(1981). The amount of carboxylase was measured as the carboxylation rate in endogenous + exogenous substrate (F L E E L, 10 mM) and is expressed in pCi per min and per mg protein. The amount of endogenous substrate is given as the total amount of radiolabeled product (pCi per mg protein) that had formed in TCA precipitable material after 2 h at 25°C.*

tional number of variables were assayed in the two systems and the results of these experiments are given in Table I. The K_m values of vitamin K quinone and hydroquinone were of the same order of magnitude, as were the K_M values of the exogenous substrates F L E E L and Fragment Su. The amount of carboxylase per mg microsomal protein in the human system was somewhat higher than in the bovine system. Similarly, the warfarin concentration, required for 50% inhibition of the vitamin K-hydroquinone driven reaction was slightly higher in the human system. The only substantial difference between the two systems was that human carboxylase contained endogenous substrate and bovine carboxylase did not. This amount of endogenous substrate was comparable with that present in carboxylase from warfarin-treated cows (Figure 3B and C) and much higher than the carboxylase level from normal cows (Figure 3A). The high amount of endogenous substrate in human liver microsomes was not due to the physical condition of the donors: the results shown in Figure 3C were obtained with livers obtained from healthy donors between 20 and 30 years old who all had gone through a traffic accident, whereas the results in Figure 3D were obtained with livers from donors who had been hospitalized for more than one week and who had died from various diseases. Both types of carboxylase contained a high amount of endogenous substrate. With the aid of Sepharose-bound monospecific and highly purified antibodies, we were able to demonstrate that 45% of the endogenous substrate was a precursor of blood clotting factor X and 21% could be indentified as a prothrombin precursor. So it seems that in human liver, a pool of clotting factors occurs and that the posttranslational processes may have a regulatory function in their excretion in plasma.

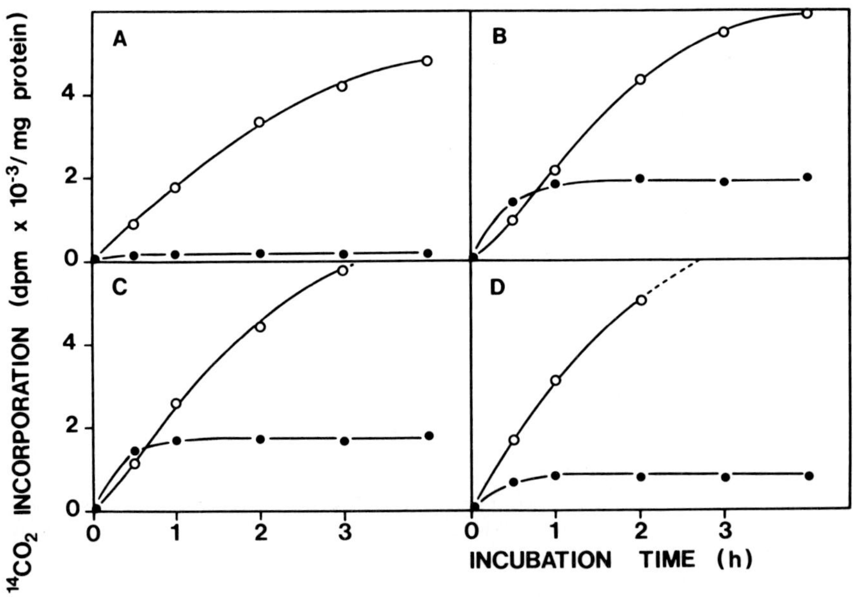

FIGURE 3. Comparison between bovine and human carboxylase. The carboxylation of endogenous substrate (closed circles) and F L E E L (open circles) were measured in solubilized liver microsomes from normal cows (A), warfarin-treated cows (B), human donors, who were killed by traffic-accidents (C) and human donors, who died from various diseases (D).

Characterization of Bovine-liver Carboxylase

For our studies concerning the bovine carboxylating system, we exclusively used Sp-carboxylase. In the presence of vitamin K hydroquinone, the carboxylation of endogenous substrate in Sp-carboxylase starts without delay, but the carboxylation of exogenous substrates exhibits a lag-phase, which may last up to half an hour (Figure 4). During this lag-phase the enzyme is presumably carboxylating the factor X-precursor. Another observation was that Sp-carboxylase contained very little vitamin K-reductase activity: the carboxylation reaction could be started with vitamin K hydroquinone but with either vitamin K quinone or vitamin K epoxide very poor carboxylation was observed.

Omitting the Sepharose bead, the antifactor X and the factor X-precursor, Sp-carboxylase consists of 40% phospholipids and 60% proteins. The phospholipids were analyzed with the aid of thin layer chromatography and were identified as more than 95% phosphatidylcholine. The presence of this phospholipid moiety was an absolute requirement for carboxyl-

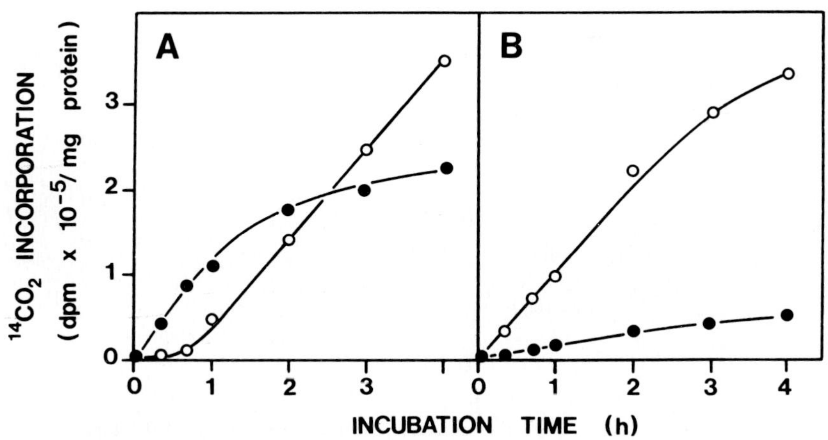

FIGURE 4. Time-course of the carboxylation by Sp-carboxylase. The carboxylation of endogenous substrate (closed circles) and F L E E L (open circles) was measured under standard conditions.

ase activity, which was demonstrated in the following experiment. Sp-carboxylase was incubated with various phospholipases (either phospholipase A_2, C or D) as described in Materials and Methods. The Sepharose beads were washed subsequently with a buffer containing 0.15 M NaCl, 20 mM Tris-HCl, pH 7.5, 0.5 mM EDTA and 2 mg/ml bovine albumin. This washing procedure efficiently removed the phospholipases as well as their digestion products (de Metz *et al.*, 1981b). After this treatment, no carboxylase activity could be detected, unless exogenous phospholipids were added to the system. Various phospholipids were assayed for their ability to restore the carboxylase activity (Table II). Only neutral phospholipids (especially phosphatidylcholine) were effective in this respect. Negatively charged phospholipids as well as

TABLE II. Reconstitution of Phospholipid-depleted Sp-carboxylase

	Carboxylase activity (pCi per min per mg protein) in:	
Phospholipid added[a]	*Sp-carboxylase*	*PL-depleted sp-carboxylase*
None	*980*	*29*
Crude microsomal extract	*1029*	*666*
Neutral PL:		
phosphatidylcholine	*1098*	*853*
sphingomyelin	*804*	*568*
phosphatidylethanolamine	*421*	*206*
lyso-PC	*960*	*784*
Negatively charged PL:		
phosphatidylinositol	*20*	*0*
phosphatidylserine	*11*	*0*
phosphatidylglycerol	*28*	*0*
phosphatidic acid	*24*	*0*

[a]*Phospholipid is abbreviated as PL and lyso-PC stands for the split product that is formed when phosphatidylcholine is digested with phospholipase A_2. The various phospholipids were added to the enzyme as mixed micelles with cholate in a 1:1 ratio (w/w). The final phospholipid concentration was 1 mg/ml in all cases.*

phosphatidic acid did not restore the enzyme activity. The phospholipid composition of the native Sp-carboxylase could also be changed by a direct exchange of phospholipids. This was accomplished by washing Sp-carboxylase with a buffer containing the phospholipid of choice (1 mg/ml), which was present in the form of mixed micelles with cholate in a 1:1 ratio (w/w). As is shown in Table II, the treatment of Sp-carboxylase with various neutral phospholipids did not dramatically affect the enzyme activity, but after the native phospholipids had been exchanged with negatively charged phospholipids, carboxylase activity could hardly be detected. The presence of cholate does not seem to be required for the enzyme activity but solely for the transport of phospholipids into Sp-carboxylase. Removal of cholate by extensive washing and dialysis had no effect on the activity of the reconstituted enzyme preparation.

The protein part of Sp-carboxylase was analyzed on polyacrylamide gels in SDS. Samples of solubilized microsomes as well as Sp-carboxylase (both containing similar amounts of active enzyme) were incubated with 0.02 mCi [^{3}H]-vitamin K for 2 h at 25°C and the proteins were eluted from Sp-carboxylase with 6 M urea and 2% SDS. Before electrophoresis all samples were dialyzed overnight against a buffer containing 0.1 % SDS, 0.1 M glycine and 0.05 M Tris-HCl, pH 8.0. After staining the gels, the Sp-carboxylase only contained a small number of distinct proteins (Figure 5). It is not known how many of these proteins belong to the carboxylase complex. Three vitamin K-binding proteins could be detected in solubilized microsomes, two of which (45 K and 30 K) were also present in Sp-carboxylase. It seems plausible, therefore, that at least these two proteins are part of carboxylase.

Stimulators of Sp-carboxylase

Sp-carboxylase is stimulated by some divalent metal ions (Mn^{2+}, Ca^{2+}) and by certain organic solvents (dimethylsulphoxide (DMSO), ketones). The divalent metal ions only stimulate the carboxylation reaction and not the formation of vitamin K epoxide (Table III). Hence, they induce a better coupling of these two reactions. The organic solvents, on the other hand, stimulate both carboxylation and epoxide formation. A number of experiments were carried out in which we studied the nature of this stimulation in greater detail. In the first place, we measured the influence of the aliphatic chain length of the ketones. In Figure 6 it is shown that the maximal stimulation is 4-fold in all cases, but that

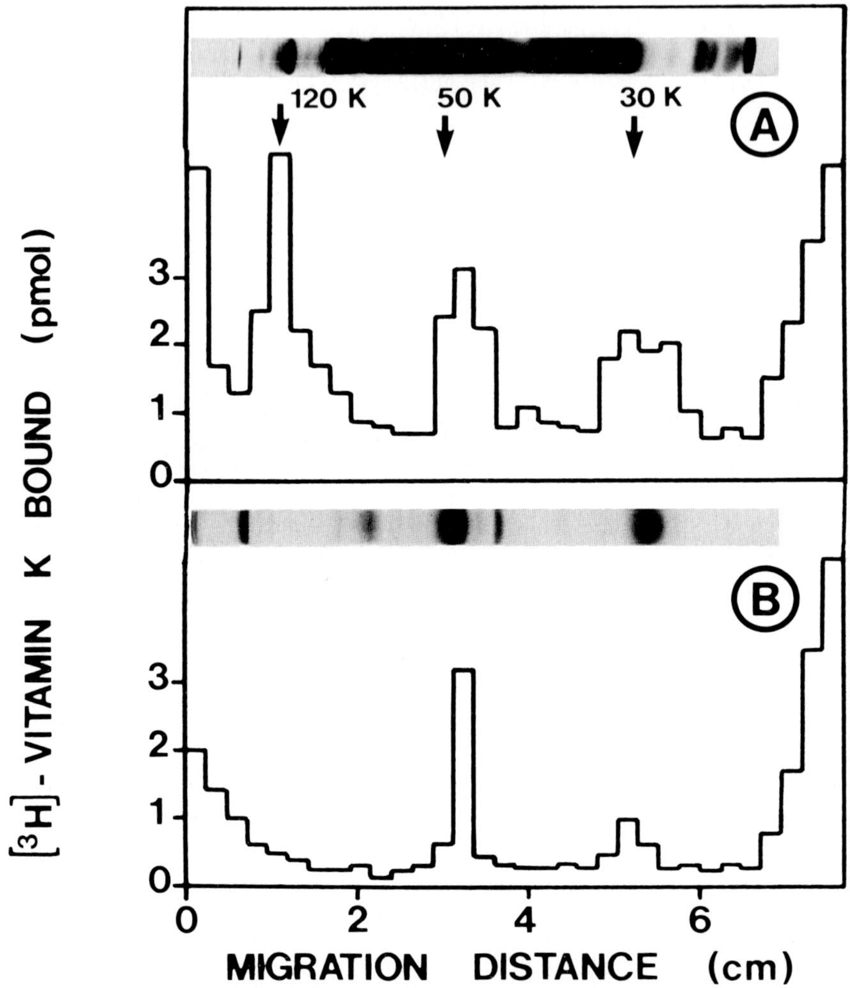

FIGURE 5. Polyacrylamide gel electrophoresis of carboxylase. Samples containing similar amounts of enzymatic activity were taken from either solubilized microsomes (top) or Sp-carboxylase (bottom), incubated with [^3H]-vitamin K and analyzed as described in the text. After electrophoresis the gels were sliced and counted. Photographs of stained gels, which were run in parallel are shown on top of each curve.

TABLE III. Stimulation of Carboxylation and Vitamin K Epoxide Formation[a]

	CO_2 incorporation (nmol per mg protein)	Formation of vitamin K epoxide (nmol per mg protein)
None	11	96
$MnCl_2$ (20 mM)	28	105
$CaCl_2$ (20 mM)	29	102
DMSO (20%, v/v)	54	480
Acetone (10%, v/v)	49	346
$MnCl_2$ (20 mM) + DMSO (20%, v/v)	134	502

[a]*The carboxylation reaction was performed in standard reaction mixtures (1 ml), containing 0.1 mM warfarin. Non-labeled CO_2 had been removed from the reaction mixtures as described in de Metz et al., 1982b. After incubation for 2 h at 25°C, 0.4 ml was used for determining the incorporated label and 0.4 ml was used for measuring the amount of epoxide formed. Dimethylsulphoxide is abbreviated as DMSO.*

with an increasing chain length, the concentration required for the optimal reaction rate is reduced.

In an attempt to find out whether the stimulation by organic solvents is caused by a change of the physical state (gel or liquid crystalline) of the phospholipids in carboxylase, we took reaction mixtures containing 10% acetone (v/v) as a model system. Sp-carboxylase was prepared and its native phospholipids were exchanged with exogenous phospholipids of known composition and well defined melting points. The following phospholipids were used: di-C12:0 PC (m.p. 0°C), di-C14:0 PC (m.p. 23°C), di-C16:0 PC (m.p. 42°C), di-C18:0 PC (m.p. 58°C), di-C18:1 cis PC (m.p. -20°C) and di-C18:1 trans PC (m.p. 9.5°C). Carboxylating reaction mixtures containing these types of carboxylase were incubated above and/or below these melting points, in the presence as well as in the absence of acetone. These data are processed in the form of Arrhenius plots in which the reaction rate is plotted against the reciprocal temperature. An example of these plots (obtained with di-C14:0 PC) is shown in Figure 7. The activation energy of the carboxylation reaction (closed circles) is obtained from the slope of the curve and, if the physical state of the phospholipid influences the carboxylation reaction, this should result in a biphasic curve. In

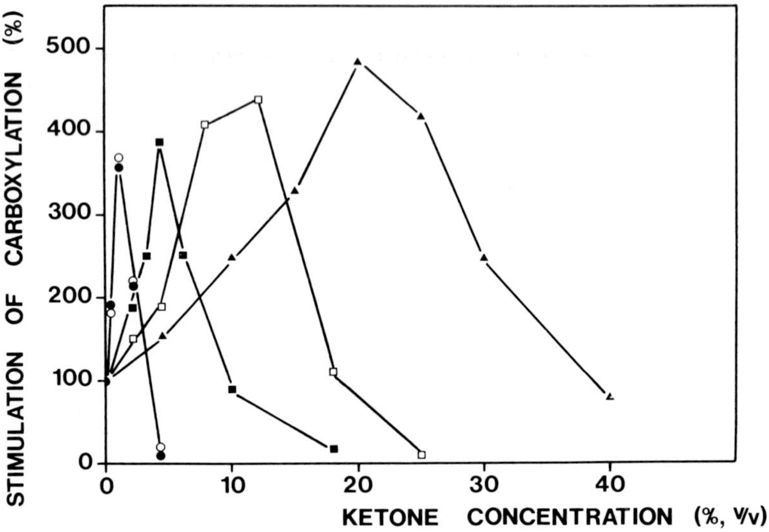

FIGURE 6. Sp-carboxylase and ketones: influence of chain length. Reaction mixtures were incubated in the presence of varying amounts of either 3-pentanone (closed circles), 2-pentanone (open circles), 2-butanone (closed squares), acetone (open squares) or DMSO (closed triangles). The stimulation of the carboxylation rate is expressed on a percentage of the rate in the absence of organic solvents.

none of the cases, however, such a biphasic curve was observed, which suggests that the physical state of the phospholipid does not influence the reaction. On the other hand, the presence of acetone dramatically influences the Arrhenius plot (open circles). At about 35°C the carboxylation rate is high, both in the presence and in the absence of acetone. But at lower temperatures it is obvious that the slope of the curve is less steep in the presence of acetone than in its absence. Hence, acetone decreases the activation energy of the carboxylation reaction. Therefore, the lower the incubation temperature is, the more pronounced will be the effect of acetone. This was demonstrated once more in experiments in which we used crude, solubilized microsomes. At 25°C the carboxylation rate was 27 pCi per min per mg protein and no effect of acetone was observed. At 5°C the reaction rate was reduced to 67% in the presence of acetone and to only 11% in its absence.

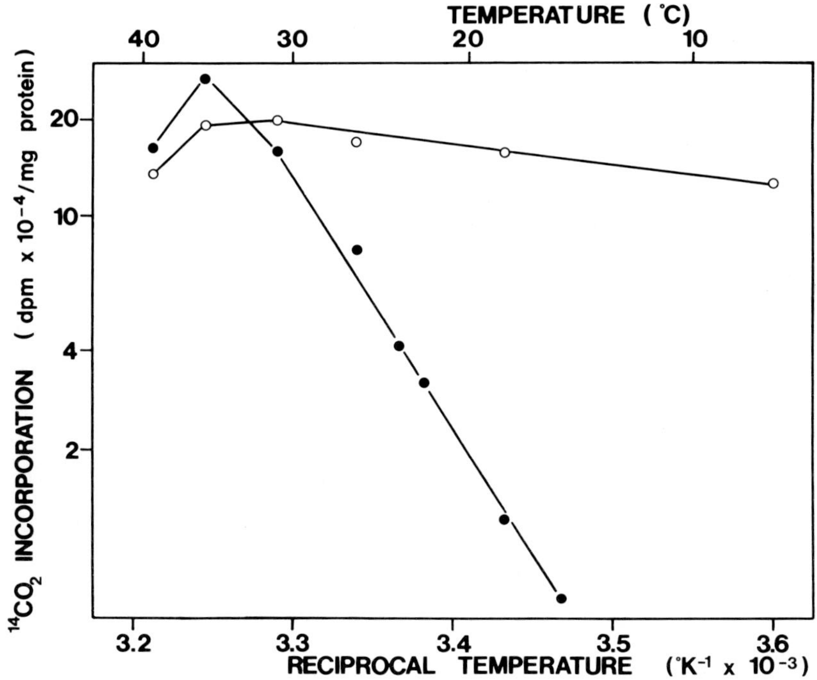

FIGURE 7. The effect of temperature on Sp-carboxylase activity. Sp-carboxylase was reconstituted with di-C14:0 PC and the carboxylation rate was measured at various temperatures and in the absence (closed circles) or presence (open circles) of acetone (10%, v/v). The data were plotted according to Arrhenius.

Finally, we investigated the influence of acetone, using Sp-carboxylase in the presence of various coenymes: three forms of vitamin K (hydroquinone, quinone and epoxide), and 3-DTT-MK-0 hydroquinone (vitamin K hydroquinone in which the phytyl side chain has been replaced by dithiothreitol). As is shown in Table IV the stimulating effect of acetone was only observed in case we used a coenzyme with a long hydrophobic side chain. This was demonstrated more clearly when we prepared Arrhenius plots of the reactions driven by vitamin K hydroquinone and 3-DTT-MK-0 hydroquinone. Both reactions were studied in the presence and in the absence of acetone (Figure 8). In the presence of 3-DTT-MK-0 hydroquinone (open and closed circles), the activation energy of the carboxylation reaction was similar to that of the car-

TABLE IV. The Effect of Various Coenzymes on Sp-carboxylase Activity.

	$^{14}CO_2$ incorporation (pCi/min, mg)[a]	
Coenzyme added	minus acetone	plus acetone
None	12	15
Vitamin K hydroquinone	1109	4880
Vitamin K quinone	115	578
Vitamin K epoxide	101	402
3-DTT-MK-0 hydroquinone	222	310

[a]The carboxylation rate was measured in standard reaction mixtures (0.25 ml). Acetone (if added) was present at a concentration of 10% (v/v).

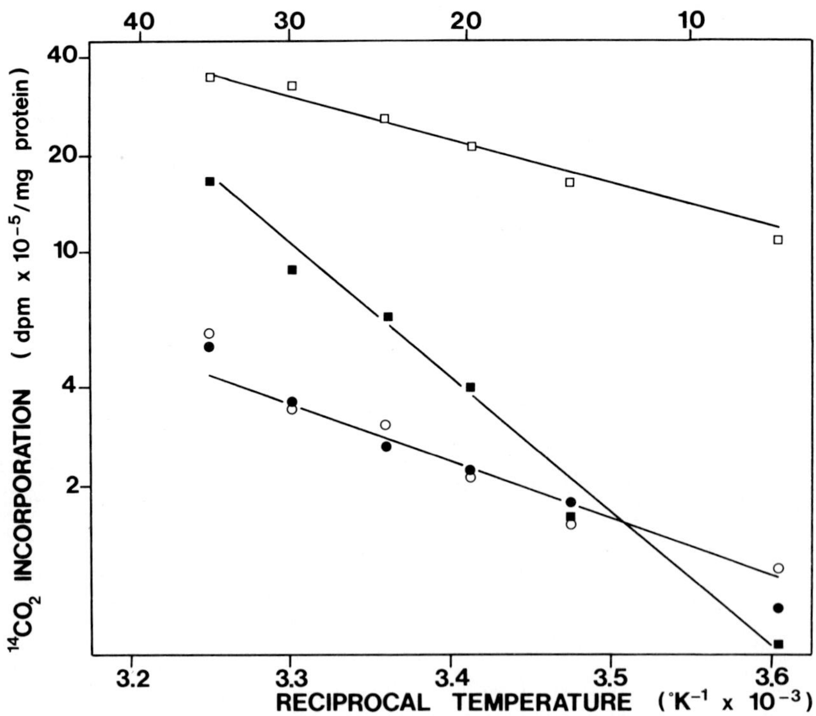

Figure 8

boxylation in the presence of vitamin K hydroquinone + acetone (closed squares); but in the absence of acetone (open squares), the activation energy of the vitamin K hydroquinone-driven reaction was substantially higher. Therefore, the most probable site of action of acetone is in solubilizing the phytyl side chain of vitamin K. Since the organic solvents do not change the K_m of vitamin K, the nature of their stimulating effect remains a matter of speculation.

DISCUSSION

When we compared the carboxylating enzyme systems from bovine and human liver, the two systems seemed to be closely similar. The only striking difference was that human liver microsomes contained substantially more carboxylatable endogenous substrate than 10 different animal species. Normal, non-warfarinized animal livers always contained extremely low amounts of endogenous substrate. Human livers were exceptional in this respect. High amounts of endogenous substrate in human livers may be explained by assuming that all our donors were vitamin K-deficient, but this seems unlikely, especially since no endogenous substrate was found in microsomes from human placenta (Soute *et al.*, 1982). The main difference between the animal liver preparations and the human ones was that our experimental animals were healthy and that their death was instantaneous, whereas the processes that had lead to the death of the human donors had not been under our control. Therefore, we think that, during the process of dying, either ribosomal protein synthesis was accelerated or the vitamin K-dependent carboxylation was slowed down in such a way that endogenous substrates accumulated in the liver. Another explanation for the high amount of endogenous substrate in human liver may be that this post-translational process has a regulatory function in the excretion of

---

FIGURE 8. The effect of various coenzymes on the activation energy of the carboxylation reaction. The carboxylation rate was measured in the presence of either 3-DTT-MK-0 hydroquinone (closed circles), 3-DTT-MK-0 hydroquinone + acetone (10% v/v, open circles), vitamin K hydroquinone (closed squares) or vitamin K hydroquinone + acetone (10% v/v, closed squares). Arrhenius plots were constructed from the data obtained at various incubation temperatures.

these plasma proteins. A pool of clotting-factor precursors would then be present in the liver cells and only be carboxylated and excreted when induced by certain signals (e.g. a low plasma concentration of clotting-factors). None of these possibilities can be verified, however, until biopsies from human volunteers are used for the production of carboxylase.

When we tried to characterize bovine-liver Sp-carboxylase, we could demonstrate that this preparation consisted of phosphatidylcholine (40%) and a number of proteins (60%). The phospholipids were required for the enzymatic activity of carboxylase, but it is not sure whether their function lies solely in maintaining a hydrophobic environment around carboxylase. They might, for instance, also have a function in the binding or the transport of vitamin K and vitamin K-metabolites before, during or after each carboxylation event. The protein part of Sp-carboxylase was analyzed on SDS gels and 5-6 proteins could be detected. Two of these proteins had a high affinity towards vitamin K since, in 2% SDS and 6 M urea as well as during electrophoresis, vitamin K remained bound to these proteins. It remains to be investigated, however, which of the various proteins belong to the carboxylase complex.

During our efforts to elute carboxylase from the Sepharose beads of Sp-carboxylase, we observed that certain organic solvents like ketones and DMSO are able to stimulate the carboxylation reaction 4 to 5 fold. No evidence could be obtained that the organic solvents influence the enzyme activity via an interaction with the phospholipid moiety. We also investigated whether these solvents facilitate the exchange of vitamin K from solvent micelles to the carboxylase complex. This would be expressed as a decrease in the K_M of vitamin K by organic solvents. No such decrease was observed. On the other hand, the stimulating effect was only observed when we used forms of vitamin K_1 as coenzyme for carboxylase. When we used derivatives of menadione, the carboxylation reaction could not be stimulated by organic solvents. We think, therefore, that these solvents interact with the (probably hydrophobic) binding site of vitamin K on the protein part of carboxylase, thereby facilitating the binding and/or mobility of the vitamin during various stages of the carboxylation reaction. This increase in mobility leads to a decrease in the activation energy of the carboxylation reaction, which is demonstrated in Arrhenius plots. From these plots, it can be seen that reactions driven by various derivatives of menadione have a low activation energy and the same holds true for vitamin K_1 + organic solvents.

In the absence of these solvents, however, the vitamin K_1-driven reaction has a high activation energy. Obviously, the difference between these two types of reaction is most clearly expressed at low reaction temperatures. At 5°C, for instance, the reaction rate of the vitamin K_1-driven reaction was 6 times higher in the presence of acetone than in its absence. This might be important in those cases where the carboxylation reaction has to be studied at low temperatures (Jones *et al.*, 1977). All these data are consistent with a model in which the long phytyl side chain of vitamin K_1 has a strong interaction with the hydrophobic protein part of carboxylase and would function only in carrying the reactive naphthoquinone group to the active site(s) of the enzyme. Short chain vitamin K analogs, which have a less hydrophobic character, might bypass this carrier function.

ACKNOWLEDGEMENTS

The authors wish to thank professor H.C. Hemker for his stimulating discussions and for kindly reading this manuscript. The technical assistance of Mr. H. Hendrix and Mrs. M. Ulrich are greatfully acknowledged.

REFERENCES

Carlisle, T.L., Shah, D.V., Schlegel, R, and Suttie, J.W. (1975). *Proc. Soc. Exp. Biol. Med. 148*, 140.
Comfurius, P., and Zwaal, R.F.A. (1977). *Biochim. Biophys.* Acta 488, 36.
Jones, J.P., Gardner, E.J., Cooper, T.C., and Olson, R.E. (1977). *J. Biol. Chem. 21*, 7738.
Laemmli, R.J. (1970). *Nature 227*, 680.
Lowry, O.H., Rosebrough, N.J., Farr, A.L., and Randall, R.J. (1951). *J. Biol. Chem. 137*, 265.
Mack, D.O., Wolfensberger, M., Girardot, J.M., Miller, J.A., and Johnson, B.C. (1979). *J. Biol. Chem. 254*, 2656.
De Metz, M., Vermeer, C., Soute, B.A.M., van Scharrenburg, G.J.M., Slotboom, A.J., and Hemker, H.C. (1981). *FEBS Lett. 123*, 215.
De Metz, M., Vermeer, C., Soute, B.A.M., and Hemker, H.C. (1981). *J. Biol. Chem. 256*, 10843.
De Metz, M., Soute, B.A.M., Hemker, H.C., Fokkens, R.,

Lugtenburg, J., and Vermeer, C. (1982). *J. Biol. Chem.* *257*, 5326.

De Metz, M., Soute, B.A.M., Hemker, H.C., and Vermeer, C. (1982). *FEBS Lett.* *137*, 253.

Singleton, W.S., Gray, M.S., Brown, N.L., and White, J.C. (1965). *J. Am. Oil. Chem. Soc.* *42*, 53.

Soute, B.A.M., Vermeer, C., de Metz, M., Hemker, H.C., and Lijnen, H.R. (1981). *Biochim. Biophys. Acta* *676*, 101.

Soute, B.A.M., de Metz, M., and Vermeer, C. (1982). *FEBS Lett.* *146*, 365.

Vermeer, C., Soute, B.A.M., de Metz, M., and Hemker, H.C. (1982). *Biochim. Biophys. Acta* *714*, 361.

Vermeer, C., Hendrix, H., and Daemen, M. (1982). *FEBS Lett.* accepted for publication.

Wallin, R., and Suttie, J.W. (1982). *Arch. Biochem. Biophys.* *214*, 155.

Zwaal, R.F.A., Roelofsen, B., Comfurius, P., and van Deenen, L.L.M. (1975). *Biochim. Biophys. Acta* *406*, 83.

DISCUSSION

Johnson: We reported at Wisconsin that the methylthio ether of menadione was essentially as active as K1 hydroquinone. This would not involve any long side chain at all, not even as long as the DTT side chain, which I agree is an active compound.

Vermeer: With this short side chain, vitamin K is rather active. In our hands it works quite OK. By this short side chain, it might overcome the restrictions that are made by the interaction between the phytyl side chain and the hydrophobic part of the carboxylase.

Olson: About six years ago, we published a series of trials like the ones you did with various vitamin K homologues and showed in microsomal membranes that the activation energy was much lower with MK-1 than MK-4 (Jones *et al.*, 1976). Now you have reconstructed sort of a complex system with the solid phase carboxylase and get similar results. How do you explain this? As regards your mechanism, Dr. Suttie has proposed that the vitamin K hydroperoxy anion is the reagent that plucks a proton from the gamma position of glutamate. You have protonated your vitamin K-hydroperoxy anion and are assuming that the forces are still sufficient to labilize the γ-proton of glutamate. Do you think that this idea is reasonable chemically?

Vermeer: This mechanism was put forward in close cooperation with organic chemists and they think it is reasonable. It is just a kind of visualization of how it may work. Any suggestions to alter and improve it are welcome.

Canfield (Texas A & M): I have a question about your binding protein studies. Were you starting with hydroquinone in your assays and then putting that over gels?

Vermeer: We had radioactive-labeled vitamin K hydroquinone that was used to incubate the mixture with and then we put in SDS and then the combination was put on gels.

Canfield: Have you ever looked for this binding with any sort of equilibrium, sedimentation or sucrose density, or anything like that, to see if you could find binding proteins?

Vermeer: No, we just used SDS gels.

VITAMIN K-DEPENDENT CARBOXYLASE: RECENT STUDIES OF THE RAT LIVER ENZYME SYSTEM[1]

J.W. Suttie
P.C. Preusch
J.J. McTigue

Department of Biochemistry
College of Agricultural and Life Sciences
University of Wisconsin-Madison
Madison, Wisconsin

Although vitamin K was shown in the late 1930's to be essential for blood coagulation, it was some time before this response was attributed to a failure to synthesize specific plasma proteins, and over 40 years before the molecular basis for this requirement was established. A number of indirect lines of evidence available in the mid 1960's and early 1970's suggested (Suttie, 1974) that prothrombin and clotting Factors VII, IX, and X were synthesized from a liver precursor pool and that vitamin K was involved as a cofactor in some posttranslational event. The nature of this event remained elusive until the independent discovery by Stenflo *et al.* (1974) and Nelsestuen *et al.* (1974) that the amino-terminal portion of prothrombin (Figure 1) contained a number of residues of a previously unidentified amino acid, γ-carboxyglutamic acid (Gla).

[1]*Previously unpublished data reported here was supported by the College of Agricultural and Life Sciences of the University of Wisconsin-Madison, and in part by grants AM-14881 and AM-21472 from the National Institutes of Health, Bethesda, Maryland.*

ISBN 0-12-387560-9

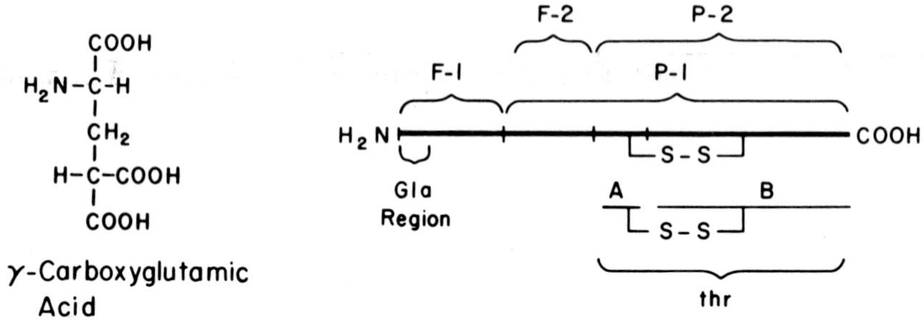

FIGURE 1. Structure of γ-carboxyglutamic acid (Gla) and a diagrammatic representation of the prothrombin molecule. Specific proteolysis of prothrombin by thrombin and Factor Xa will cleave prothrombin into the specific large peptides shown: fragment-1 (F-1), fragment-2 (F-2), prethrombin-1 (P-1), prethrombin-2 (P-2), and thrombin (thr). For details of the activation of prothrombin to thrombin, see Jackson and Nemerson (1980). The Gla residues in bovine prothrombin are located at residues 7, 8, 15, 17, 20, 21, 26, 27, 30, and 33, and they occupy homologous positions in the other vitamin K-dependent plasma proteins.

GENERAL PROPERTIES OF THE VITAMIN K-DEPENDENT CARBOXYLASE

The discovery of Gla residues in prothrombin led to the demonstration (Esmon *et al.*, 1975b) that crude rat liver microsomal preparations contain an enzymatic activity that promotes a vitamin K-dependent incorporation of $H^{14}CO_3^-$ into Glu residues of microsomal precursors of these vitamin K-dependent proteins to form Gla residues. The enzyme activity was soon shown to be active in a number of detergent-solubilized systems, and it was demonstrated (Suttie *et al.*, 1976) that the pentapeptide Phe-Leu-Glu-Glu-Val would serve as a substrate for this enzyme. Most subsequent studies have utilized this or similar peptide substrates, rather than the endogenous substrates, to study the enzyme.

Early studies of the requirements of this enzymatic system have been reviewed (Olson and Suttie, 1978; Suttie, 1978, 1980a; Johnson, 1981), and these reviews as well as the proceedings of a 1979 symposium (Suttie, 1980b) dealing with vitamin K action and vitamin K-dependent proteins provide access to much of the older literature in the field. The initial report of the vitamin K-dependent carboxylation reaction utilized a rat liver postmitochondrial supernatant preparation, and no reductant was added. Studies with washed microsomes demonstrated that the activity had a requirement for NAD(P)H and/or a reduced pyridine nucleotide-generating system. It was subsequently shown that the chemically-reduced form of the vitamin, the hydroquinone (vitamin KH_2), could substitute for these cytoplasmic factors. Crude systems contain several vitamin K reductase activities, and the enzymatic activity is now studied with [NADH + vitamin K] or with vitamin KH_2 as a vitamin source. Early studies also demonstrated that the carboxylation reactions did not require ATP, and all of the available data are consistent with the view that the energy to drive this carboxylation reaction is derived from the reoxidation of the reduced form of vitamin K.

The carboxylase has a requirement for O_2, and there is no evidence that any other compound which might accept electrons from vitamin KH_2 will substitute for O_2 in this reaction. A number of studies appear to have ruled out the involvement of biotin in the system, and the available data are consistent with the hypothesis that CO_2 is the active form of CO_2/HCO_3^- utilized by the enzyme. Studies of vitamin K analogs have suggested that the only important structural feature in a detergent-solubilized system is a 2-methyl-1,4-naphthoquinone substituted at the 3-position with a rather hydrophobic group (Mack *et al.*, 1979).

The direct vitamin K antagonist 2-chloro-3-phytyl-1,4-naphthoquinone is an effective inhibitor of the carboxylase, but the commonly used 4-hydroxycoumarin based anticoagulants, such as Warfarin, are inhibitors only at high concentrations. Their physiological site of action appears to be at a separate site, the vitamin K epoxide reductase. Sulfhydryl poisons, spin-trapping agents, chelating agents, and CN^- have been reported to inhibit the reaction, while pyridoxal phosphate and Mn^{2+} stimulate the reaction. The significance of some of these observations in the crude detergent-solubilized microsomal systems which have been used is difficult to assess at this time.

CURRENT RESEARCH EFFORTS

As indicated in Figure 2, the vitamin K-dependent carboxylase is only one of a number of vitamin K related enzymatic

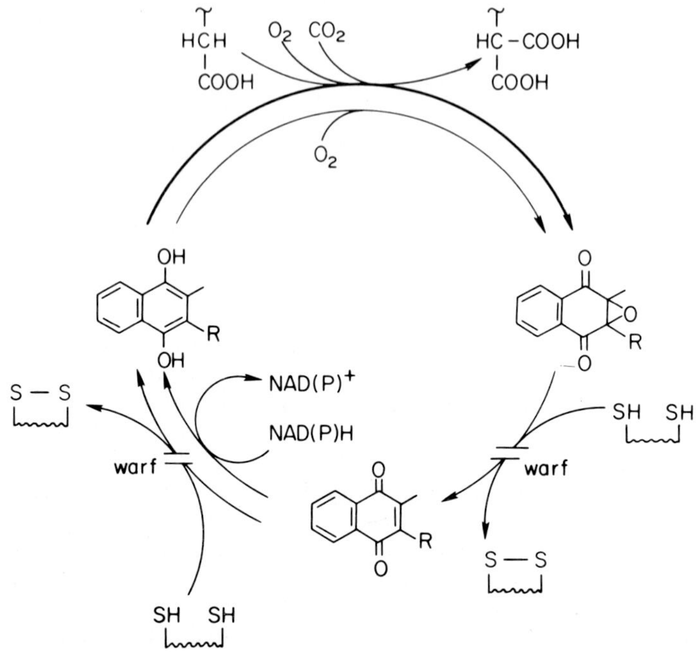

FIGURE 2. Vitamin K-dependent reactions catalyzed by crude rat liver microsomes. The possibility that the carboxylation and epoxidation activities may be catalyzed by the same enzyme is discussed in the text. The dithiol-dependent reductions of the epoxide and of vitamin K quinone are extremely sensitive to the action of coumarin anticoagulants such as Warfarin. These reactions are dithiothreitol-dependent in in vitro *systems, but the physiologically-active reductant has not been identified. It is possible that other as yet unidentified metabolites of vitamin K are formed during the carboxylation reaction.*

activities associated with rat liver microsomal preparations. The relationship of the vitamin K 2,3-epoxidation reaction to the carboxylase activity will be discussed below, and the action of the vitamin K-epoxide reductase and the various microsomal quinone reductases appear essential to an effective recycling of the microsomal vitamin K pool. Recent studies in our laboratory have centered around three major areas: (1) a determination of the specificity of the glutamyl substrate binding site of the enzyme through the use of low-molecular-weight substrates of specific sequence, (2) attempts to elucidate the molecular role of vitamin K in the enzymatic reaction, and (3) investigation of the properties of some of the other microsomal enzymes associated with the role of the vitamin K-dependent carboxylase.

SUBSTRATE SPECIFICITY OF THE VITAMIN K-DEPENDENT CARBOXYLASE

The natural substrates for the vitamin K-dependent carboxylase are of course the high-molecular-weight microsomal precursors of the specific proteins which are to be carboxylated. These precursors accumulate in the livers of animals made hypoprothrombinemic by vitamin K insufficiency or by anticoagulant treatment, and two forms of the prothrombin precursor have been isolated from rat liver (Esmon *et al.*, 1975a; Grant and Suttie, 1976). Additional forms of this protein can be distinguished by isoelectric focusing (Graves *et al.*, 1980), but the structural features responsible for these variants have not been elucidated. There is apparently a tight association between these precursor proteins and the membrane-associated carboxylase. DeMetz *et al.* (1981) have shown that 60-70% of the vitamin K-dependent carboxylase activity in bovine liver can be retained on a column of Sepharose linked bovine Factor X antibody, and data from our laboratory (Table I) indicate that about 20% of the carboxylase activity in rat liver microsomes can be shown to be associated with that material which reacts with an antibody directed against rat prothrombin. These solid-phase preparations will carboxylate an added low-molecular-weight peptide substrate as well as the bound specific precursors. The carboxylase does not readily dissociate from its precursors, even after they are carboxylated, suggesting that there is some protein/protein association in addition to that directed at the Glu-binding site.

Although low-molecular-weight peptide substrates with apparent K_m values in the range of a few mM have a much lower affinity for the enzyme than the natural substrates, they have been useful as probes of the binding site specificity. Inspection of the amino acid sequences of the Gla region of the known vitamin K-dependent clotting factors (Jackson and Nemerson, 1980) fails to reveal any unique sequence that might be a signal for the carboxylase. A collaborative effort between our laboratory and Dr. Dan Rich of the University of Wisconsin School of Pharmacy has resulted in the syn-

TABLE I. Immunoadsorption of Prothrombin Precursor-carboxylase Complexes[a]

	Vitamin K-dependent activity (dpm/0.5 g liver)	
Immunoadsorbent	*Peptide Carboxylation*	*Protein Carboxylation*
Non-immune agarose	*482000*	*60200*
Antiprothrombin agarose	*374000*	*46700*
% Adsorbed	*22%*	*22%*

[a]Microsomes from a Warfarin-treated rat were solubilized in Triton X-100 and passed over an antiprothrombin agarose column or a control column linked to non-immune IgG. The activity adsorbed to the column was then calculated. The column material carried out an active vitamin K-dependent carboxylation of both peptide substrates and bound prothrombin precursors (Swanson and Suttie, 1982).

thesis and assay of a large number of potential peptide substrates. The peptide Phe-Leu-Glu-Glu-Leu has been used as a standard for assay; and, in the case of this reference peptide, it has been demonstrated (Finnan and Suttie, 1980) that only the first of the two adjacent Glu residues is carboxylated to any significant extent. This observation has not been verified in other peptides of similar structure. The data in Table II indicate that, in general, peptides with Glu-Glu sequences are better substrates than those with single Glu residues, and that modification of the residue preceding the Glu-Glu sequence has a profound influence on the substrate activity. Peptides containing Asp, D-Glu, Gln, or Homo-Glu residues have been demonstrated not to be carboxylated to any significant extent. We have reinvestigated this restriction following the recent report of Hamilton *et al.* (1982) that the enzyme can form β-carboxyaspartic acid. Under appropriate conditions, Boc-Asp-OBzl is carboxylated to about 10% the extent of Boc-Glu-OBzl, but peptides containing Asp are essentially inactive.

The relatively low apparent affinity of the low-molecular-weight peptide substrates suggests that some secondary or tertiary structure might be important in providing tight binding to the enzyme, and a peptide homologous to prothrombin precursor sequence 18-23 whose structure is constrained by a disulfide loop and a Pro-residue has been synthesized

TABLE II. Substrate Specificity of the Vitamin K-dependent Carboxylase[a]

Structure	*Activity*	*Structure*	*Activity*
Phe-Leu-Glu-Glu-Leu	*100*	*Boc-Glu-Glu-Leu-OMe*	*107*
Phe-Ala-Glu-Glu-Leu	*72*	*Boc-Glu-Asp-Leu-OMe*	*11*
Phe-Gly-Glu-Glu-Leu	*4*	*Boc-Asp-Asp-Leu-OMe*	*<1*
Leu-Glu-Glu-Leu	*8*	*Boc-Gln-Gln-Leu-OMe*	*<1*
Phe-Glu-Leu-Glu-Leu	*1*	*Boc-Aad-Aad-Leu-OMe*	*<1*
Phe-Glu-Ala-Leu-Glu-Ser-Leu	*6*	*Boc-D-Glu-Glu-Leu-OMe*	*<1*

[a]*All peptides were assayed at a concentration of 1 mM and the activity expressed relative to the activity of Phe-Leu-Glu-Glu-Leu. Aad is amino adipic or homoglutamic acid, and Phe-Glu-Ala-Leu-Glu-Ser-Leu is homologous to residues 29-35 of bovine prothrombin (Rich* et al., *1981).*

(Rich *et al.*, 1983) to probe this possibility. The data in Table III indicate that it is not a good substrate, and that a homologous peptide containing the Ala rather than a Pro residue is actually a better substrate for the carboxylase. It is likely that the Ala containing peptide has a *trans* amide bond between residues 21-22 while this bond is *cis* in the Pro-containing peptide. These data suggest that the secondary and tertiary structure of the precursor, or perhaps some spatially removed recognition sequence on the protein, are of much more importance than the primary sequence in determining whether a specific Glu site is carboxylated.

The peptide corresponding to prothrombin precursor residues 13-29, which has six Glu residues, has been isolated by Soute *et al.* (1981) and found to have an apparent K_m of only 1 μM. These data suggest that peptides with either a multiple Glu-Glu sequence or a significant secondary structure would be good substrates. Synthesis of disulfide containing peptides of this size is difficult, but other low-molecular-weight substrates that have lower K_m's than those previously available have been synthesized. As part of an effort to produce better substrates and inhibitors of the vitamin K-dependent carboxylase, a peptide joined to the 3-position of menadione was synthesized (Lehrman *et al.*, 1981). Although this compound is an inhibitor of the carboxylase at high concentrations, probably through an interaction at the vitamin K-binding site, the data in Figure 3 indicate that at low concentrations, it is an effective enzyme substrate with a

TABLE III. Carboxylase Activity of Peptides Analogous to Prothrombin Sequence 18-23

Substrate	*Activity*[a]
H-Ala-Leu-Glu-Glu-Pro-Ala-OH	*16%*
S—S *H-Cys-Leu-Glu-Glu-Pro-Cys-OH*	*7%*
S—S *H-Cys-Leu-Glu-Glu-Ala-Cys-OH*	*48%*

[a]*The activity of all peptides at 1 mM was compared to 1 mM Phe-Leu-Glu-Glu-Leu (Rich* et al.*, 1983).*

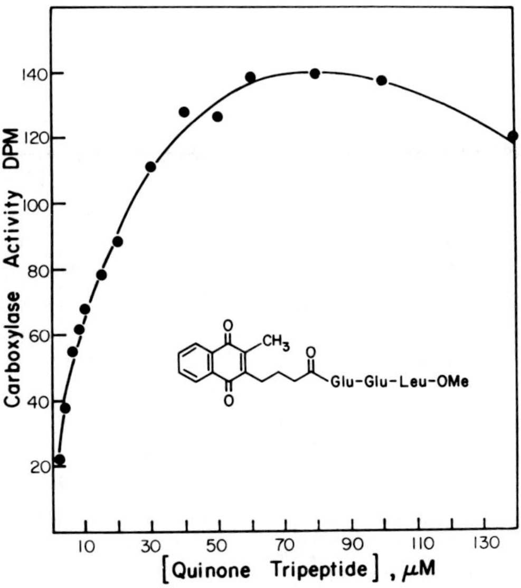

FIGURE 3. Carboxylation of a 2-Me-1,4-naphthoquinone linked tripeptide substrate. The compound illustrated is an effective inhibitor of both endogenous protein and added peptide substrate carboxylation at mM concentrations (Lehrman et al., 1981), but is an effective substrate at low concentrations.

apparent K_m of 10-15 μM. The basis for this increased enzyme affinity has not yet been resolved, but it is hoped that continued efforts in this direction will result in the synthesis of peptides which are carboxylated at concentrations approaching those of the natural substrates. Such compounds would be useful in mechanism studies and might serve as affinity probes in attempts to purify the enzyme.

ENZYMATIC MECHANISM OF THE CARBOXYLASE

Failure to reproducibly obtain highly purified preparations of the liver microsomal vitamin K-dependent carboxylase has hampered efforts to elucidate the mechanism of action of this enzyme, but some progress has been made. The vitamin could presumably act to promote the activation or transfer of CO_2, or it could be directly involved in labilizing the γ-hydrogen of a Glu residue prior to CO_2 fixation. Friedman *et al.* (1979) have demonstrated that release of a tritium atom at the γ-position of a peptide-bound Glu residue to water is dependent on the vitamin KH_2 and oxygen concentrations but not dependent on CO_2 concentration. This observation would appear to establish γ-hydrogen abstraction as the major vitamin K-dependent step. The data do not, however, preclude an additional interaction of some form of the vitamin with CO_2. Neither do the data establish whether the reaction is an ion-mediated or radical-mediated event, and both possibilities have been considered. Liver microsomes contain an activity, "vitamin K epoxidase", first described by Willingham and Matschiner (1974) that converts vitamin K hydroquinone to the 2,3-epoxide of the vitamin (Figure 2). Any hypothesis of the role of vitamin K in the carboxylation reaction must consider the possible relationship between these two activities, and much of our recent work has been directed toward attempts to establish whether or not both activities are catalyzed by the same enzyme. Both reactions are located in the same tissues (Friedman and Smith, 1977), and both activities are enriched in the rough endoplasmic reticulum (Carlisle and Suttie, 1980). Both enzymatic activities utilize the reduced vitamin as a substrate (Sadowski *et al.*, 1977) and are active with the same forms of the vitamin (Friedman and Smith, 1979). With the exception of low CO_2 concentrations, which limit carboxylation but have no effect on epoxidase activity, a number of conditions which stimulate or inhibit one reaction have been reported to have the same effect on the other (Suttie *et al.*, 1978). Recent studies have added further evidence that these two reactions are catalyzed by the same enzyme. When both activities are assayed at varying O_2 concentrations, it was found that the O_2 concentration needed for half-maximal activity was very similar (Figure 4).

It has been demonstrated (Suttie *et al.*, 1980) that increasing the concentration of a peptide substrate for the carboxylase, which increases the number of carboxylation events in the system, also increases epoxidase activity. This observation would be difficult to explain if there were not some interrelationship between the two systems but does not indicate what the stoichiometry of this interrelationship might be. We have now demonstrated (Larson *et al.*, 1981) that increasing the peptide substrate concentration or adding Mn^{2+}, which is a stimulator of the carboxylase activity, to the system causes a parallel increase in both carboxylation and epoxidation (Table IV). Continued studies of this response indicated that at saturating concentrations of CO_2 there is an apparent equivalent stoichiometry between epoxide

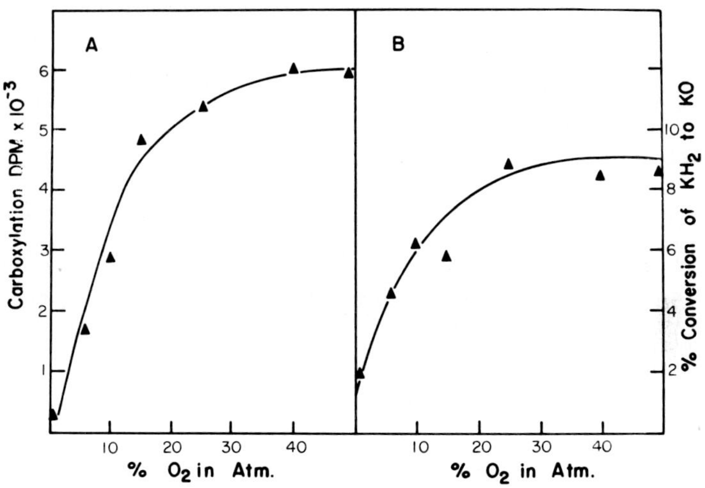

FIGURE 4. Effect of O_2 concentration on vitamin K-dependent carboxylation and vitamin K epoxidation. Carboxylase activity (panel A) and epoxidase activity (panel B) were measured in the presence of 1 mM peptide substrate and 4 mM $NaHCO_3$ utilizing a microsomal preparation from vitamin K-deficient rats.

TABLE IV. Effect of Substrate Concentration and Manganese on the Stoichiometry of Vitamin K Epoxide and Gla Formation[a]

Additions	*nMoles CO_2 into peptide/ml*	*nMoles CO_2 into protein/ml*	*nMoles K oxide/ml*	*KO/CO_2*
1 mM peptide	*1.85*	*1.95*	*4.1*	*1.08*
10 mM peptide	*5.4*	*1.87*	*6.5*	*0.89*
10 mM $MnCl_2$	*9.9*	*1.83*	*9.5*	*0.82*

[a]*The peptide substrate used was Phe-Leu-Glu-Glu-Leu, and CO_2 fixed into peptide and protein was calculated from a determination of the specific radioactivity of the CO_2/HCO_3 pool at the end of the incubation. The incubations were conducted under conditions of saturating CO_2 (Larson* et al.*, 1981).*

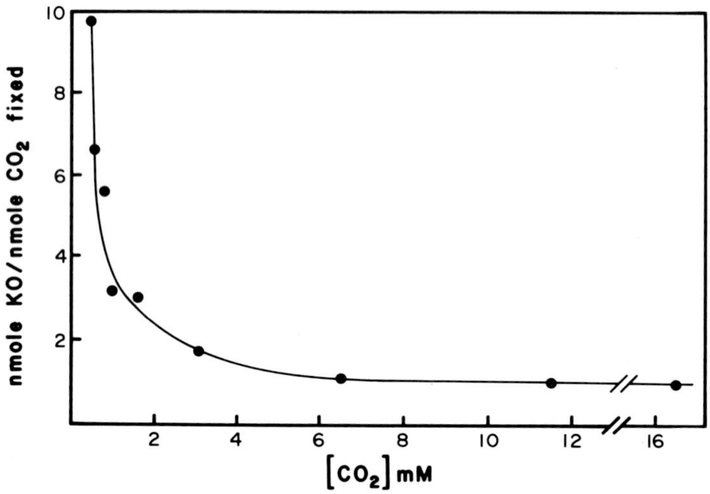

FIGURE 5. The ratio of the amount of vitamin K epoxide formed/CO_2 fixed into Gla as a function of CO_2 concentration. For details see Larson et al., *1981.*

formation and Gla formation, but as the CO_2 concentration was lowered, a large excess of vitamin K epoxide was produced (Figure 5). These observations lend further support to the hypothesis that CO_2 fixation is secondary to the primary vitamin K-dependent event. How these two reactions are coupled has not been elucidated at this time, but one possibility would be through an oxygenated intermediate that would be on the pathway to the epoxide and would also be involved in hydrogen abstraction from the γ-position of the Glu substrate. Glutathione peroxidase has been shown to act on and reduce a number of hydroperoxides to the corresponding alcohols, and Larson and Suttie (1978) have shown that the activity of the vitamin K epoxidase and vitamin K-dependent carboxylase in detergent-solubilized microsomes can be decreased by the addition of increasing amounts of glutathione peroxidase. More recent studies (Larson *et al.*, 1981) have indicated that tritium release from the γ-position is influenced in a similar fashion. These data suggest that a hydroperoxide of the vitamin, a logical intermediate in epoxide formation, is involved in the carboxylation reaction; and a mechanism which is consistent with the available data is shown in Figure 6.

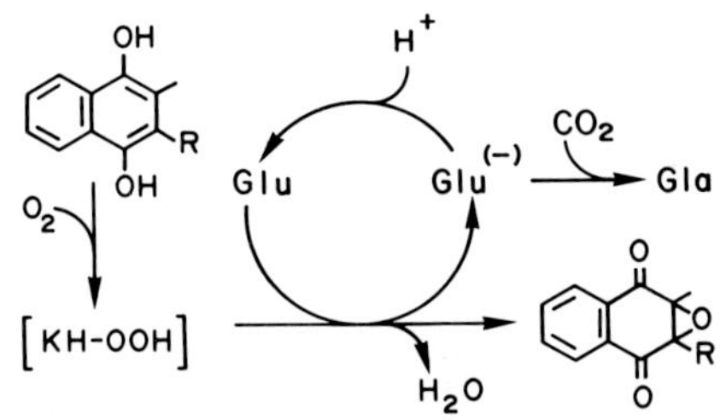

FIGURE 6. Proposed mechanism for coupling of carboxylation and epoxidation. Some oxygenated intermediate would be formed from attack of oxygen on the reduced vitamin. This could conceivably be a hydroperoxide anion, a hydroperoxide radical, or a hydroperoxide. This intermediate would then activate the γ-position of the Glu residue and be converted to epoxide. The observed stoichiometry of the reaction indicates a tight coupling at this point. In the absence of a suitable Glu-containing substrate, the oxygenated intermediate is either not formed at all or it is not converted to the epoxide. In the absence of CO_2, the activated substrate is not carboxylated, but presumably is able to revert to the unactivated form. The activated glutamyl residue could be either a carbanion, as shown, or a carbon radical. In the latter case, additional radical mechanisms would be required to form the appropriate adduct from CO_2. This mechanism is consistent with the available data but is to a large extent hypothetical.

The results of some experiments recently carried out in our laboratory would appear to be consistent with this general model of carboxylase action. In an attempt to develop competitive inhibitors of the Glu-binding site of the enzyme, peptides containing phosphoserine residues have been synthesized. A peptide containing phosphoserine residues has a negative charge at the same distance from the polypeptide backbone as a Glu residue, but no available methylene carbon. Such peptides are apparent competitive inhibitors of the carboxylation of another low-molecular-weight peptide (Figure 7); and, as predicted by the model shown in Figure 6, they also inhibit the epoxidation reaction (Figure 8). As with the observed stimulation of the enzyme by additional substrate, such an observation is difficult to explain unless the reactions are closely coupled. It has also been suggested (Olson and Suttie, 1977) that the epoxide is not a

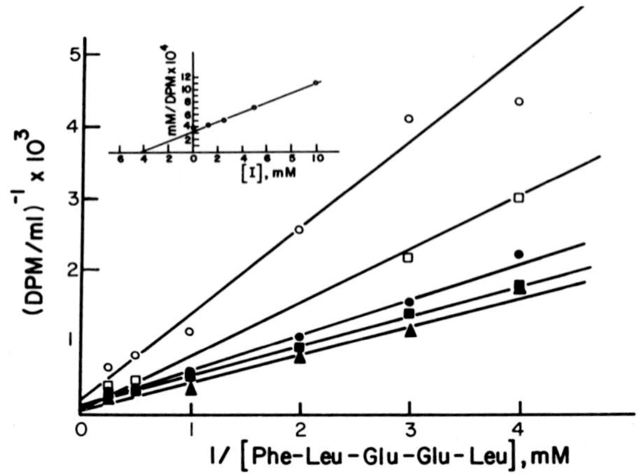

FIGURE 7. Effect of a peptide containing phosphoserine residues on vitamin K-dependent carboxylase activity. Double reciprocal plots of the effect of Boc-phosphoSer-phosphoSer-Leu-OMe on the vitamin K-dependent carboxylation of Phe-Leu-Glu-Glu-Leu. Inhibitor concentrations used were: 1.25 mM, ■——■ ; 2.5 mM, ●——●; 5.0 mM, □——□ ; and 10.0 mM, ○——○ . A replot of the slope is shown in the insert (Suttie et al.*, 1982).*

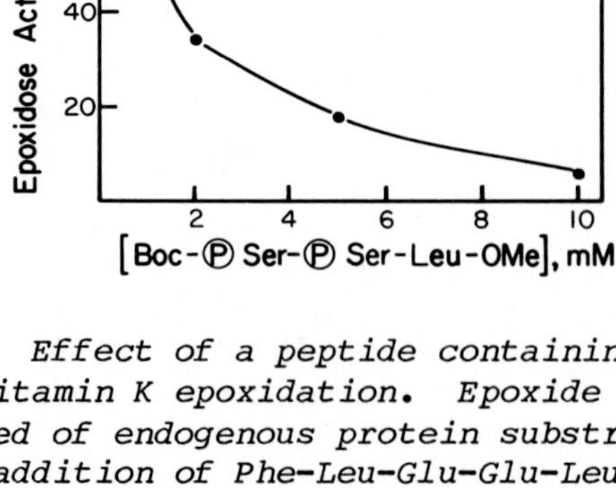

FIGURE 8. Effect of a peptide containing phosphoserine residues on vitamin K epoxidation. Epoxide formation in a system depleted of endogenous protein substrates was stimulated by the addition of Phe-Leu-Glu-Glu-Leu and the effect of Boc-phosphoSer-phosphoSer-Leu-OMe on the epoxide formation was measured.

direct product of this reaction, but rather that oxidation of the reduced vitamin by the carboxylase results in the formation of H_2O_2, and that the epoxide arises from the well known (Tischler *et al.*, 1940) oxidation of the vitamin to its epoxide by H_2O_2. The data shown in Figure 9 indicate that production of vitamin K epoxide by the intact animal or in a microsomal incubation is stereospecific and suggest that a chemical oxidation route of epoxide formation is unlikely (Preusch and Suttie, 1983).

OTHER ENZYMES RELATED TO THE CARBOXYLASE

The metabolic cycling of vitamin K by the enzyme activities indicated in Figure 2 suggest that these enzymes may function with the carboxylase in a membrane-associated morphological complex and that efficient recycling of the epoxide may be essential for carboxylase activity. Simple calculations of the dietary requirement for vitamin K relative to the daily formation of Gla, as measured by its excretion, also indicate that each mole of vitamin must be extensively reutilized.

The vitamin K epoxide reductase has been known for some time to be sensitive to the coumarin anticoagulants, and on

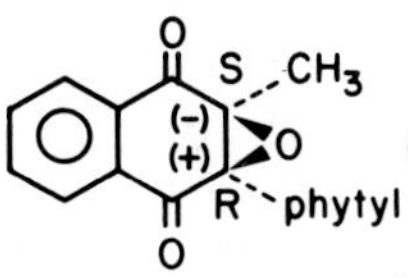

(+)-2S,3R-Vitamin K_1 epoxide

	$[\theta]_{342}$ of Vitamin K_1 epoxide
H_2O_2 and Vitamin K quinone	0 (racemic)
isolated from Rat liver	+ 3800° cm^2/dmole
<u>in vitro</u> microsomal incubation	+ 3900° cm^2/dmole
phase-transfer epoxidation	~+ 150° cm^2/dmole

FIGURE 9. Stereochemistry of vitamin K epoxide. The epoxide has been produced by a standard chemical oxidation and by a partially stereospecific oxidation. The metabolite isolated from either the intact rat or form an in vitro *microsomal incubation is highly optically active. The naturally occurring (+) isomer has been tentatively assigned the 2S,3R configuration on the basis of comparison to simple naphthoquinone epoxides of known conformation.*

TABLE V. In Vitro *Concentration of 4-Hydroxycoumarin Giving 50% Inhibition of Various Vitamin K-Dependent Activities*[a]

Activity	Warfarin		Difenacoum	
	Normal rat liver	Warfarin-resistant rat liver	Normal rat liver	Warfarin-resistant rat liver
Vit. K-dependent carboxylase				
Vitamin KH_2-initiated	3 mM	3 mM	1.4 mM	1.4 mM
Vitamin K + NADH-initiated	2 mM	2 mM	1.0 mM	1.0 mM
Epoxidase	4 mM	4 mM	2.0 mM	2.0 mM
DT-diaphorase				
Soluble assay	33 μM	30 μM	49.0 μM	67.0 μM
Liposomal assay	840 μM	540 μM	17.0 μM	17.0 μM
Epoxide reductase	4 μM	50 μM	2.5 μM	4.0 μM

[a]*Difenacoum is a 4-hydroxycoumarin which is an effective rodenticide for control of the Warfarin-resistant strain of rats. The epoxide reductase is not only the enzyme which is most sensitive to Warfarin in the normal rat but it is the only one whose sensitivity is substantially altered in the preparation from the Warfarin-resistant rat. Difenacoum is a potent inhibitor in the normal rat, and its inhibition of the epoxide reductase is altered only slightly in the preparation from the Warfarin-resistant rat (Hildebrandt and Suttie, 1982).*

the basis of the altered coumarin sensitivity of this enzyme in the tissues of the Warfarin-resistant rat (Whitlon *et al.*, 1978), this activity has also been assumed to be the physiologically-important site of coumarin action. We have recently investigated the sensitivity of other vitamin K-dependent enzymes that might conceivably be involved in this response (Table V), and it can be seen that the epoxide reductase appears to be the most likely site of action of these important drugs. One activity which was not probed in the study reported in Table V was the DTT-dependent vitamin K quinone reductase activity, however this activity has also been reported (Fasco and Principe, 1982) to be Warfarin sensitive. A recent collaborative study (Fasco *et al.*, 1982) has now shown that the Warfarin sensitivity of this enzyme, as well as the epoxide reductase, is altered in tissues from Warfarin-resistant rats. The present data do not allow a conclusion as to whether both of these DTT-dependent activities are catalyzed by the same enzyme or if some other explanation to the apparent dual genetic defect is possible.

Vitamin K epoxide can be chemically reduced to vitamin K quinone by dithiols in an organic base catalyzed reaction, and a detailed study (Preusch *et al.*, 1982) of these reactions has led us to postulate that the enzymatic mechanism might be similar to that shown in Figure 10. Studies of an enzyme preparation obtained from the Warfarin-resistant rat have again proved valuable in assessing this model. As indicated in Figure 11, microsomal preparations from normal rats are able to convert vitamin K epoxide to vitamin K quinone and, after a short lag period, to vitamin KH_2. A different response is observed with preparations from Warfarin-resistant rats. A new product, which has been identified as the 3-hydroxy derivative of vitamin K, appears at the same time as the quinone (Fasco *et al.*, 1983) and, depending on incubation conditions, may account for over 50% of the total product. This product could result if the mutated enzyme was altered such that the half enol-3-hydroxy derivative of the vitamin, which is shown as an intermediate of the enzyme reaction in Figure 10, dissociated from the enzyme and spontaneously rearranged to the isolated hydroxy derivative of the quinone. This metabolite is not a substrate for either the vitamin K-dependent carboxylase or epoxidase; and, if it were also formed *in vitro*, its formation could explain the increased vitamin K requirement observed in the Warfarin-resistant rat (Hermodson *et al.*, 1969). These observations have also furnished a possible clue to the inhibitory action of the 4-hydroxycoumarins as vitamin K antagonists. Although these compounds do not appear to be good structural

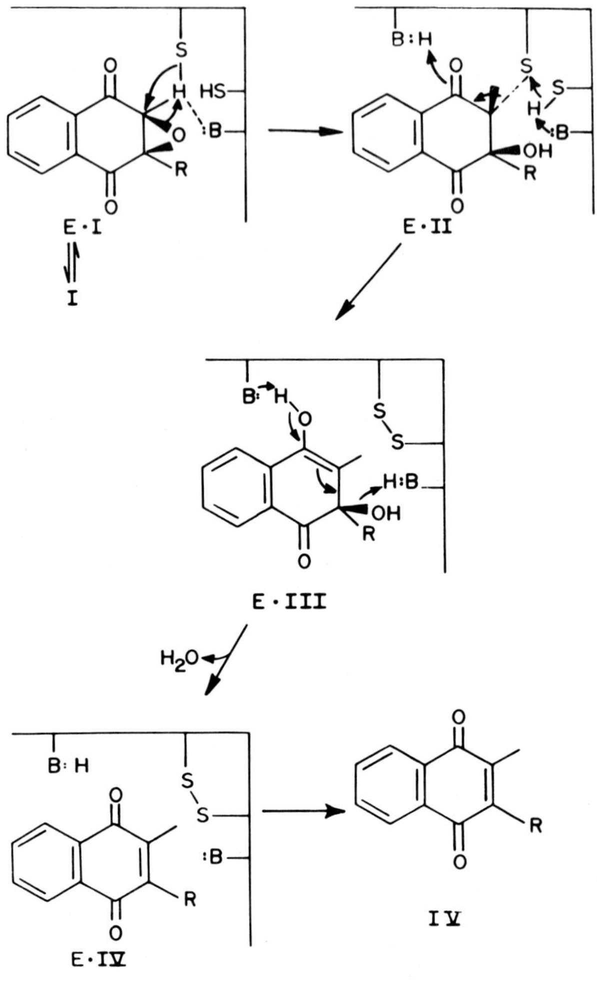

FIGURE 10. Proposed model for the enzymatic reduction of vitamin K epoxide. This tentative model is based on a kinetic analysis of intermediates which have been characterized from a thiol-mediated chemical reduction of the epoxide. It assumes an enzyme-bound thiol adduct (E·II) and a half enol, 3-hydroxy derivative of the vitamin as a transition state (E·III) on the way to the formation of vitamin K quinone (IV). It is assumed that the role of DTT in the in vitro *system commonly used is to reduce the enzyme to allow another cycle of reduction.*

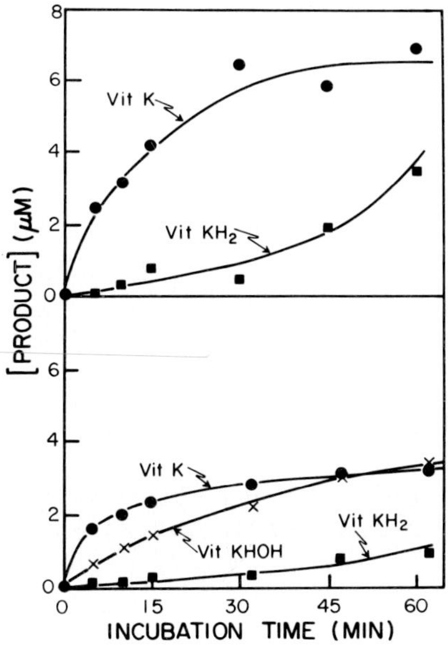

FIGURE 11. Metabolism of vitamin K epoxide by rat liver microsomes. Reduction of vitamin K epoxide with DTT as a reductant by a preparation from normal rat liver microsomes (top panel) results in the formation of vitamin K quinone (vit K) and, after a short lag period, formation of vitamin K hydroquinone (vit KH_2). When the incubation is carried out with a microsomal preparation from Warfarin-resistant rats (bottom panel), an additional metabolite, 3-hydroxy vitamin K (vit KHOH), is formed at the same time as vitamin K quinone (Fasco et al., *1983).*

analogs of either vitamin K or its epoxide, they do seem (Figure 12) to be good analogs of the transition state the half enol form, of the vitamin which was postulated in Figure 10. If they do in fact bind at this site, the altered binding affinity which results in the release of the intermediate and formation of the 3-hydroxy derivative would also explain the lowered affinity of the mutant enzyme for its inhibitor.

SUMMARY AND PERSPECTIVE

The rat liver microsomal vitamin K-dependent carboxylase appears to represent a unique carboxylation reaction which

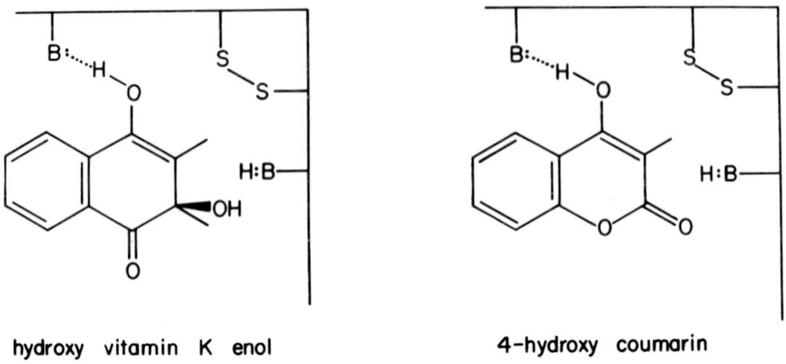

FIGURE 12. Proposed interaction of 4-hydroxycoumarin anti-coagulants with the vitamin K epoxide reductase. Structural similarity suggests that the coumarin anti-coagulants are much better analogs of the hydroxy half enol intermediate proposed in Figure 10 than of the substrate vitamin K epoxide.

carries out an important posttranslational modification of a limited number of proteins. In contrast to the simple primary sequence recognition sites involved in some posttranslational modifications, the available data would suggest that some macromolecular recognition site is required to insure that only specific glutamyl residues of these proteins become carboxylated. The nature of these interactions, if they exist, has not been identified. The available data strongly suggest that the formation of vitamin K 2,3-epoxide from the reduced vitamin is the enzymatic step which is associated with the abstraction of the γ-hydrogen of the substrate Glu residue. How this is accomplished and what is the chemical nature of the resulting Glu residue are questions that are essential to an understanding of the mechanism but which cannot be answered at the present time. These points remain both the most interesting and most difficult to experimentally attack with the systems currently available. It seems increasingly likely that the series of vitamin K transformations indicated in Figure 2 might form some closely associated membrane complex. It is hoped that continued investigations of these associated activities may offer a clearer explanation of the physiological responses which are seen in response to vitamin K insufficiency or to Warfarin antagonism.

REFERENCES

Carlisle, T.L., and Suttie, J.W. (1980). *Biochemistry 19*, 1161.

De Metz, M., Vermeer, C., Soute, B.A.M., Van Scharrenburg, G. J.M., Slotboom, A.J., and Hemker, H.C. (1981). *FEBS Lett. 123*, 215.

Esmon, C.T., Grant, G.A., and Suttie, J.W. (1975a). *Biochemistry 14*, 1595.

Esmon, C.T., Sadowski, J.A., and Suttie, J.W. (1975b). *J. Biol. Chem. 250*, 4744.

Fasco, M.J., and Principe, L.M. (1982). *J. Biol. Chem. 257*, 4894.

Fasco, M.J., Hildebrandt, E.F., and Suttie, J.W. (1982). *J. Biol. Chem. 257*, 11210.

Fasco, M.J., Preusch, P.C., Hildebrandt, E., and Suttie, J.W. (1983). *J. Biol. Chem.* in press.

Finnan, J.L., and Suttie, J.W. (1979). *In* "Vitamin K Metabolism and Vitamin K Dependent Proteins" (J.W. Suttie, ed.), p. 509. University Park Press, Baltimore.

Friedman, P.A., and Smith, M.W. (1977). *Biochem. Pharmacol. 26*, 804.

Friedman, P.A., and Smith, M.W. (1979). *Biochem. Pharmacol. 28*, 937.

Friedman, P.A., Shia, M.A., Gallop, P.M., and Griep, A.E. (1979). *Proc. Natl. Acad. Sci. USA 76*, 3126.

Grant, G. A., and Suttie, J.W. (1976). *Biochemistry 15*, 5387.

Graves, C.B., Grabau, G.G., Olson, R.E., and Munns, T.W. (1980). *Biochemistry 19*, 266.

Hamilton, S.E., Tesch, D., and Zerner, B. (1982). *Biochem. Biophys. Res. Commun. 107*, 246.

Hermodson, M.A., Suttie, J.W., and Link, K.P. (1969). *Am. J. Physiol. 217*, 1316.

Hildebrandt, E.F., and Suttie, J.W. (1982). *Biochemistry 21*, 2406.

Jackson, C.M., and Nemerson, Y. (1980). *Annu. Rev. Biochem. 49*, 765.

Johnson, B.C. (1981). *Mol. Cell. Biochem. 38*, 77.

Larson, A.E., and Suttie, J.W. (1978). *Proc. Natl. Acad. Sci. USA 75*, 5413.

Larson, A.E., Friedman, P.A., and Suttie, J.W. (1981). *J. Biol. Chem. 256*, 11032.

Lehrman, S.R., Rich, D.H., Goodman, H.L., and Suttie, J.W. (1981). *In* "Peptides: Synthesis, Structure and Function" (D. Rich, and E. Gross, eds.), p. 513. Pierce Chemical, Rockford.

Mack, D.O., Wolfensberger, M., Girardot, J.M., Miller, J.A., and Johnson, B.C. (1979). *J. Bio. Chem. 254,* 2656.

Nelsestuen, G.L., Zytkovicz, T.H., and Howard, J.B. (1974). *J. Biol. Chem. 249,* 6347.

Olson, R.E., and Suttie, J.W. (1978). *Vitamins and Hormones 35,* 59.

Preusch, P.C., Patterson, J.L., Mctigue, J.J., and Suttie, J.W. (1982). *Fed. Proc. 41,* 466 (abstr.).

Preusch, P.C., and Suttie, J.W. (1983). *J. Biol. Chem. 258,* 714.

Rich, D.H., Lehrman, S.R., Kawai, M., Goodman, H.L., and Suttie, J.W. (1981). *J. Med. Chem. 24,* 706.

Rich, D.H., Kawai, M., Goodman, H.L., and Suttie, J.W. (1983). *J. Med. Chem.,* in press.

Sadowski, J.A., Schnoes, H.K., and Suttie, J.W. (1977). *Biochemistry 16,* 3856.

Soute, B.A.M., Vermeeer, C., De Metz, M., Hemker, H.C., and Lijnen, H.R. (1981). *Biochim. Biophys. Acta 676,* 101.

Stenflo, J., Fernlund, P., Egan, W., and Roepstorff, P. (1974). *Proc. Natl. Acad. Sci. USA 71,* 2730.

Suttie, J.W. (1974). *Vitamins and Hormones 32,* 463.

Suttie, J.W. (1978). *In* "The Fat-soluble Vitamins" (H.F. DeLuca, ed.), p. 211. Plenum Press, New York.

Suttie, J.W. (1980). *CRC Crit. Rev. Biochem. 8,* 191.

Suttie, J.W. (1979). *"Vitamin K Metabolism and Vitamin K Dependent Proteins,"* p. 592. University Park Press, Baltimore.

Suttie, J.W., Hageman, J.M., Lehrman, S.R., and Rich, D.H. (1976). *J. Biol. Chem. 251,* 5827.

Suttie, J.W., Larson, A.E., Canfield, L.M., and Carlisle, T.L. (1978). *Fed. Proc. 37,* 2605.

Suttie, J.W., Geweke, L.O., Martin, S.L., and Willingham, A.K. (1980). *FEBS Lett. 109,* 267.

Suttie, J.W., Wood, G., Kawai, M., Lehrman, S.R., and Rich, D.H. (1982). *Fed. Proc. 41,* 467 (abstr.).

Swanson, J.C., and Suttie, J.W. (1982). *Biochemistry 21,* 6011.

Tischler, M., Fieser, L.F., and Wendler, N.L. (1940). *J. Am. Chem. Soc. 62,* 2866.

Whitlon, D.S., Sadowski, J.A., and Suttie, J.W. (1978). *Biochemistry 17,* 1371.

Willingham, A.K., and Matschiner, J.T. (1974). *Biochem. J. 140,* 435.

DISCUSSION

Graves (Iowa S. U.): Would you state once again what you think are the important structural features that are required for good carboxylation to occur? Do you have a model for this?

Suttie: There does not appear to be a lot of specificity in the primary sequence. I find it significant that you can form β-carboxy aspartate if you start with a very simple compound. But once you put it into a peptide bond, and get a little bulk there, it doesn't seem to act as a substrate. At the present time, I don't see a lot of significance in the primary sequences. I think the important structural feature will turn out to be that the protein contain a macromolecular recognition site for the enzyme. There is some work in press from other laboratories based upon sequencing of cDNA which would suggest that the vitamin K dependent plasma proteins are synthesized with a signal sequence of amino acids followed by a basic section which might be anywhere from 4-5 up to 10-12 residues, depending upon where the signal peptidase cleaves. This "pro" piece might be very important in specificity. Following that, the structure of the vitamin K dependent protein begins.

Graves: I am curious that you tried the D-analogues and that they did not serve as substrates. Have you tried them as inhibitors? We have found that D-peptides are not phosphorylated by protein kinases, but serve as rather good competitive inhibitors.

Suttie: Every peptide that we synthesize is tested both as a substrate and as a potential inhibitor against a good substrate. The peptides containing D residues were not very good inhibitors.

Graves: What about the nitro-analogue?

Suttie: The nitro-analogue is very interesting. A peptide containing the nitro-analogue is an inhibitor, but not a very good one. On the other hand, we went one step further and took it up to pH 10, generated the supposedly stable carbanion, and got a result that we simple do not understand; that is, it was not carboxylated, but it stimulated the carboxylation of another substrate.

Wold: I would like to propose that the critical issue in

whether a residue is going to be a substrate or not is simply reflected by its pK. To make it very simple, we can say that only a glutamate or aspartate that is protonated can be carboxylated. It seems to me that that is consistent with clusters of glutamic acid where the last one to titrate will have a very high pK. The last proton off will really be held on quite strongly. It might also be consistent with the effect of organic solvents in changing the dielectric constant and thus increasing the pK. One test might actually be to put some of your more complex aspartate peptides in acetone or dimethylsulfoxide and see if they now become reasonable substrates.

Suttie: This is something we haven't done yet. They have some curious properties that we have noted in the past: these aspartyl containing peptides (and we have made a number of them) that are not substrates are not inhibitors either, but do some of the stimulatory effects on other peptides which I just mentioned.

Johnson: How does that fit with the fact that for every native protein that is carboxylated, no glutamyl residue beyond about amino acid 33 from the amino terminal end is carboxylated? Only the first few glutamyl residues not beyond amino acid residue 32-33 are carboxylated.

Suttie: That is presumably a problem solved by macromolecular recognition sites. It is not as straight forward as it might once have seemed. Everybody that has looked at the structure of the F1 region of prothrombin, and certainly the amino terminal region, has probably constructed a model. If you do that, it all looks so simple because except for the disulfide bridge (which causes a bend, of course), you synthesize a perfect amphiplatic helix where all the Gla residues lie on one flat plane. The other side of that helix is very hydrophobic, suggesting that this greasy part lays against a greasy part in the protein and all the Gla's are available in one stretch. There is now a four-angstrom resolution structure of the F-1 portion of prothrombin and there is absolutely no helical region in this portion. The 10 Gla residues in the model are, however, sitting at the end of a domain, and are more or less planer. That is about all you can say at the present time. That does not indicate that before carboxylation, the Gla residues may have been in a helical segment.

Olson: If one has activated oxygen on the surface of the carboxylase, regardless of when the epoxide is formed, one could get a stereo-specific epoxide. Only if the epoxide is formed chemicaly from H_2O_2 and vitamin K, would it be racemic. Could you also comment on the rate of CO_2 as well as other agents in affecting the epoxidase/carboxylase ratios?

Suttie: Obviously, oxygen has to interact with the vitamin. I would suggest that the semiquinone of the vitamin interacts with the oxygen to form an initial oxygen adduct which then forms the epoxide. You can also certainly postulate that the epoxide is formed chemically from an activated oxygen or hydrogen peroxide, and it is possible that it interacts on the surface of the protein, therefore it forms stereo selectively. I would argue that this would be an enzymatic reaction, but that's another point.

Olson: It seems to me that the epoxide formation could be at one of several stages in the sequence of this complicated reaction and not be coexistent with carboxylation. Do you believe there is a single active site?

Suttie: I'm sure that epoxide formation is not co-existent with carboxylation. If that were true, I do not think you would get carboxylation without epoxidation. I think that what you can conclude from the evidence at the present time is that epoxide formation is apparently coupled with glutamyl activation. The one thing that we need better numbers on are the relationships between epoxide formation and hydrogen release. In terms of CO_2 affecting epoxidation, the fact is that it doesn't affect it all that much. The rate of epoxidation is relatively independent of carboxylation. It varies, but you are talking about variation within one order of magnitude, whereas by varying the CO_2 concentration, you obviously can vary the carboxylation reaction over a couple orders of magnitude.

Olson: Are we talking about orders of magnitude? That would be 10-fold.

Suttie: We don't see large changes.

Olson: Cyanide ions increase epoxidation and reduce carboxylation. It doesn't mean that we are not dealing with a single protein, but it does raise questions about how it happens?

Suttie: I don't know how it happens. That's what I would like to find out.

Canfield (Texas A & M): Have you looked at the role of the cytosol in your expoxide reductase in heart? Also, in your rate of formation of hydroquinone, do you interpret that as meaning that it is being formed by the same enzyme or does your system contain two enzymes and you think one is acting on the intermediate of the other?

Suttie: We run the epoxide reductase reaction under conditions where there is essentially none of the so-called ERSA. We do not specifically try to get rid of it all. We really have not addressed the question of whether or not the presence of this protein might have given the enzyme some more stereo-selectivity. I really do not know. The second point, as to whether there is one enzyme or two enzymes, is interesting. If you look at crude preparations of the epoxide reductase, there is no doubt that the quinone is formed first and then the hydroquinone. There is no reason to believe that both reactions are catalyzed by the same enzyme, except that the same dithiol-dependent systems will reduce the quinone to the hydroquinone. Both reactions are inhibited by warfarin and the warfarin reductase, is altered in the warfarin resistant rat. You are left now with a limited number of possibilities. Perhaps there have been two simultaneous mutations of two completely different enzymes (I tend not to believe that). Another possibility is that the same enzyme does in fact catalyze both reactions. It is possible that one polypeptide contains two active sites and the mutation has affected both of them, or that the mutation is somehow in the system that utilizes the soluble dithiol to reduce the enzyme-bound dithiol. I think there are a number of possibilities. At the present time I would prefer to believe that perhaps the same enzyme catalyzes both reaction, but that also raises some problems.

Wallin (U. of Tromso): How important do you think the hydroquinone reductases are in this carboxylation?

Suttie: I think they are rather important. We have recently been looking at the hypoprothrombinemic effect of salicylates and in contrast to what some people have reported recently, it doesn't appear that salicylates are inhibitors of the vitamin K epoxide reductase. However, if you maintain a rat on a barely marginal level of vitamin K, you can knock down clotting factor synthesis with salicylate. Salicylates are potent inhibitors of DT diaphorase, and there are other lines of evidence to suggest that under certain conditions, DT diaphorase can be an important system. Given marginal levels of vitamin K, DT diaphorase must be important, because if you remove this activity you affect the whole system. We should also keep in mind that physiologically, it is probably true that epoxide reductase is normally the important enzyme. We play around with a lot of numbers, and it turns out that if you make some reasonable assumptions, every mole of vitamin must be used a few hundred to a few thousand times. The vitamin is almost acting as a bound prosthetic group, and it seems likely then, that there is a complex of these proteins and the vitamin is continually cycled.

Knauer (OMRF): What is the relationship of the molecular events that you outline and the molecular events that, say chloro-K undergoes, and does that shed any light on your mechanistic process? Where does chloro-K fit?

Suttie: I don't think anybody knows. Without ever having done the experiments all that critically, I think people assume that chloro-K is just a competitive inhibitor of the vitamin binding site. There is some data that suggests that it is. People have looked for epoxide chloro-K and have not found it. But again, if you look at that structure, it is hard to believe that the epoxide is stable, so I don't think anybody really knows.

Knauer: But no intermediates would fit in there?

Suttie: No intermediates have been detected.

Girardot: You said that aspirinate would inhibit the vitamin K reductase. In our system we have inhibition with aspirinate which reacts acetyl salicylate at 600 μM. Could you comment on that, using the K hydroquinone to initiate a reaction?

Suttie: Salicylate will affect DT-diaphorase at about 1 μM and at high levels, it will affect the epoxide reductase.

UTILIZING SUBSTRATES WITH MODIFIED GLUTAMYL RESIDUES TO ELUCIDATE THE REACTION MECHANISM OF THE VITAMIN K-DEPENDENT γ-GLUTAMYL CARBOXYLASE[1]

Paul A. Friedman
David L. Anton
Stephen K. Kwan

Department of Pharmacology
Harvard Medical School
and
Charles A. Dana Research Institute
Harvard-Thorndike Laboratory of Beth Israel Hospital
Department of Medicine
Beth Israel Hospital and Harvard Medical School
Boston, Massachusetts

INTRODUCTION

In efforts to clarify the detailed reaction mechanism of the vitamin K dependent gamma-glutamyl carboxylase investigators have been hampered by the lack of a highly purified preparation of the enzyme. Although information presented in these proceedings would suggest that such preparations may now be available, it has nevertheless been possible in the past few years using crude carboxylase preparations to reach certain conclusions about the mechanism. Beyond the most basic description of the enzyme system (i.e., that to gamma carboxylate certain glutamyl residues of appropriate proteins this intrinsic membrane protein requires oxygen, carbon dioxide or bicarbonate, and the hydroquinone form of vitamin K as cofactor) (for review, see Suttie, 1980; and Johnson, 1981), radiolabeled vitamins K as well as certain synthetic

[1]*This research was supported by grants HL25066 and HL27150 from the National Institutes of Health.*

ISBN 0-12-387560-9

oligopeptide substrates for the carboxylase have proven most useful in extending our knowledge of this reaction.

For instance, subsequent to the observations of Matschiner and coworkers (Matschiner *et al.*, 1970; Willingham and Matschiner, 1974) that tritiated phylloquinone was metabolized by crude rat liver microsomal preparations to phylloquinone 2,3-epoxide, the carboxylase activity was described (Esmon *et al.*, 1975) and found to require oxygen (Esmon *et al.*, 1975; Friedman and Shia, 1976). It was proposed that formation of the epoxide might be coupled to carboxylation; indeed, it was then demonstrated that both reactions are located in the same tissues (Friedman and Smith, 1977), both activities are enriched in the rough endoplasmic reticulum (Carlisle and Suttie, 1980), both activities utilize the reduced vitamin as a substrate (Sadowski *et al.*, 1977), both are active with the same forms of the vitamin (Friedman and Smith, 1979), and the extent of both carboxylation and formation of the epoxide are increased in solubilized crude vitamin K dependent gamma glutamyl carboxylase preparations to which peptide substrates have been added (Suttie *et al.*, 1980). The culmination of this series of experiments was the demonstration that when the ratio of vitamin K epoxide formed to carboxylation events was determined (Larson *et al.*, 1981), a molar ratio of 1 was attained at concentrations of CO_2 saturating for the carboxylation event (5-10mM). Thus, under physiological conditions, these events would be coupled.

Significant progress had been made in understanding the role of the vitamin in the reaction, but, because the carboxylated substrates were proteins, further investigation of the carboxylation event was hampered. Yet, once it had been demonstrated that certain oligopeptides containing Glu Glu sequences were substrates for the carboxylase (Suttie *et al.*, 1976), it became apparent that these Glu residues could be modified in ways that might contribute to our understanding of the carboxylation reaction. In this presentation we will review some of our earlier experiments using peptides containing tritium labeled glutamic acid, and we will describe the synthesis of and initial studies with several new substrates with modified glutamyl residues.

METHODS

Preparation of the Vitamin K Dependent Gamma Glutamyl Carboxylase

Microsomal pellets, obtained from homogenates of the livers of either vitamin K-deficient or warfarin-treated rats

(Friedman and Shia, 1976), were treated in one of three ways. For some experiments microsomes were solubilized with Triton X-100 and used directly in assays. In other experiments, in which it was required to lower the concentration of bicarbonate in the microsomes, the solubilized microsomes were lyophilized and then reconstituted with water which had been boiled and degassed (Larson *et al.*, 1981). Finally, several experiments were done with an acetone powder prepared from the microsomal pellets as we have previously described (Friedman and Shia, 1980). The particular preparation used for an experiment is indicated in the text.

Conditions for Assays

The conditions for measuring carboxylation and epoxidation are as have been described (Friedman and Shia, 1976; Friedman and Smith, 1977). Volumes of reactions for individual experiments vary, and, therefore, are indicated for each experiment in the text. Also, since the particular peptide utilized for different experiments varied, it too is indicated for each experiment. When ^{14}C was assayed for in the presence of tritium labeled peptide, this was done using a Searle Mark III liquid scintillation counter with a program that measured only 0.2% of the ^{3}H radioactivity present while retaining 70% counting efficiency for ^{14}C.

Tritium release into water was determined by bulb to bulb distillation using a Thunberg vessel as has been described (Friedman *et al.*, 1979). Counting efficiency for tritium was 40%.

Determination of CO_2 Content of Samples

The titration method described by Rose (Rose, 1976) was used with the following modifications. Incubation and titration were carried out in a sealed vial under nitrogen that had been bubbled through KOH. $Ba(OH)_2$ was 10 mM; HCl was 10 mM. These modifications allowed detection of as little as 50 nmoles of CO_2 with a reproducibility of $\pm$10%.

Preparation of Peptides

The preparation of the tritium labeled pentapeptide used in these studies has been described elsewhere (Friedman *et al.*, 1979). The tritium labeled L-glutamic acid, which was incorporated into the peptide was purchased from New England Nuclear and it had a specific activity of 50 Ci/mmole and was tritium labeled at both the beta and gamma carbons (55% beta and 45% gamma as determined by tritium NMR analysis), with a random distribution between the R and S positions at each carbon.

Syntheses of the other peptides are described in the text.

RESULTS AND DISCUSSION

Following incubation of the tritium labeled pentapeptide with the acetone powder carboxylase preparation, it could be demonstrated that the quantity of tritiated water recovered after bulb to bulb distillation of the reaction mixture was dependent on reduced vitamin K, native enzyme, time, and oxygen (Table 1A). Furthermore, in experiments in which carboxylation and tritium release were measured in the same reaction mixtures, it could be demonstrated that carboxylation and tritium release were stimulated to identical degrees by increasing concentrations of vitamin, enzyme, and oxygen, as well as by increasing time of incubation (data not shown). With these data in hand, an important question about the mechanism of the carboxylation reaction could be asked. Must vitamin K dependent tritium release (a reflection of gamma C-H bond cleavage in Glu) be coupled to carboxylation or can the two events be uncoupled when the concentration of bicarbonate in the reaction becomes limiting (Table 1B)? Clearly it is shown in this experiment that when the concentration of bicarbonate is increased over a twenty-fold range there results a ten-fold increase in the amount of CO_2 fixed. In that same experiment, however, over the twenty-fold bicarbonate concentration range there is essentially no change in the amounts of vitamin K dependent tritium released (Friedman *et al.*, 1979). To calculate from the amounts of tritiated water recovered the actual numbers of gamma C-H bonds cleaved requires a number of assumptions: 1) no tritium is released at the beta position; 2) that all tritium released is recovered as water; 3) that a negligible isotope effect exists at the gamma carbon; and 4) that the enzyme is acting chirally. with these assumptions it was calculated that in these reactions between 1950 to 2300 pmoles of Glu had undergone gamma C-H bond cleavage. Thus, vitamin K dependent gamma C-H bond cleavage, exceeding twenty times the amount of CO_2 fixed at the lowest bicarbonate concentration, is a demonstration that this cleavage indeed can be uncoupled from carboxylation. It should be noted that this twenty-fold excess would be increased further if a significant isotope effect existed or if there were tritium transfer other than to water. (We have not detected tritium in compounds other than water). On the other hand, even in the extremely unlikely event that this vitamin K dependent enzyme possessed a mechanism which resulted in catalysis of all the tritium from the Glu substrate, gamma C-H bond cleavage of 500 to 600 pmole of Glu still would have occurred. Experiments described below (see Table III) strongly suggest that no significant tritium release from the beta position of this pentapeptide occurred during our incubations.

From these data it could be concluded that there is not a concerted reaction mechanism in which an active form of the vitamin, the glutamyl residue and CO_2 interact with the enzyme. In addition vitamin K cannot be functioning solely to activate or transfer CO_2 in a manner analogous to biotin. Vitamin K is required for cleavage of the gamma C-H bond of Glu whether or not a carboxylation ensues.

Since it had been shown previously that low CO_2 concentrations which limit carboxylation have no effect on vitamin K epoxidase activity (Larson *et al.*, 1981; Larson *et al.*, 1979), a logical extension of the above experiments was to examine the relationship of vitamin K epoxide formation to tritium release in appropriate reaction mixtures (Table II). In these experiments bicarbonate is limiting (i.e., bicarbonate is at a concentration at which uncoupling of CO_2 fixation from both epoxide formation and tritium release is seen). At such limiting CO_2 concentrations increasing the peptide substrate concentration results, as expected, in a larger increase in epoxidation than in CO_2 fixation. As Table II shows, such an increase in peptide substrate concentration stimulated tritium release and epoxidation to the same extent.

Although these results do not indicate a unique mechanism, they support a general one in which an active, oxygenated form of vitamin K (perhaps a vitamin K hydroperoxide) is required for cleavage of the substrate glutamyl gamma C-H bond. Concurrent with the cleavage is formation of vitamin K epoxide.

There exist several limitations to using the tritium labeled pentapeptide. Subsequent to our synthesis of that peptide, data were presented showing that the majority of carboxylations of the pentapeptide occurred on the glutamyl residue closer to the amino terminal phenylalanine (Finnan and Suttie, 1979). It seemed likely that tritium release would also occur mainly from that residue. Thus, the second

TABLE 1A. Vitamin K dependent tritium release[a]

Condition	cpm
Complete system	1694
No enzyme	94
No vitamin	140
No incubation	92
No oxygen	120
Boiled enzyme	97

[a]Incubations were for 30 min in room air at 27°C in a total volume of 130 µl. The reaction mixture contained 100 µl of resuspended acetone powder, 2 mM NADH, 2 mM pyridoxal 5'-phosphate, 1.6 mM tritiated pentapeptide (5×10^5 cpm), and either 5 µl of ethanol or 5 µg of menaquinone 3 in 5 µl of ethanol (penultimate addition). $NaHCO_3$ (0.5 µmol) was added last, and the tubes were capped during the incubation. Reactions were terminated by addition of 270 µl of 10% perchloric acid, and the mixtures were centrifuged to remove precipitated protein. An aliquot of the supernatant solution (200 µl) was assayed for tritiated water. Results are expressed as cpm for the entire 130 µl incubation mixture.

Table 1B. Dissociation of vitamin K dependent tritium release from carboxylation[b]

HCO_3^-, mM	CO_2 fixed, pmol	[^3H] water recovered, cpm	γC-H bonds cleaved, pmol
0.2	92	1669	1985
0.5	221	1869	2222
1	390	1816	2160
2	654	1922	2285
4	1028	1642	1952

[b]Incubations were as described in A with the following exceptions: prior to the addition of bicarbonate, the test tubes were sealed with serum caps and the gas phase was made N_2/O_2(80:20); reactions were initiated by injecting the indicated amounts of $NaHCO_3$ (containing 1.3×10^6 cpm of $NaH^{14}CO_3$) through the serum caps. Bicarbonate concentrations at zero time were determined by measuring total CO_2. Results are the average of triplicate determinations from which controls without vitamin K have been subtracted. (Friedman et al., 1979).

TABLE II. Coupling of epoxidation to γ-hydrogen removal

Peptide concentration	Vitamin K epoxide	γC-H bonds cleaved	Vitamin K epoxide/ γC-H bonds cleaved
	nmol/ml	nmol/ml	ratio
0.75 mM peptide	10.4	9.3	1.12
5.75 mM peptide	30.7	29.0	1.06

Vitamin K epoxide values represent peptide-dependent epoxidation. The latter is calculated by subtracting the amount of epoxide formed in the absence of any peptide substrate from the amounts formed in incubations with peptide substrate. Lyophilized microsomes containing 0.3 mM CO_2 (as bicarbonate) were used and vitamin KH_2 was 80 μg/ml. 10 mM $MnCl_2$ was added for maximal activity. The vitamin K epoxide/ CO_2 fixed ratios were 16.5 for low peptide and 27.4 for high peptide. (Larson et al.*, 1981).*

tritium labeled glutamyl residue was of little use in studying the reaction and likely was contributing to a higher background of tritiated water in our controls. Secondly, there was tritium at both beta carbons of the glutamyl residues which could potentially confound interpretation of results. Thirdly, synthesis of the radiolabeled pentapeptide and its purification was not a trivial undertaking; thus, this peptide was not readily available. In addition, since uncoupling of vitamin K dependent tritium release from carboxylation can occur, it is also possible that during the purification of the carboxylase these two activities could be separated or physically uncoupled. Use of the tritiated substrate would be necessary to detect these situations. For these reasons we decided to synthesize a simpler substrate which was tritium labeled at only the gamma carbon of that glutamyl residue which is predominantly activated.

It had been reported some years ago that when glutamic acid is heated in deuterated water containing 6N DCl only the gamma carbon becomes labeled with deuterium (Rattner, Rittenberg, and Schoenheimer, 1940). We were able to confirm this result by ^{1}H NMR (data not shown). When we carried out this reaction with tritiated water then, we specifically labeled the gamma carbon with tritium. This glutamic acid was esterified with benzyl alcohol at the gamma position, the amino group was blocked with Boc, and the resulting protected amino acid was coupled using dicyclohexylcarbodiimide to Glu

(OBZl)-Leu-OMe. Hydrogenolysis of this peptide yielded the potential substrate Boc-[γ-^{3}H]Glu-Glu-Leu-OMe. Indeed, this compound proved to be an excellent substrate for the tritium release reaction (Table III). In the experiment described, when equivalent concentrations of the peptides possessing the same amount of radioactivity were compared, 4.6 times more vitamin K dependent tritium release occurred from the tripeptide than from the pentapeptide. The pentapeptide used here was synthesized over three years ago; the rate of tritium release from this peptide in the absence of vitamin K is about 5 times greater than when it was first tested. The reason for this is not clear. However, even when first tested the ratio of tritium released in the presence of K versus that released in its absence was never greater than about 10. With the tripeptide the ratio is greater than 200. This new substrate then not only is simpler to synthesize and is radiolabeled solely on the relevant carbon, but it also provides for a much more sensitive assay for vitamin K dependent gamma C-H bond cleavage. The apparent Km for tritium release of the tripeptide is about 2.9 mM while that for carboxylation is about 2.5 mM.

Since the tripeptide is labeled only in the N-terminal Glu residue with the tritium only in the gamma position and since only 45% of the tritium in the pentapeptide is at the gamma position, it can be calculated that the tripeptide has 4.4 times the specific activity of the pentapeptide in the gamma hydrogens of the N-terminal Glu. This corresponds well to the ratio of vitamin K dependent tritium release of 4.6 found in Table III. This suggests that all the tritium is released from the gamma-position with none or negligible release from the beta-position.

An important aspect of the reaction mechanism that remains unknown is whether the activated glutamyl residue exists as a carbanion or as a carbon radical. To try to differentiate between these two possibilities, we have synthesized tripeptide substrates which contain a beta-fluoroglutamyl residue in place of the glutamyl residue normally activated. If the carboxylation reaction proceeds via a carbanion, one should see vitamin K dependent elimination of fluoride from the beta-position. We have synthesized beta-fluoroglutamic acid from protected beta-hydroxyglutamic acid using the fluorodehydroxylation reaction described by Kollonitsch for other appropriately hydroxylated compounds (Kollonitsch *et al.*, 1979). The beta-fluoroglutamic acid can be introduced into the tripeptide after appropriate protection.

When these tripeptides are incubated with solubilized microsomes, the peptides are not carboxylated, and we have not been able to detect vitamin K dependent fluoride elimina-

TABLE III. Comparison of Tritiated Peptides

		^{3}H release cpm	Vitamin K dependent ^{3}H release cpm	% ^{3}H release
Boc [γ-^{3}H] Glu Glu Leu OMe	*-K**	*76*		
(711,333 cpm/rxn)	*+K*	*16315*	*16239*	*2.3*
Phe Leu [β, γ-^{3}H] Glu [β, γ-^{3}H] Glu Leu	*-K*	*3273*		
(739,539 cpm/rxn)	*+K*	*6985*	*3712*	*0.5*

Incubations were for 2 hr at 17°. Final volume was 125 µl. The reaction mixtures contained solubilized microsomes (100 µl), 2mM pyridoxal -5'-phosphate, 2mM peptide, and finally isopropanol (5 µl) or 25 µg vitamin K_1 hydroquinone in 5 1 of isopropanol. Reactions were stopped by the addition of 270 µl 10% perchloric acid, precipitated protein was removed by centrifugation, and the supernatant was assayed for tritiated water as described in Methods.

**K = Vitamin K hydroquinone.*

tion. At least two further experimental problems will have to be addressed before results with these peptides will yield firm conclusions. First, using crude microsomes we are near the limit of detection for F- measured with an ion specific fluoride electrode. Thus, a more active preparation of the carboxylase will facilitate these experiments enormously. Second, while we can show that the fluorinated peptide inhibits carboxylation of the pentapeptide substrate, such data suggest that the fluorinated peptide can bind to the carboxylase but do not indicate whether the gamma C-H bond of the beta-fluoroglutamyl residue can be activated by the enzyme. Studies to determine this are underway.

REFERENCES

Carlisle, T.L., and Suttie, J.W. (1980). *Biochemistry 19,* 1161.

Esmon, C., Sadowski, J., and Suttie, J. (1975). *J. Biol. Chem. 250,* 4744.

Finnan, J.L., and Suttie, J.W. (1979). *In* "Vitamin K Metabolism and Vitamin K-Dependent Proteins" (J.W. Suttie, ed.), p. 509. University Park Press, Baltimore.

Friedman, P., and Shia, M. (1976). *Biochem. Biophys. Res. Commun. 70,* 647.

Friedman, P.A., and Shia, M.A. (1980). *Biochim. Biophys. Acta 616,* 362.

Friedman, P.A., Shia, M.A., Gallop, P.A., and Griep, A.E. (1979). *Proc. Natl. Acad. Sci. USA 76,* 3126.

Friedman, P.A., and Smith, M.W. (1977). *Biochem. Pharmacol. 26,* 804.

Friedman, P.A., and Smith, M.W. (1979). *Biochem. Pharmacol. 28,* 937.

Johnson, B.C. (1981). *Mol. and Cell. Biochem. 38,* 77.

Kollonitsch, J., Marburg, S., and Perkins, L.M. (1979). *J. Org. Chem. 44,* 771.

Larson, A.E., Friedman, P.A., and Suttie, J.W. (1981). *J. Biol. Chem. 256,* 11032.

Larson, A.E., Whitlon, D.S., and Suttie, J.W. (1979). *Fed. Proc. 38,* 876 (Abtr.).

Matschiner, J., Bell, R., Amelotti, J., and Knauer, T. (1970). *Biochim. Biophys. Acta 201,* 309.

Rattner, S., Rittenberg, D., and Schoenheimer, R. (1940). *J. Biol. Chem. 135,* 357.

Rose, J.G. (1976). *Anal. Biochem. 75,* 358.

Sadowski, J.A., Schnoes, H.K., and Suttie, J.W. (1977). *Biochemistry 16,* 3856.

Suttie, J.W. (1980). *CRC Crit. Rev. Biochem 8,* 191.
Suttie, J.W., Geweke, L.O., Martin, S.L., and Willingham, A.K. (1980). *FEBS Lett. 109,* 267.
Suttie, J.W., Hageman, J.M., Lehrman, S.R., and Rich, D.H. (1976). *J. Biol. Chem. 251,* 5827.
Willingham, A., and Matschiner, J. (1974). *Biochem. J. 140,* 435.

DISCUSSION

Olson: There is a slight disparity between your first CO_2 experiment (Friedman *et al.* 1979) and the later epoxidase experiments (Larson *et al.* 1981) in terms of the stoichiometry between protons released and moles of epoxide formed. I think in the first CO_2 experiment, there were about twice as many moles of hydrogen released as CO_2 fixed; at bicarbonate levels, that should give you 1:1 equivalents. In the second experiments, however, you found that epoxide formed paralelled the hydrogen release. Is this true?

Friedman: I think that you are describing something that isn't real. The acetone powder preparation is a more uncoupled preparation than the lyophilized microsome preparation. You can demonstrate uncoupling both with respect to tritium release and with respect to epoxide formation as you limit CO_2 using the lyophilized, solubilized microsome preparation. The system, for reasons that I do not understand, is better coupled even at those lower CO_2 concentrations than the acetone powder. It suggests the possibility that in the acetone powder, we already may have modified a site on the enzyme that facilitates the fixing of CO_2.

Olson: Can you conclude from the fluoride experiments that you have a glutamic acid radical and not a carbanion?

Friedman: No, I don't think so. That is the problem with a negative experiment.

Olson: It is extraordinary that you do not find tritium exchange in another negative experiment. We have also tried unsuccessfully to enrich the pentapeptide with tritiated water during carboxylation. Are you astonished by this result and can you explain it?

Friedman: No. For example, you could propose a mechanism in which a peroxy anion or radical is the active species; either a proton or a hydrogen radical could be removed from the substrate and you have a KOOH. The latter, in breaking down to give the epoxide could generate OH- which picks up an H+ from elsewhere. If that were the mechanism, and that is compatible with the data, it would be very difficult, I think, to get any tritium back in to the substrate.

Olson: But if you are not carboxylating, you have an exposed negative charge.

Friedman: Or a radical.

Olson: Or a radical, with protons all around.

Friedman: Well, we don't know if there are protons all around.

Olson: I think in enzymes and even in lipids there are protons around.

Friedman: All I can say is that the experiment was one that did not allow the tritium back into the substrate. It may be that the γ-carbon is sequestered and not sitting there in the media with lots of protons around it.

Suttie: Was there epoxide formation in the presence of the γ-fluoride?

Friedman: We have not yet looked for epoxide formation in the presence of the fluorine-containing peptide substrate.

Kun: I have seen the elaborate attempts of putting fluoride atom into this peptide. If you would use the keto acid homologue in this peptide, perfluorylchloride would be a very good reagent - it would introduce fluorinate quantitatively into the carboxyl C. This is just a suggestion.

Friedman: I am aware of a synthesis in which the keto derivative can be put into the peptide and then fluorinate it. I think that under those conditions the peptide bonds would break.

Kun: Very mild conditions.

Friedman: Then we could do that. However, you would introduce two fluorines. It might change the pK of the gamma carboxyl group to make it even more unfavorable than it might already be, but that is a possibility.

Girardot: This is the problem of concentration of the proton at the active site of the enzyme. Is there any way you know that 10^{-7} M H+, even if you drop the CO_2 concentration, is enough H+ at the enzyme CO_2 active site?

Friedman: Why do you say that?

Girardot: Proton concentration is very low in comparison to the CO_2 concentration so that it is difficult to say that you've got to get water protons back into glutamic acid.

Friedman: I think we have made these calculations as to the amount of tritium specific activity that is available and we should have seen some.

Girardot: In fact, we did not see sufficient tritium back either when we did the experiment; this however may still be purely an isotope (mass) effect.

Friedman: I know. There must be a reason for that.

Vermeer: Did you ever try to measure the exchange of tritium and CO_2 in solid phase carboxylase?

Friedman: We haven't done that. We have looked at the solid phase carboxylase and we get some activity. We don't get anywhere near the activity that you get. All we have looked at there is carboxylation.

Canfield (Texas A&M): Do you have any idea what things might be incorporated in besides CO_2; do you see anything at all?

Friedman: No, but we speculate that if you ended up with a γ-hydroxy glutamate residue, for example, you would have a very hard time writing a carbanion mechanism if that is what you have in mind.

Canfield: I bring it up because there is some loss of peptide. If you start quantitating peptide that you put in, and peptide plus carboxylated products you get back, they do not add up. You lose a peptide to something else, and I don't know what it is.

Friedman: How do you measure that?

Canfield: I measure it with a fluorometric assay, so I know the nmoles I put in and the total nmoles I get back, and what's in the count doesn't match. Have you seen anything at all that you could suggest which might be worth looking for?

Friedman: No. I haven't observed that, but I haven't looked for it.

Suttie: These systems are not all that clean and some peptide is getting hydrolyzed in the system, so in addition to some possible meaningful reaction occurring, there are a lot of nonmeaningful reactions that are getting rid of things. that may be why it is difficult to balance the total reaction.

VITAMIN K-DEPENDENT CARBOXYLASE: A HEME PROTEIN?

Robert E. Olson
Anne L. Hall
Florence C. Lee
William K. Kappel
Robert G. Meyer
William J. Bettger

Departments of Biochemistry
St. Louis University School of Medicine
St. Louis, Missouri
and
University of Pittsburgh School of Medicine
Pittsburgh, Pennsylvania

The vitamin K-dependent glu-peptide carboxylase represents a new class of biotin-independent carboxylases (Olson and Suttie, 1977). This enzyme, however, has proven to be a very difficult protein to purify. A number of laboratories around the world, many of them represented at this meeting today, have attempted over the past five years to purify this enzyme and thus far the best results represent about a 400-fold purification from the level found in liver microsomes, far from homogeneity (Houser *et al.*, 1978; Canfield *et al.*, 1980; Hall *et al.*, 1979; Price *et al.*, 1980; Wallin *et al.*, 1980; Wallin and Suttie, 1982). The fact that the enzyme is an intrinsic membrane-bound protein has not simplified the problem.

A number of interesting advances have been reported in this meeting but there are still some very serious questions that remain. Three of these are: 1) What is the nature of the enzyme in terms of size, shape, subunit structure and

[1]*Thanks are due the National Institutes of Health, Bethesda, Maryland for grants AM-09992, AM-32365, AM-27937 and HL-07050 to R.E. Olson.*

ISBN 0-12-387560-9

prosthetic groups? Vitamin K is not classically a coenzyme for this enzyme since it is not tightly associated with the enzyme and does not undergo complete regeneration during the action of the enzyme. Vitamin K hydroquinone is, in fact, a substrate for the reaction; 2) Do the epoxidase and carboxylase activities of the presently available preparations of the enzyme reside in the same peptide chain? It is granted by all observers that these two activities are closely associated but the ratio of epoxidase to carboxylase activities can be made to vary widely by addition of bicarbonate, cyanide and manganese ions (Larson *et al.*, 1981; Lee *et al.*, 1982); 3) What is the mechanism of action of this enzyme? It requires reduced vitamin K, oxygen and CO_2 and hence is either an oxygenase or an oxidase (as well as a carboxylase) depending on whether or not vitamin K epoxide is a mandatory metabolite. Most investigators agree that the last step in this multistep reaction is the fixation of CO_2. The alpha proton of a carboxylic acid is usually quite unreactive and must be activated by some chemical mechanism. Vitamin K is clearly involved in the activation process which is the main energy requiring step. Whether the mechanism involves radicals or ions is not clear. This activation may involve proton abstraction or induction of proton loss by electronic shifts in the carboxyl group of glutamic acid.

In fact, most of the work on mechanism that has been described in the last few years has been directed toward generation of hypotheses about chemical changes in the substrate (vitamin K) and not in the enzyme (Olson and Suttie, 1977; Larson *et al.*, 1978; Friedman *et al.*, 1979; Hall *et al.*, 1982). The enzyme, of course, must contain reactive groups which are important in the overall vitamin K-dependent carboxylation reaction. An accessible and essential sulfhydryl group has been established, but little else is known about the enzyme. The stoichiometry of the reaction catalyzed by this enzyme is also not entirely clear. Does it permit the oxidation of KH_2 to vitamin K and hydrogen peroxide with vitamin K epoxide formation as a side reaction, or does it oxidize KH_2 with the formation of one mole of vitamin K epoxide plus one mole of water during CO_2 fixation? These alternatives are shown in Figure 1.

REQUIREMENTS OF THE VITAMIN K-DEPENDENT CARBOXYLASE

The requirements of this enzyme system are given in Table I. The substrate can vary from a simple substituted glutamic acid (N-t-BOC-glutamic acid-α-benzyl ester) to preprothrombin

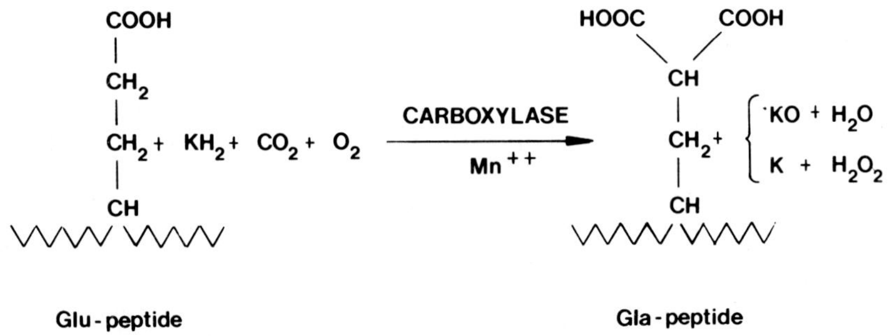

FIGURE 1. The reaction(s) catalyzed by the vitamin K-dependent glu-peptide carboxylase.

TABLE I. Requirements for the Vitamin K-Dependent Peptide Carboxylase

1.	*Vitamin KH_2*
2.	*CO_2*
3.	*O_2*
4.	*Glu-Peptide Substrate (n = 1 to 584)*

Reaction is stimulated by Mn^{++}, DTT, and Pyridoxal-5^1-Phosphate.

Reaction is inhibited by: p-CMB, Cl-K and Warfarin (variably).

Reaction is not inhibited by puromycin, cycloheximide, metyrapone, avidin, imido-ATP, FCCP, A-23187, X-537A.

containing 582 amino acids. Pentapeptides developed by Suttie *et al.* (1976) and by Houser *et al.* (1977), Phe-leu-glu-glu-leu, Phe-leu-glu-glu-val and Phe-leu-glu-glu-ile have been used successfully. Rich *et al.* (1979) have studied a large number of related peptides and have observed a range of carboxylation activities. Oxygen is required at pO_2 of 10 mm Hg for half maximal activity (Jones *et al.*, 1976) which is higher than that for cytochrome oxidase, but similar to that reported for cytochrome P-450-dependent cholesterol 7-α-hydroxylase (Boyd *et al.*, 1973). Carbon dioxide is required and appears to be the proximal reactant rather than bicarbonate (Jones *et al.*, 1977). ATP and biotin are not required and avidin has no effect on the reaction. Divalent cations, particularly Mn^{2+}, stimulate pentapeptide carboxylation as does pyridoxal-5'-phosphate. Inhibitors of protein synthesis such as cycloheximide and actinomycin D have no effect on carboxylation but inhibitors of sulfhydryl groups (PCMB or copper) have marked inhibitory effects (Johnson *et al.*, 1979).

PURIFICATION OF THE VITAMIN K-DEPENDENT CARBOXYLASE

Attempts to purify the enzyme by a variety of methods have not been fruitful. Ion exchange resins, chaotropic ions, sulfhydryl binding columns, thrombin affinity columns (which are specific for the physiological substrate, preprothrombin), hydrophobic affinity columns, and gel filtration in various combinations have not been highly successful. Affinity columns employing vitamin K homologues and analogues have not been useful. Attempts to purify the enzyme by reacting immunospecific adsorbents with its substrate preprothrombin have been partially successful since the enzyme and its substrate are closely associated and purify together. It has been extraordinarily difficult to remove the substrate from the enzyme when this enzyme-substrate complex is partially purified.

A four-step purification method resulting in a 300-fold purified enzyme from rat liver microsomes is presented in Table II. The procedure involves isolation of the rough reticulum where the enzyme and substrate are both concentrated, and a partial purification of the enzyme-substrate complex on a heparin affinity column. The action of the heparin column is to bind the substrate preprothrombin to which the enzyme also binds. The enzyme-substrate complex is then eluted with hypertonic salt solutions as shown in Figure 2. The fourth inefficient step is to remove the substrate from the enzyme by a prothrombin antibody affinity column, which yields a small amount of free enzyme, still showing several bands on SDS-PAGE. By subjecting the partially purified enzyme to isoelectric focusing it is possible to identify one peak which contained all the carboxylase activity, as shown in Figure 3, but unfortunately the isolation of this protein was not possible in the presence of ampholytes.

TABLE II. *Partial Purification of Vitamin K-Dependent Carboxylase from Rat Liver Microsomes*

Material	Volume (ml)	Total protein (mg)	Total activity (units)[a]	Specific activity (units/mg)	Recovery %	Fold purified
Whole microsomes	250	3,248	37,037	11	100	1
RER[b]	70	194	25,200	130	68	11
Heparin column	4	3.8	4,200	1,100	11	96
Antibody column	2	0.2	650	3,250	2	285

[a]1 unit of activity = 1 picomole of CO_2 fixed in 30 min. at 18°C.
[b]RER = rough endoplasmic reticulum.

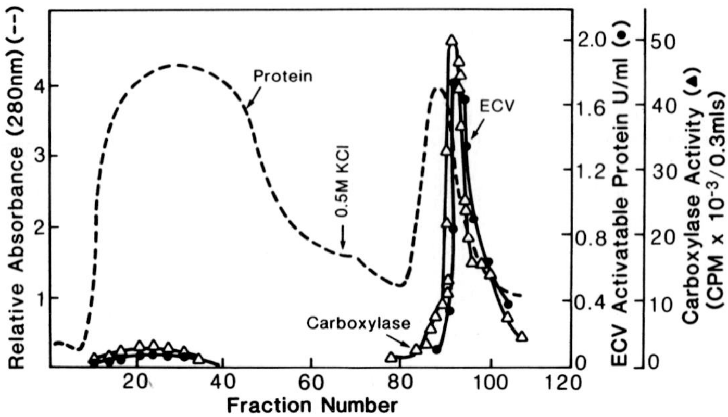

FIGURE 2. Heparin-affinity chromatography for separation of the preprothrombin-carboxylase complex from solubilized rat liver microsomes. The Triton-solubilized microsomes in imidazole buffer pH 7.2 containing 0.1 M KCl were loaded on the column and washed with the same buffer. Each fraction is 2.5 ml. Elution with 0.5 M KCl was initiated at the 70th fraction. ECV = Echis carinatus *venom assay for preprothrombin. Carboxylase was assayed for as previously described (Hall* et al.*, 1982).*

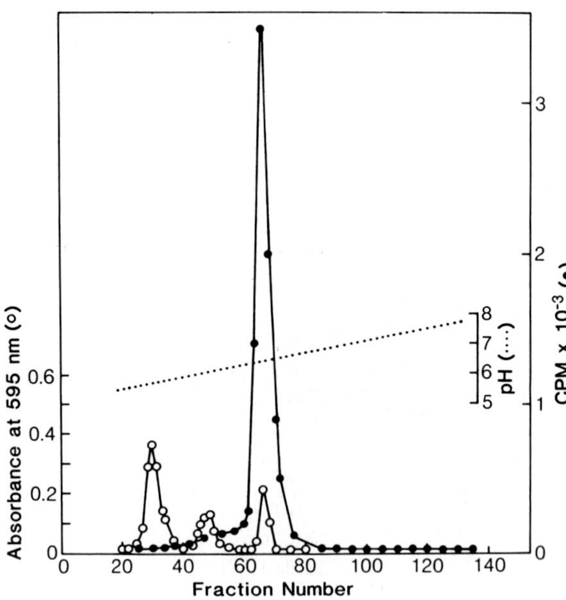

Figure 3

IS THE VITAMIN K CARBOXYLASE AN IRON METALLOPROTEIN?

The main objective of this presentation is to consider the evidence bearing on the view that the vitamin K-dependent glu-peptide carboxylase is an iron or iron-porphyrin (heme) enzyme. There are three lines of evidence. The first is the fact that most oxygenases or oxidases require a metal or a flavin for activity (Malmstrom, 1982). The evidence for the carboxylase being a flavoprotein is not convincing (Preusch, 1981). The second is that iron deficiency causes a fall in carboxylase levels in both Sprague-Dawley and warfarin-resistant rats. The third is that $^{59}FeSO_4$ when given to vitamin K-deficient rats, yields isotopic iron in the liver reticulum, which co-purifies with the enzyme through the steps described in Table II. The effects of four weeks of iron deficiency on blood hemoglobin content and carboxylase and iron levels in rat liver are shown in Table III. All values declined more than 50% in the iron-deficient rats. The vitamin K-dependent carboxylase levels were restored in 24 hours by subcutaneous injection of 1 mg of $FeSO_4$/100 g rat. Little or no effect on hemoglobin levels was noted as expected from the regeneration time of erythrocytes.

When 100 µCi of ^{59}Fe was injected into warfarin-resistant rats over a four day period of vitamin K-deficiency prior to the isolation of the enzyme, ^{59}Fe was incorporated into the hepatic reticulum. As shown in Figure 4, the radioactivity was associated with both the crude protein in the void volume and the enzyme fraction. The enzyme fraction was concentrated via Amicon ultrafiltration and applied in imidazole buffer, pH 7.2 to a prothrombin antibody affinity column to remove preprothrombin. As shown in Figure 5, the ^{59}Fe and the carboxylase now eluted together in the void volume. No residual activity was observed in fractions representing the elution of retained protein with 6 M urea. When the carboxylase was again concentrated by Amicon filtration and applied to a second heparin affinity column, whose thrombin binding sites had been previously saturated with prothrombin, most of the enzyme activity and ^{59}Fe co-eluted in the void volume as

---

FIGURE 3. Isoelectric focusing of partially purified carboxylase. Condition as described in Vesterberg, O. (1971).

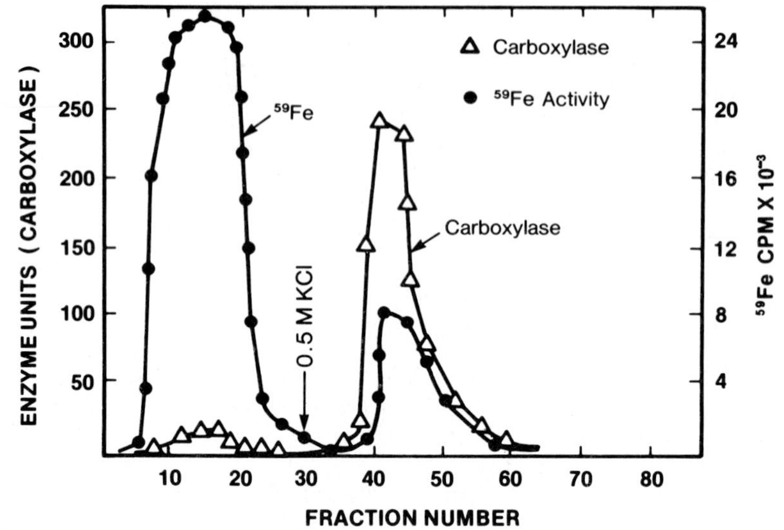

FIGURE 4. Heparin affinity chromatography of ^{59}Fe-labeled, Triton-solubilized rat liver microsomes. Conditions same as Figure 2. One unit of carboxylase activity is one picomole of CO_2 fixed per 30 minutes at 18°C. The ^{59}Fe appearing in the void volume is 10 times the value shown.

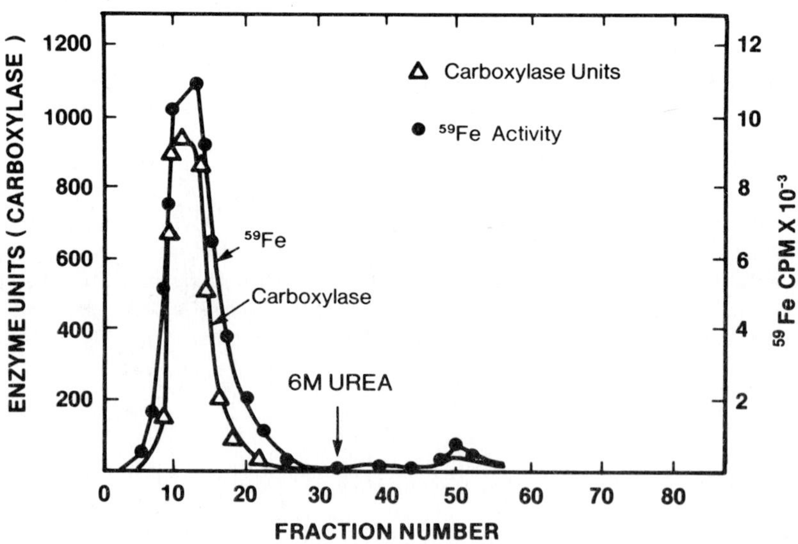

Figure 5

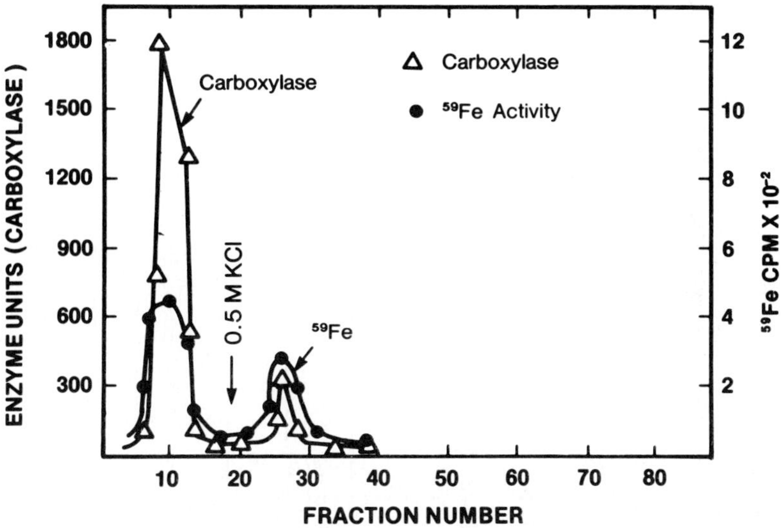

FIGURE 6. Second heparin affinity chromatography of ^{59}Fe and carboxylase. Conditions same as Figure 4 except the thrombin binding capacity of the column was saturated by previous treatment with prothrombin.

shown in Figure 6. A small amount of both enzyme and ^{59}Fe was retained. These data are summarized in Table IV which shows that the ^{59}Fe and carboxylase co-purify during the three final steps in the procedure that led to 150-fold purification of the enzyme. Similar experiments with ^{54}Mn did not show the same degree of copurification with the carboxylase.

When the ^{59}Fe-labeled eluate from the heparin column was subjected to acrylamide gel electrophoresis under normal (disc) and denaturing conditions (SDS), the results shown in Figure 7 were obtained. Under standard conditions most of the ^{59}Fe migrated with a sizeable slow moving peak, and some with a lesser faster moving peak, both well behind hemoglobin. Under denaturing conditions all of the ^{59}Fe migrated with heme.

FIGURE 5. Prothrombin antibody affinity chromatography of ^{59}Fe and carboxylase. Conditions as described by Graves et al., *(1980).*

TABLE III. *Effect of Iron Status on Liver Vitamin K-Dependent Carboxylase Activity, Liver Iron and Blood Hemoglobin Levels in Rats*

	Carboxylase units/mg protein		Hemoglobin g/100 mls.		Liver iron ppm wet wt.	
Group	W.R.	S.D.	W.R.	S.D.	W.R.	S.D.
Iron adequate	40	34	13.3	14.2	301	395
Iron deficient	23[a]	17[a]	3.4[a]	3.6[a]	98[a]	135
Iron deficient + Fe[b]	42	42	3.3[a]	3.4[a]	147[a]	196[a]
Iron adequate + Fe[b]	39	36	13.4	14.0	331	411

[a] $p<0.05$ compared with iron adequate group; $n = 8$ for all groups.
[b] 1 mg Fe/100 gm body weight s.c. followed by sacrifice in 24 hours.
WR = Warfarin-resistant rats.
SD = Sprague-Dawley controls.

TABLE IV. *Partial Purification of Vitamin K-Dependent Carboxylase from Rat Liver Microsomes Labeled with ^{59}Fe*

Step	Protein (mg)	Carboxylase (units)	^{59}Fe total activity cpm x 10^{-4}	Carboxylase specific activity units/mg protein	^{59}Fe specific activity cpm/mg protein x 10^{-4}
(1) Microsomal extract	331.0	51,360	4180	155	272
(2) Heparin affinity column	29.0	48,220	32	1,660	1.1
(3) Prothrombin antibody column	12.3	38,904	20.5	3,208	1.7
(4) 2nd Heparin affinity column	1.0	24,000	8.5	24,000	8.5

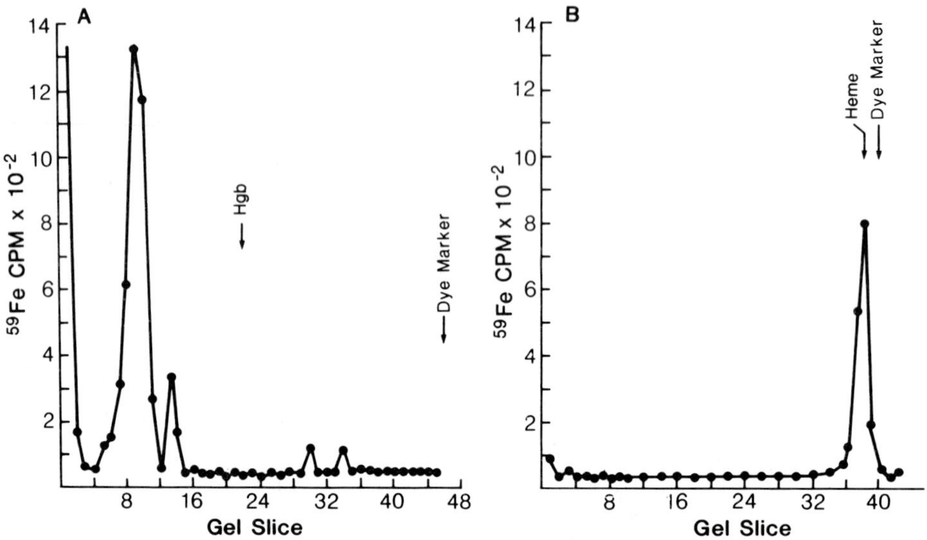

FIGURE 7. Electrophoresis of ^{59}Fe-labeled proteins from heparin-affinity column eluate. Left panel (A) standard disc gel electrophoresis of labeled protein. The position of a hemoglobin standard is marked. Right panel (B) SDS-PAGE of denatured solid-phase proteins. All the ^{59}Fe migrated with heme.

PREPARATION OF SOLID-PHASE VITAMIN K-DEPENDENT CARBOXYLASE

We have prepared a solid-phase rat liver carboxylase using antibodies to rat plasma prothrombin raised in rabbits. Our method is similar to that of DeMetz *et al.* (1981a) who used beef liver carboxylase precipitated with Sepharose-bound antibodies to factor X. Our antibody was also linked to Sepharose and the gel reacted with the enzyme- substrate complex from Triton-solubilized microsomes at 4°C for 72 hours (Olson *et al.*, 1982). The effect of this procedure on enzyme and preprothrombin from crude and heparin-purified carboxylase is shown in Table V. Preprothrombin levels were determined with *Echis carinatus* venom.

Virtually all of the precursor protein was bound by the antibody gel. The percent of carboxylase enzyme activity bound by the antibody gel from four different enzyme preps ranged from 40% to 69% with a mean value of 50%. It seems possible from these results that the enzyme(s) is present in

TABLE V. Preparation of Solid Phase Vitamin K-Dependent Carboxylase

Enzyme source	*Crude extract*	*Heparin eluate*
Initial activity		
carboxylase (units)[a]	*8,250*	*65,000*
ECV activity (NIH) units)[b]	*30*	*100*
% Bound		
carboxylase	*50%*	*60%*
ECV	*100%*	*100%*
% Free		
carboxylase	*50%*	*40%*
ECV	*0*	*0*

[a] *1 unit = 1 pmole $^{14}CO_2$ fixed/mg protein/30 minutes at 18°C.*

[b] Echis carinatus *activatable protein = NIH thrombin units.*

excess of the prothrombin precursor protein. It is known that about 50% of the precursor protein in solubilized rat microsomes precipitates with anti-factor X antibodies (Willingham *et al.*, 1979). Whether the enzyme which carboxylates factor X precursor is the same as the one that carboxylates preprothrombin is not known. The vitamin K-carboxylases may be a family of isozymes.

With 5 mM DDT present, the carboxylase activity of the solid phase enzyme toward exogenous pentapeptide is linear over a two hour period (Figure 8A). This contrasts with the report of DeMetz *et al.* (1981b) who reported a 30 minute lag in developing pentapeptide carboxylase activity with the beef enzyme precipitated with antibodies to factor X. The activity of the solid phase enzyme was linear over a protein range of 6 μg to 24 μg as shown in Figure 8B. The optimum temperature for peptide carboxylation for both the solid phase and soluble enzyme appeared to be around 18°C.

The solid phase enzyme was assayed for one hour at 18°C under various conditions illustrated in Figure 9. Triton was present in each assay to the extent of 0.2%. Phosphatidylcholine (0.2 mg) stimulated carboxylase activity approximately 25%. It also proved to be a stabilizing factor. $MnCl_2$ (10 mM) stimulated carboxylase activity 200%. Dithiothreitol (DDT) enhanced activity another two-fold by maintaining the

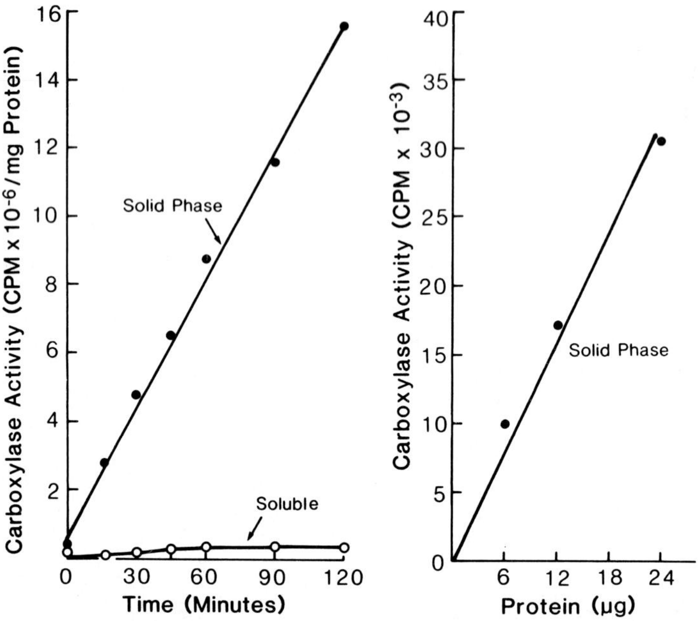

FIGURE 8. Relationship of carboxylase activity to time and protein concentration. Solid circles (●) represent solid-phase enzyme from rat liver. Open circle (o) represent Triton-solubilized enzyme whose specific activity is 10% of that shown in graphs. Left panel (A) shows relationship of activity to time. Right panel (B) shows relationship of activity to protein concentration. 24 μg of protein is equivalent to 0.2 ml of Sepharose-antibody gel.

vitamin KH_2 in a reduced form for a longer period. Rat serum albumin (0.2 mg) stimulated activity probably as a result of its free fatty acid binding properties. With all ligands present, the specific activity of the solid-phase enzyme was 240 times that of the soluble preparation.

One surprise was the total lack of stimulation of pentapeptide carboxylation of the solid-phase enzyme by 1 mM pyridoxal-5'-phosphate (Figure 10). Peptide carboxylation by the Triton-solubilized enzyme is stimulated two- to three-fold by pyridoxal-5'phosphate (Suttie *et al.*, 1979). It is likely that the PLP receptor site is blocked by antibodies in the solid-phase enzyme. $CuCl_2$ (2 mM), a blocker of -SH groups, inhibited the solid-phase enzyme which was completely reversed with 5 mM DTT.

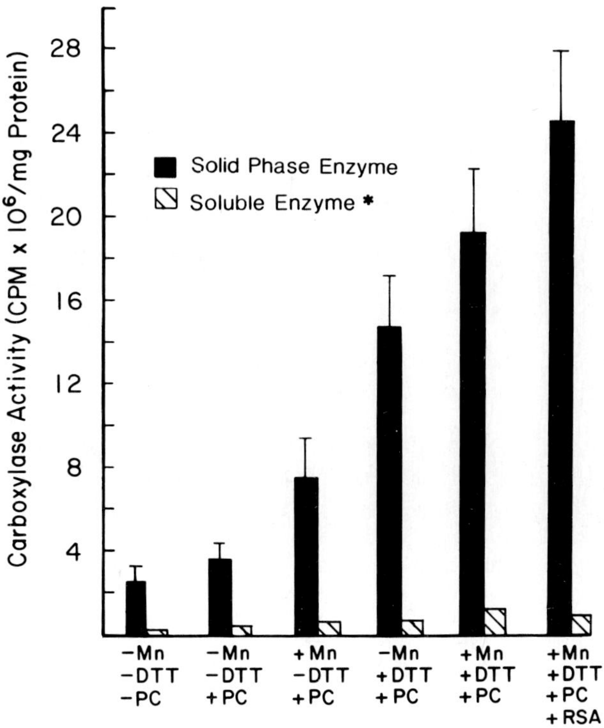

FIGURE 9. Effect of various factors on the carboxylase activity preparations of solid-phase and soluble enzyme. The specific activity of the soluble enzyme was 10% of that shown. Mn = Mn^{2+}, DTT = Dithiothreitol; PC = phosphatidylcholine, RSA = rat serum albumin. The bars shown for the solid-phase enzyme are the standard error of the mean of 3-4 experiments.

As shown in Figure 11 the carboxylated preprothrombin comigrated with prothrombin at an apparent molecular weight of 75,000. No other antibody precipitated carboxylated protein was present except for a very minor protein with an apparent molecular weight of 55,000. This carboxylation of the endogenous precursor occured rapidly over a period of one to two minutes and then ceased. The enzyme-substrate complex was thus functional in the antibody precipitated solid-phase state.

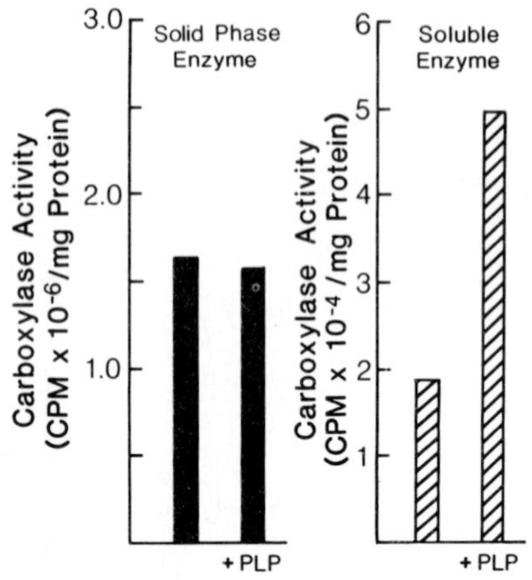

FIGURE 10. The effect of pyridoxal phosphate (PLP) on carboxylase activity of solid-phase and soluble enzyme preparations. The left panel shows the non-response of the solid-phase carboxylase to 1 mM PLP. The right panel shows the typical response of the soluble enzyme to PLP.

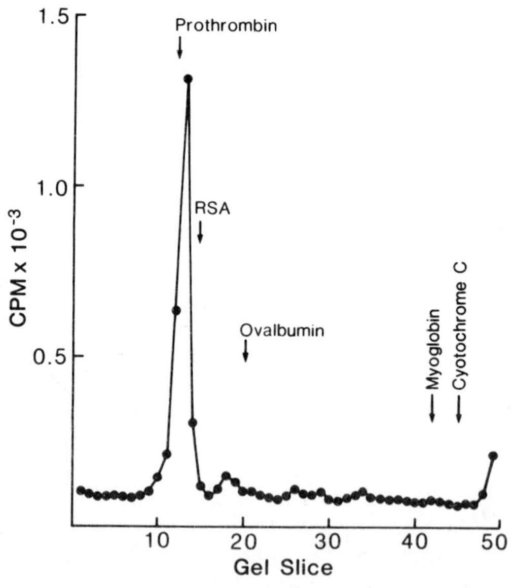

FIGURE 11. SDS electrophoresis of preprothrombin carboxylated by the solid-phase enzyme. The enzyme was incubated with $^{14}CO_2$ under standard conditions for 10 minutes at 18°C.

The solid-phase enzyme exhibited a mixed "cytochrome type" difference spectrum in the presence of dithionite showing a minimum at 410 nM and a maximum at 426 nM. ^{59}Fe studies similar to those described above showed incorporation of ^{59}Fe into the solid-phase enzyme preparation at about the same specific activity as noted in Step 4 of Table IV.

DISCUSSION AND CONCLUSIONS

The activity and stability of the solid-phase carboxylase is impressive. It is truly remarkable that an antibody directed to the native substrate of the carboxylase will form a stable and biologically active system with the enzyme-substrate complex. This has been shown by DeMetz *et al.* (1981a) for beef carboxylase with antibodies to factor X and by us for rat carboxylase with antibodies to prothrombin. It is also difficult to understand why the carboxylase-substrate complex resists dissociation with high salt and even with repeated exposure of antibodies to the substrate. Answers to these questions await further research.

The action of DTT in increasing carboxylase activity appear to be two-fold. First, it reduces the rate of decay of KH_2 which, under the usual conditions of assay, has a half-time of about two minutes (Hall *et al.*, 1982). In addition, it stabilizes an essential sulfhydryl group in the enzyme. The action of divalent cations in stimulating carboxylase with pentapeptide appear to be due to a reduction in the Km for the pentapeptide (Kappel and Olson, 1982) but the mechanism is obscure. The role of pyridoxal-5-phosphate, which stimulates the activity towards pentapeptide of the soluble but not the solid-phase enzyme, remains obscure. PLP forms a Schiff base with a key lysine in the enzyme (Kappel and Olson, 1982). The derivative, in turn, may augment the electron withdrawing power of the metalloenzyme and hence promote activation of the gamma-proton of peptide-bound glutamate.

Finally, we have adduced arguments that the vitamin K-dependent carboxylase is a heme enzyme. There is no proof for this hypothesis since it can be tested adequately only after the enzyme has been purified to homogeneity and its prosthetic group(s) identified. Nonetheless, for reasons already cited it represents an exciting possibility.

Even though cytochrome P-450 does not copurify with the carboxylase and inhibitors of cytochrome P-450 do not block glu-peptide carboxylation, there are, nonetheless, some interesting parallels in the action of cytochrome P-450 and the vitamin K-dependent carboxylase. Cytochrome P-450 acts as a mixed function oxygenase, forms epoxides, can be uncoupled by organic hydroperoxides, and contains heme (White and Coon, 1980). This led us to speculate about the possibility that the vitamin K-dependent carboxylase might be an analogous system, modified to accommodate vitamin K hydroquinone as its substrate and in the process generate a ferric complex which could activate the gamma-proton of glutamic acid.

This hypothesis, shown in Figure 12, postulates that reduced vitamin K is a substrate which not only forms the ini-

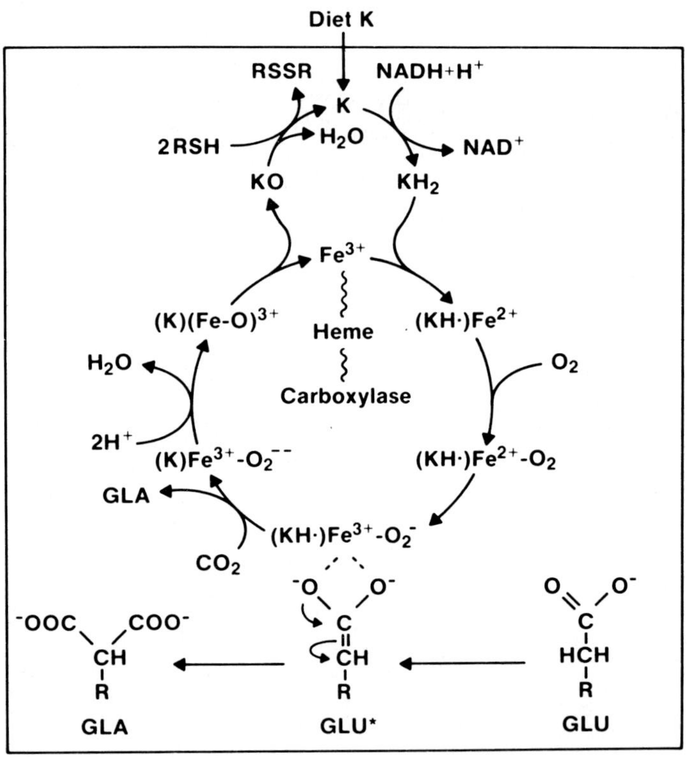

FIGURE 12. Hypothetical mechanism for a vitamin K-dependent carboxylase containing a heme group. The vitamin K cycle permitting reutilization of vitamin K epoxide is depicted at the top. The activation of the gamma-methylene proton of GLU resulting in the formation of GLA is shown at the bottom.

tial enzyme-substrate complex with the carboxylase, but also supplies the electrons in separate steps which reduce the iron and eventually oxygen. In the P-450 system, these electrons are derived from NADPH via a flavin reductase. In our model the first electron from KH_2 reduces the iron from Fe^{3+} to Fe^{2+} and forms a complex with the vitamin K semi-quinone. Oxygen then adds to generate a hemoglobin type $KH.Fe^{2+}O_2$ complex which then generates a ferric iron moiety and a superoxide anion. This ferric iron is electron attracting and would permit the activation of the gamma-carbon of peptide bound glutamate as shown. It is next visualized that the CO_2 fixation will occur by nucleophilic attack of the GLU carbanion on CO_2 forming GLA. At this point the second electron from vitamin K would form the peroxide level iron complex in the heme protein which, in analogy with the P-450 system can dissociate to form hydrogen peroxide, or dismutate to water in the presence of two protons derived from GLU, one from the carboxyl and one from the alpha carbon atom. This would generate the ternary complex $[K{\cdot}Fe^{3+}{\cdot}O]$ in which oxygen has a formal valence of zero. This oxene oxygen is a very reactive species that could then combine with vitamin K to generate the vitamin K-2,3-epoxide. At the top of the diagram one sees the now established K-salvage cycle in which K-oxide is reduced to vitamin K by vitamin K epoxide reductase and the vitamin K is then further reduced by a pyridine nucleotide catalyzed step to KH_2. This model thus is a bicycle, in which one wheel represents recycling of substrate and the other regeneration of the enzyme after carboxylation.

Finally, Figure 13 shows a model of the molecular biology of the biosynthesis of prothrombin (and other vitamin K-dependent proteins). In this model a portion of the rough reticulum of the rat liver cell is shown. The ribosome with its attached mRNA is shown generating the nascent peptide, preprothrombin, which enters the bilayer, probably through a leader sequence (MacGillivray *et al.*, 1979). Scheinbuks (1979) has shown in our laboratory that during vitamin K-dependent carboxylation of washed membranes, nascent chains, which are releasable by puromycin, become enriched with $^{14}CO_2$. We believe that under physiological conditions GLA synthesis occurs on the nascent chain. Our experiments show that the precursor peptide, even when it accumulated in vitamin K deficient rats and detached from ribosomes is firmly bound by the membrane and is not released by concentrations of desoxycholate that release vesicular and peripheral proteins. We imagine that vitamin K is embedded in the membrane with its functional group pointing outward and its alkyl side chain buried in the hydrocarbon portion of the bilayer (Jones *et al.*, 1976). We know that reducing equivalents from NADH

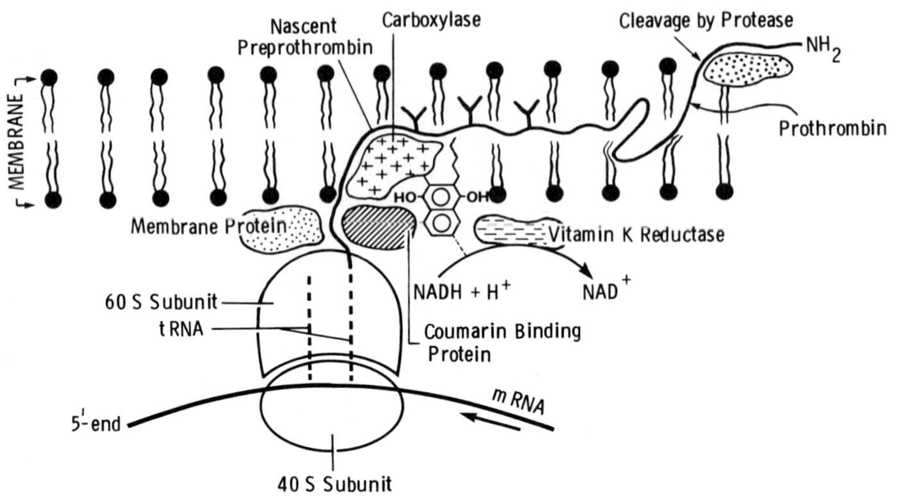

FIGURE 13. Rough reticulum of a hepatocyte synthesizing prothrombin. One leaf of the endoplasmic membrane is shown. Vitamin K is depicted as inserted on the cytoplasmic side whereas the signal peptide cleavage enzyme is shown on the lumenal side. The carboxylase may be a transmembrane protein although it is shown as a central (intrinsic) membrane protein.

and a vitamin K reductase are required for carboxylation and according to Wallin *et al.* (1978), the responsible enzyme is principally cytosolic, i.e., DT-diaphorase, although there are other NADH dependent reductases that could play this role (Wallin and Hutson, 1982). The coumarin binding protein is presented in close apposition to the carboxylase system and is near the surface of the membrane. It is not established whether the inhibitory coumarin binding protein is vitamin K epoxide reductase (Hildebrandt and Suttie, 1982) or warfarin is an allosteric effector of vitamin K-dependent carboxylase (Searcey *et al.*, 1978).

ACKNOWLEDGEMENT

The authors are indebted for expert technical assistance to Terry Lin, Karen Smail and Judith Montibellor.

REFERENCES

Bettger, W.J., Hall, A.L., and Olson, R.E. (1981). *Fed. Proc. 40*, 1665.

Boyd, G.S., Grimwade, A.M., and Lawson, N.E. (1973). *Eur. J. Biochem. 37*, 334.

Canfield, L.M., Sinsky, T.A., and Suttie, J.W. (1980). *Arch. Biochem. Biophys. 202*, 515.

DeMetz, M., Vermeer, C., Soute, B.A.M., Van Scharrenburg, G.J.M., Slotboom, A.J., and Hemker, H.C. (1981a). *FEBS Lett. 123*, 215.

DeMetz, M., Vermeer, C., Soute, A.M., and Hemker, H.C. (1981b). *J. Biol. Chem. 256*, 10843.

Friedman, P.A., Shia, M.A., Gallop, P.M., and Griep, A.E. (1979). *Proc. Natl. Acad. Sci. USA 76*, 3126.

Graves, C.B., Grabau, G.G., Olson, R.E., and Munns, T.W. (1980). *Biochem. 19*, 266.

Hall, A.L., Turner, P.M., Dankle, B.F., Wang, D.A., and Olson, R.E. (1979). *In* "Vitamin K Metabolism and Vitamin K-Dependent Proteins", Ed. J.W. Suttie, p. 433. University Park Press, Baltimore, Maryland.

Hall, A.L., Kloepper, R., Zee-Cheng, R.K-Y., Lee, F.C., and Olson, R.E. (1982). *Arch. Biochem. Biophys. 214*, 45.

Hildebrandt, E.F., and Suttie, J.W. (1982). *Biochem. 21*, 2406.

Houser, R.M., Carey, D.J., Dus, K.M., Marshall, G.R., and Olson, R.E. (1977). *FEBS Lett. 75*, 226.

Houser, R.M., Hall, A.L., and Olson, R.E. (1978). *Fed. Proc. 37*, 1588.

Johnson, B.C., Mark, D.O., Delaney, R., Wolfenberger, M.R., Esman, C., Price, J.A., Suen, E., and Girodot, J.M. (1979). *In* "Vitamin K Metabolism and Vitamin K-Dependent Proteins", Ed. J.W. Suttie, p. 455. University Park Press, Baltimore, Maryland.

Jones, J.P., Fausto, A., Houser, R.M., Gardner, E.J., and Olson, R.E. (1976). *Biochem. Biophys. Res. Commun. 72*, 589.

Jones, J.P., Gardner, E.J., Cooper, T.G., and Olson, R.E. (1977). *J. Biol. Chem. 252*, 7738.

Kappel, W.K., and Olson, R.E. (1982). *Fed. Proc. 41*, 500.

Larson, A.E., and Suttie, J.W. (1978). *Proc. Natl. Acad. Sci. USA 75*, 5413.

Larson, A.E., Friedman, P.A., and Suttie, J.W. (1981). *J. Biol. Chem. 256*, 11032.

Lee, F.C., Espinosa, I.A., and Olson, R.E. (1982). *Fed. Proc. 41*, 467.

MacGillivray, R.T.A., Chung, D.W., and Davie, E.W. (1980). *In* "Vitamin K Metabolism and Vitamin K-Dependent Proteins", Ed. J.W. Suttie, p. 546. University Park Press, Baltimore, Maryland.

Malmstrom, B.G. (1982). *Annu. Rev. Biochem. 51,* 21.

Olson, R.E., and Suttie, J.W. (1977). *Vitamins and Hormones 35,* 59.

Olson, R.E., Hall, A.L., Huang, J.S., and Huang, S.S. (1982). *Clin. Res. 30,* 560A.

Preusch, P.C., and Suttie, J.W. (1981). *J. Nutr. 111,* 2087.

Price, J.A. and Johnson, B.C. (1980). *In* "Vitamin K Metabolism and Vitamin K-Dependent Proteins", Ed. J.W. Suttie, p. 500. University Park Press, Baltimore, Maryland.

Rich, D.H., Lehrman, S.R., Kawai, M., Goodman, H.L., and Suttie, J.W. (1979). *In* "Vitamin K Metabolism and Vitamin K-Dependent Proteins", p. 471. University Park Press, Baltimore, Maryland.

Scheinbuks, J. (1979). *In* "Vitamin K Metabolism and Vitamin K-Dependent Proteins", Ed. J.W. Suttie, p. 279. University Park Press, Baltimore, Maryland.

Searcey, M.T., Graves, C.B., and Olson, R.E. (1977). *J. Biol. Chem. 252,* 6260.

Suttie, J.W., Hageman, J.M., Lehrman, S.R., and Rich, D.H. (1976). *J. Biol. Chem. 251,* 5827.

Suttie, J.W., Lehrman, S.R., Geweke, L.O., Hageman, J.M. and Rich, D.H. (1979). *Biochem. Biophys. Res. Commun. 86,* 500.

Vesterberg, O. (1971). *Methods in Enzymology 22,* 389.

Wallin, R., Gebhardt, O., and Prydz, H. (1978). *Biochem. J. 169,* 95.

Wallin, R., Canfield, L.M., Sinsky, T.A. and Suttie, J.W. (1980). *In* "Vitamin K Metabolism and Vitamin K-Dependent Proteins", Ed. J.W. Suttie, p. 490. University Park Press, Baltimore, Maryland.

Wallin, R., and Hutson, S. (1982). *J. Biol. Chem. 257,* 1583.

Wallin, R., and Suttie, J.W. (1982). *Arch. Biochem. Biophys. 214,* 155.

White, R.E. and Coon, M.J. (1980). *Annu. Rev. Biochem. 49,* 315.

Willingham, A.K., Martin, S.L., Graves, C.B., Graham, G.G. and Munns, T.W. (1979). *In* "Vitamin K Metabolism and Vitamin K-Dependent Proteins", Ed. J.W. Suttie, 553. University Park Press, Baltimore, Maryland.

DISCUSSION

Weiner (U. of Colorado): It seems to me that even though you do not have pure enzyme, you could make some estimates as to the stoichiometry of the iron in the enzyme, perhaps by knowing how much prothrombin is attached. Have you tried to make those estimates to see if there is any reasonable stoichiometry there?

Olson: I'm not sure that it is possible to do meaningful stoichiometry on the impure ^{59}Fe-containing enzyme-substrate complex purified with the aid of antibodies. My estimate of purity of the enzyme at 200 fold purification from microsomes is 5% but it may be less. The specific activity of the ^{59}Fe used in our experiments was 13.26 mCi/mg which translated into 20 cpm/picrogram, corrected for half life. Pure preprothrombin contains 3000 N.I.H. thrombin units per milligram or 3/μg. In the preparation of the solid-phase enzyme 39,000 carboxylase units and 100 Echis units of preprothrombin were coprecipitated with 138,00 cpm of ^{59}Fe. In this complex then, there were 13.8 nanograms of radioactive iron, 300 μg of preprothrombin, and an unknown amount of enzyme. If the enzyme had a molecular weight of 100,000 and was present in stoichiometric proportions to preprothrombin, 428 micrograms should have been present. 428 micrograms of enzyme corresponds to 4.28 nanomoles of enzyme and 6.9 nanograms of iron corresponds to 0.24 nanomoles. We do not know how much non-radioactive iron was present in the complex, and we do not know how much the injected ^{59}Fe was diluted in body iron pools before being synthesized into the various heme proteins of the reticulum. We are not sure that the enzyme and substrate are in a 1:1 ratio. The total non-immunoglobin protein which this amount of complex represents is 3.0 milligrams. If all the assumptions were true, the enzyme-substrate complex would be 24% pure.

Floyd (OMRF): If you inactivate by allowing it to react and then become inactive, what happens if you try to isolate it? How does it differ versus an active enzyme? Are there metabolites of phospholipids covalently bound?

Olson: I do not know how one would isolate an inactive enzyme because we are so dependent on the activity as a guide.

Floyd: You could carry it through the same procedure that you did with your active enzyme and follow it by SDS-PAGE electrophoresis or whatever is necessary and see if it runs in the same place. One could see if there is an attachment, perhaps of a phospholipid, or something like that.

Olson: The preparations are dependent on phospholipid. We have not looked at inactive preparations, but even our active preparations have several bands in SDS-PAGE.

Friedman: You mentioned earlier something about the data being compatible with a mixed function oxidase mechanism. The one point that I don't think was brought up by anybody is the fact that glutathione peroxidase, as well as horse radish peroxidase, would inhibit carboxylation, epoxidation, and tritium release to the same extent in any given system where the only variable is the amount of peroxidase enzyme that is put into that system. While we are dealing with a crude system, we believe that the glutathione peroxidase is in some way destroying an activated oxygenated vitamin intermediate. It is sort of hard to conceive of it being a classical mixed function oxidase mechanism, I think.

Olson: As you know, we have been unable to repeat the glutathione peroxidase inhibition of the carboxylase with homogeneous glutathione peroxidase purified according to Tappel (Hall *et al.*, 1982). We have not been able to inhibit carboxylase more that 10% using that enzyme. We have been able to inhibit carboxylation using the horse radish peroxidase, which is a marvelous oxidant for reduced vitamin K. Horse radish peroxidase will oxidize 200 μg/ml of KH_2 in a few seconds. This in turn reduces substrate levels to the point where epoxidation, carboxylation and tritium release are all inhibited. I think that we need to get a pure enzyme before we can talk convincingly about kinetics and mechanism.

Suttie: What relationship do you see between those reactions and the formation of the vitamin K epoxide?

Olson: The mechanism I proposed dissociates the activation of the V-proton from epoxide formation. This mechanism is more analogous to many other biochemical mechanisms resulting in activation of α-carbon protons. One way is to make a thioester, which is the way that coenzyme A works. Another is to inhibit the ionization of the carboxyl which Dr. Wold suggested. There are examples in which carboxyl groups chelate with iron, and in different valent states will move electrons into the iron center or out of it, which in turn will affect the acidity of the γ-proton.

Suttie: I can understand that. Are you attempting to relate observations on this particular carboxylase and epoxidation?

Olson: I'm not attempting to relate them directly because, in analogy to cytochrome P-450, the oxidation of the heme iron from ferrous to ferric is dissociated from carboxylation of substrate because the two reactions, epoxidation and carboxylation, can be dissociated. It is likely that they occur at different steps in the oxidation/reduction of metalloenzyme. No one denies now that vitamin K-epoxide is a very important product of the carboxylase complex. Whether both actions occur at one active center is unclear; certainly they are not linked in a concerted mechanism. I agree that the activation of the γ-proton of glutamate is a partial reaction.

Suttie: It's a half-reaction?

Olson: That's correct because, you see, downstream it is very possible that activated oxygen could pick up vitamin K and generate epoxide. This is a half-reaction.

Wallin (U. of Tromso): You claim that the carboxylase is a heme protein in microsomes?

Olson: That is my hypothesis.

Wallin: According to Omura and Sato (1964), there is only room for P_{450} and P_5 in microsomes. You claim or at least suggest that carboxylase is a heme protein; I say Omura and showed earlier that there there is only room for P_{450} and C_5 as heme in microsomes.

Olson: Based on what, the total heme?

Wallin: Yes. If they calculate the protoheme, they add up very nicely (the P_5 and P_{450}). There has to be very little of the carboxylase.

Olson: That assumes that their estimates are free from error. We do not know how potent this enzyme is, nor how many moles of enzyme are present per gram of microsomes. Until we have a pure enzyme, we cannot answer that question.

ADP-RIBOSYLATION OF PROTEINS: AN OVERVIEW

Martha Vaughan
Joel Moss

Laboratory of Cellular Metabolism
National Heart, Lung, and Blood Institute
National Institutes of Health
Bethesda, Maryland

The role of NAD as a substrate in oxidation-reduction reactions integral to a number of different metabolic pathways is, of course, well known. Many of these reactions and the enzymes that catalyze them have been extensively studied over the past several decades. It was less than twenty years ago, however, that NAD was first implicated as a substrate in reactions that result in the post-translational covalent modification of proteins by ADP-ribosylation (for review see Hayaishi and Ueda, 1977; Hilz, 1981; Mandel *et al*., 1982; Pekala and Moss, 1983). In these reactions, the high energy bond linking the ADP-ribose and nicotinamide moieties of β-NAD is cleaved with transfer of the ADP-ribose to an appropriate acceptor, either a protein or another ADP-ribose already attached to a protein. As shown in Table I, two general types of enzymes that catalyze ADP-ribose transfer have been identified, poly(ADP-ribose) synthetase and mono-ADP-ribosyltransferase (Hayaishi and Ueda, 1977; Hilz, 1981; Vaughan and Moss, 1981; Mandel *et al*., 1982).

Poly(ADP-ribose) synthetase appears to catalyze both the initial ADP-ribosylation of protein and the subsequent addition of ADP-ribose moieties linked through ribose-ribose bonds to form a homopolymer (Kawaichi *et al*., 1980). Found in animal cells, the enzyme is localized primarily in the nucleus, tightly bound to chromatin. Although the role(s) of poly(ADP-ribose) synthetase in cell function has not been precisely defined, the enzyme may be involved in DNA synthesis and cell proliferation, cell differentiation, and DNA repair. Consistent with other evidence that the synthetase

POSTTRANSLATIONAL COVALENT
MODIFICATIONS OF PROTEINS

ISBN 0-12-387560-9

TABLE I. Classification of Protein ADP-Ribosyltransferases

- I. *Poly(ADP-ribose) synthetase*
- II. *Mono(ADP-ribosyl)transferases*
  - A. E. coli *bacteriophage transferases*
  - B. *Transferases in animal cells*
    - 1. *Transferase A from turkey erythrocytes*
    - 2. *Transferase B from turkey erythrocytes*
  - C. *Bacterial toxins*
    - 1. *Choleragen and* E. coli *heat-labile enterotoxin*
    - 2. Bordetella pertussis *toxin*
    - 3. *Diphtheria toxin and Pseudomonas Exotoxin A*

functions in DNA repair is the finding that the purified synthetase requires DNA for activity and fragmented DNA enhances enzymatic activity much more effectively than does native DNA. At least three different enzymes participate in the degradation of poly ADP-ribosylprotein. The pyrophosphate bonds are susceptible to phosphodiesterase activity (Fujimura *et al.*, 1967; Reeder *et al.*, 1967; Futai*et al.*, 1968) and a poly(ADP-ribose) glycohydrolase cleaves the ribose-ribose bonds (Miyakawa *et al.*, 1972; Miwa *et al.*, 1974). The ADP-ribosylprotein linkage is hydrolyzed by ADP-ribosyl-protein hydrolase (Okayama *et al.*, 1978). The synthesis and gradation of poly(ADP-ribose), at least under certain circumstances, may account for an important fraction of NAD turnover and result in decreased cell NAD content. Poly ADP-ribosylation and its role in cellular metabolism are discussed in detail in Chapters 17 and 18.

Another class of ADP-ribosyltransferases catalyzes the synthesis of mono-ADP-ribosyl proteins (Table I) (Hilz, 1981; Vaughan and Moss, 1981). These mono-ADP-ribosyltransferases may be divided into three groups: 1) bacteriophage enzymes that will not be discussed at this conference (Goff, 1974; Zillig *et al.*, 1977), 2) enzymes endogenous to animal cells, and 3) bacterial toxins that exert their effects on animal cells by catalyzing the ADP-ribosylation of specific cellular proteins.

There is evidence for the existence of mono-ADP-ribosyltransferases in several animal tissues (Hilz, 1981; Vaughan and Moss, 1981). Only two, however, have been purified and characterized, both from turkey erythrocytes (Moss *et al.*, 1980; Yost and Moss, 1983). Although these enzymes catalyze the same model reactions, the ADP-ribosylation of arginine and other simple guanidino compounds, as well as the mono-ADP-ribosylation of proteins, they are distinctly different in physical and kinetic properties. The two erythrocyte transferases, designated A and B, are described in Chapter 17. Whereas the natural substrates and functions of the ADP-ribosyltransferases of animal cells are unknown, the mechanisms of action of several of the bacterial toxins have been defined in more or less detail.

Five bacterial toxins with ADP-ribosyltransferase activity have been identified. On the basis of their substrate specificities, these toxins may be divided into three groups (Table I). Toxins in two of the groups appear to exert their effects on cells by altering the activity of adenylate cyclase (Moss and Vaughan, 1979; Katada and Ui, 1981, 1982; Katada *et al.*, 1982). Adenylate cyclase is a membrane-bound enzyme that catalyzes the synthesis of cAMP from ATP. Its activity is physiologically regulated by hormones, neurotransmitters, and autocoids that act through specific receptors to either stimulate or inhibit the cyclase (Rodbell, 1980; Ross and Gilman, 1980). The actions of both stimulatory and inhibitory agonists on the catalytic unit of the cyclase system are dependent on GTP (Rodbell, 1980; Ross and Gilman, 1980; Cooper and Londos, 1982). It is now generally believed that they are mediated through specific regulatory proteins termed the stimulatory GTP-binding protein (Gs) and inhibitory GTP-binding protein (Gi) (Cooper and Londos, 1982). Choleragen, the toxin responsible for the symptomatology of cholera, and *E. coli* heat-labile enterotoxin, one of the agents involved in the pathogenesis of "traveler's diarrhea", exert their effects on cells by ADP-ribosylating a subunit of the stimulatory GTP-binding protein, Gs (Moss and Vaughan, 1981). This modification results in activation of adenylate cyclase leading to the increase in intracellular cAMP produced by choleragen and *E. coli* toxin. Choleragen-catalyzed ADP-ribosylation of the regulatory component of adenylate cyclase is considered in detail in Chapter 17.

Another type of toxin ADP-ribosyltransferase that has somewhat different effects on adenylate cyclase is produced by *Bordetella pertussis*. It enhances the effects of stimulatory agonists and interferes with the action of specific inhibitory ligands (Katada and Ui, 1981, 1982; Katada *et al.*, 1982). The membrane protein ADP-ribosylated by the pertussis toxin is clearly different from that ADP-ribosylated by choleragen and *E. coli* heat-labile enterotoxin and is believed to be a subunit of the inhibitory GTP-binding protein (Gi) of the cyclase complex (Katada and Ui, 1982).

Diphtheria toxin and Pseudomonas Exotoxin A are members of a third group of bacterial toxins. They cause cell death by catalyzing the ADP-ribosylation of elongation factor II and thereby inhibiting protein synthesis (Iglewski *et al.*, 1975; Pappenheimer, 1977). In contrast to the toxins of the first two groups, diphtheria toxin and Pseudomonas Exotoxin A are single polypeptides in which the determinants for cell surface binding and the structures responsible for ADP-ribosyltransferase activity are localized to discrete functional domains (Pappenheimer, 1977; Vasil *et al.*, 1977). The other toxins are complex molecules in which these two functions are subserved by structurally different subunits that are noncovalently associated in the holotoxin (Gill, 1977; van Heyningen, 1977; Moss and Vaughan, 1979).

Different groups or classes of ADP-ribosyltransferases use different amino acids (as well as different proteins) as ADP-ribose acceptors (Table II). The poly(ADP-ribose) synthetase modifies certain specific glutamic acid residues in histones H1 and H2B (Burzio *et al.*, 1979; Riquelme *et al.*, 1979; Ogata *et al.*, 1980a,b). For both avian erythrocyte transferases, *E. coli* heat-labile enterotoxin and choleragen, free arginine can serve as an ADP-ribose acceptor (Moss and Vaughan, 1977, 1978; Moss and Richardson, 1978; Yost and Moss, 1983). Presumably the sites of ADP-ribosylation in their respective protein substrates, which have not been identified, are arginine residues or possibly other types of guanidino groups resulting from post-translational modification. Diphtheria toxin and Pseudomonas Exotoxin A with remarkable specificity ADP-ribosylate a single amino acid in elongation factor II. This is a complexly modified histidine residue termed diphthamide (Van Ness *et al.*, 198)a,b). Although it has thus far been found only in elongation factor II, its role in the function of that protein is unknown. The amino acid modified by *Bordetella pertussis* toxin has not been identified.

It is perhaps premature to suggest an analogy between the several groups of ADP-ribosyltransferases with different substrate specificities for amino acids and proteins and the family of protein kinases. The latter have, of course, been much more extensively studied and characterized, along with their substrates and the phosphoprotein phosphatases that are an obligatory part of the regulatory systems in which they participate. Nevertheless, with increasing information on the ADP-ribosyltransferases, their natural substrates, the effects of ADP-ribosylation on the function of these proteins, and the enzymes that might remove ADP-ribosyl residues from proteins, a similarly important regulatory role for this type of covalent post-translational modification of proteins may be revealed.

TABLE II. Substrates for ADP-ribosyltransferases

	ADP-ribosyltransferases	*ADP-ribose acceptors*	*Natural substrates*
I.	*Poly(ADP-ribose) synthetase*	*glutamic acid*	*Synthetase, histones, other proteins*
II. A.	*Bacteriophage transferases*	*arginine*	*RNA-polymerase, other proteins*
II. B.1	*Transferase A (turkey erythrocyte)*	*arginine*	*?*
II. B.2	*Transferase B (turkey erythrocyte*	*arginine*	*?*
II. C.1	*Choleragen and* E. coli *toxin*	*arginine*	*GTP-binding protein*[a]
II. C.2	Bordetella pertussis *toxin*	*?*	*?*[b]
II. C.3	*Diphtheria toxin and Pseudomonas exotoxin A*	*diphthamide*	*Elongation factor II*

[a]*GTP-binding protein of adenylate cyclase complex associated with receptors for stimulatory ligands (Gs).*

[b]*Probably GTP-binding protein of adenylate cyclase complex that mediates action of inhibitory agonists (Gi).*

REFERENCES

Burzio, L.O., Riquelme, P.T., and Koide, S.S. (1979). *J. Biol. Chem. 254*, 3029.

Cooper, D.M.F., and Londos, C. (1982). *In* "Horizons in Biochemistry and Biophysics" (L.D. Kohn, ed.), Vol. 6, p. 309. John Wiley and Sons, Ltd., New York.

Fujimura, S., Hasegawa, S., and Sugimura, T. (1967). *Biochim. Biophys. Acta 134*, 496.

Futai, M., Mizuno, D., and Sugimura, T. (1968). *J. Biol. Chem. 243*, 6325.

Gill, D.M. (1977). *In* "Advances in Cyclic Nucleotide Research" (P. Greengard, and G.A. Robison, eds.), Vol. 8, p. 85. Raven Press, New York.

Goff, C.G. (1974). *J. Biol. Chem. 249*, 6181.

Hayaishi, O., and Ueda, K. (1977). *Annu. Rev. Biochem. 46*, 95.

Hazeki, O., and Ui, M. (1981). *J. Biol. Chem. 256*, 2856.

Hilz, H. (1981). *Hoppe-Seyler's Z. Physiol. Chem. 362*, 1415.

Iglewski, B.H., and Kabat, D. (1975). *Proc. Natl. Acad. Sci. USA 72*, 2284.

Katada, T., and Ui, M. (1981). *J. Biol. Chem. 256*, 8310.

Katada, T., and Ui, M. (1982). *Proc. Natl. Acad. Sci. USA 79*, 3129.

Katada, T., Amano, T., and Ui, M. (1982). *J. Biol. Chem. 257*, 3739.

Kawaichi, M., Ueda, K., and Hayaishi, O. (1980). *J. Biol. Chem. 255*, 816.

Mandel, P., Okazaki, H., and Niedergang, C. (1982). *Prog. Nucleic Acid Res. Mol. Biol. 27*, 1.

Miwa, M., Tanaka, M., Matsushima, T., and Sugimura, T. (1974). *J. Biol. Chem. 249*, 3475.

Miyakawa, N., Ueda, K., and Hayaishi, O. (1972). *Biochem. Biophys. Res. Commun. 49*, 239.

Moss, J., and Richardson, S.H. (1978). *J. Clin. Invest. 62*, 281.

Moss, J., and Vaughan, M. (1977). *J. Biol. Chem. 252*, 2455.

Moss, J., and Vaughan, M. (1978). *Proc. Natl. Acad. Sci. USA 75*, 3621.

Moss, J., and Vaughan, M. (1979). *Annu. Rev. Biochem. 48*, 581.

Moss, J., and Vaughan, M. (1981). *Mol. Cell. Biochem. 37*, 75.

Moss, J., Stanley, S.J., and Watkins, P.A. (1980). *J. Biol. Chem. 255*, 5838.

Ogata, N., Ueda, K., and Hayaishi, O. (1980a). *J. Biol. Chem. 255*, 7610.

Ogata, N., Ueda, K., Kagamiyama, H., and Hayaishi, O. (1980b). *J. Biol. Chem. 255*, 7616.

Okayama, H., Honda, M., and Hayaishi, O. (1978). *Proc. Natl. Acad. Sci. USA 75*, 2254.

Pappenheimer, A.M., Jr. (1977). *Annu. Rev. Biochem. 46*, 69.

Pekala, P., and Moss, J. (1983). *In* "Current Topics in Cellular Regulation" (E.R. Stadtman, ed.). Academic Press, New York, in press.

Reeder, R.H., Ueda, K., Honjo, T., Nishizuka, Y., and Hayaishi, O. (1967). *J. Biol. Chem. 242*, 3172.

Riquelme, P.T., Burzio, L.O., and Koide, S.S. (1979). *J. Biol. Chem. 254*, 3018.

Rodbell, M. (1980). *Nature (Lond.) 284*, 17.

Ross, E.M., and Gilman, A.G. (1980). *Annu. Rev. Biochem. 49*, 533.

van Heyningen, S. (1977). *Biol. Rev. 52*, 509.

Van Ness, B.G., Howard, J.B., and Bodley, J.W. (1980a). *J. Biol. Chem. 255*, 10710.

Van Ness, B.G., Howard, J.B., and Bodley, J.W. (1980b). *J. Biol. Chem. 255*, 10717.

Vasil, M.L., Kabat, D., and Iglewski, B.H. (1977). *Infect. Immun. 16*, 353.

Vaughan, M., and Moss, J. (1981). *In* "Current Topics in Cellular Regulation" (B.L. Horecker, and E.R. Stadtman, eds.), Vol. 20, p. 205. Academic Press, New York.

Yost, D.A., and Moss, J. (1983). *J. Biol. Chem.* in press.

Zillig, W., Mailhammer, R., Skorko, R., and Rohrer, H. (1977). *In* "Current Topics in Cellular Regulation" (B.L. Horecker, and E.R. Stadtman, eds.), Vol. 12, p. 263. Academic Press, New York.

NAD:ARGININE ADP-RIBOSYLTRANSFERASES: ENZYMATIC ACTIVITIES IN ANIMAL CELLS AND BACTERIAL TOXINS

Joel Moss
Martha Vaughan

Laboratory of Cellular Metabolism
National Heart, Lung, and Blood Institute
National Institutes of Health
Bethesda, Maryland

INTRODUCTION

ADP-ribosyltransferases can be distinguished based on their utilization of different amino acids as ADP-ribose acceptors. Although the transferases appear to preferentially recognize certain proteins as substrates, the site of modification in the peptide chain is restricted; only specific amino acid residues are ADP-ribosylated. Among the residues ADP-ribosylated by different transferases are glutamate (Riquelme *et al.*, 1979; Burzio *et al.*, 1979; Ogata *et al.*, 1980a,b), a post-translationally modified histidine (Van Ness *et al.*, 1980a,b) and arginine (Moss and Vaughan, 1980). The mono-ADP-ribosyltransferases that have been described that ADP-ribosylate arginine residues have the interesting property of being able to use the free amino acid as an acceptor (Moss and Vaughan, 1980). Among these enzymes are the bacterial toxins, choleragen and *E. coli* heat-labile enterotoxin (Moss and Vaughan, 1980), and two NAD:arginine ADP-ribosyltransferases that have been purified from turkey erythrocytes. The two bacterial toxins exert their effects on cells by activating adenylate cyclase; both mono-ADP-ribosylate, a regulatory subunit of the cyclase system known as G/F, or a GTP-binding protein. The toxins will also ADP-ribosylate a number of other proteins as well as arginine and other low molecular weight guanidino compounds. The erythrocyte ADP-ribosyltransferases were discovered as the result of

ISBN 0-12-387560-9

a search for animal enzymes that could use free arginine as an ADP-ribose acceptor. It is not surprising, therefore, that some of their properties resemble those of the bacterial toxins, although they are clearly different in many ways. The functions and physiological substrates of the erythrocyte transferases are unknown.

CHOLERAGEN AND *E. COLI* HEAT-LABILE ENTEROTOXIN

Interaction with the Putative Cell Surface Receptor, Ganglioside G_{M1}

Both choleragen and *E. coli* heat-labile enterotoxin exert their effects on cells through the activation of adenylate cyclase, leading to an increase in intracellular cAMP (for review, see Finkelstein, 1973; Bennett *et al.*, 1976; Gill, 1977; van Heyningen, 1977; Moss and Vaughan, 1979a, 1981; Holmgren, 1981). The toxins are homologous in structure. Each contains a single A subunit, composed of two polypeptide chains, A1 and A2, linked through a single disulfide bond, in noncovalent association with five B subunits. The B subunits are involved in the binding of the toxin to the cell receptor; the A subunit is responsible for the activation of adenylate cyclase. The specific cell surface-binding site for choleragen appears to be the monosialoganglioside G_{M1} based on experimental evidence of many types from numerous laboratories (see reviews Finkelstein, 1973; Bennett *et al.*, 1976; Gill, 1977; van Heyningen, 1977; Moss and Vaughan, 1979a, 1981; Holmgren, 1981). In particular, cultured fibroblasts deficient in gangliosides and lacking a chemically detectable G_{M1} did not bind choleragen or accumulate intracellular cAMP in response to the toxin (Fishman *et al.*, 1976; Moss *et al.*, 1976a). Following incorporation of G_{M1}, but not other gangliosides, into these fibroblasts, the cells exhibited an increase in ^{125}I-toxin binding and toxin responsiveness (Moss *et al.*, 1976a; unpublished data). Incorporation of G_{M1} into liposomal model membranes enhanced ^{125}I-toxin binding (Moss *et al.*, 1976a). G_{M1} and its oligosaccharide altered the physical properties of the toxin and its B subunits ("choleragenoid") (Mullin *et al.*, 1976; Moss *et al.*, 1977; Fishman *et al.*, 1978). Interaction of choleragen with G_{M1} thus appeared to be specific for the B subunits of toxin and the oligosaccharide moiety of GM1. Although the holotoxin structure with the B subunits and cellular G_{M1} are necessary for choleragen action on intact cells, in cell-free systems neither is required. Activation of adenylate cyclase in broken cells requires only the A subunit, specifically the A1 peptide (Gill and King, 1975), and NAD (Gill, 1975, 1976).

ADP-Ribosyltransferase and NAD Glycohydrolase Activities

Both choleragen and *E. coli* heat-labile enterotoxin activate adenylate cyclase by catalyzing the transfer of ADP-ribose from NAD to a regulatory component of the cyclase system (see review, Moss and Vaughan, 1981). This ADP-ribose acceptor protein appears to be a guanyl nucleotide-binding protein, referred to as G/F, that participates in the coupling of stimulatory hormone receptors to the catalytic unit and in the activation of the catalytic unit when GTP is present (Rodbell, 1980; Ross and Gilman, 1980). G/F has been identified in membranes incubated with [^{32}P]NAD and toxin after separation of membrane proteins by sodium dodecyl sulfate-polyacrylamide gel electrophoresis (Cassel and Pfeuffer, 1978; Gill and Meren, 1978; Johnson *et al.*, 1978; Watkins *et al.*, 1981). In some systems, only one ADP-ribosylated protein of 42,000 daltons was found (Cassel and Pfeuffer, 1978; Gill and Meren, 1978). In others, such as the S49 lymphoma cells, at least two were evident--one of 45,000 and the other of 52,000-53,000 daltons (Johnson *et al.*, 1978). The ADP-ribosylation of these substrates appears to be catalyzed by the toxins themselves and not by activation of an endogenous transferase.

The ADP-ribosyltransferase activities of the toxins were studied in the absence of cellular components by taking advantage of their ability to catalyze three model reactions. Choleragen and *E. coli* heat-labile enterotoxin catalyze the ADP-ribosylation of various purified proteins, unrelated to the cyclase system (Moss and Vaughan, 1978), and low molecular weight guanidino compounds (Moss and Vaughan, 1977; Moss *et al.*, 1979a,b) such as arginine (Reactions 1A and 1B). Presumably, it is an arginine on the protein that serves as the ADP-ribose acceptor.

$$\text{NAD}^+ + \text{protein} \rightleftharpoons \text{ADP-ribose-protein} + \text{nicotinamide} + \text{H}^+ \qquad (1A)$$

$$\text{NAD}^+ + \text{guanidino-R} \rightleftharpoons \text{ADP-ribose-guanidino-R} \qquad (1B)$$

$$R = -(CH_2)_3CH(NH_2)(COOH)$$

ADP-ribosylation of guanidino compounds such as arginine is stereospecific (Oppenheimer, 1978; Moss *et al.*, 1979a,b);

the toxins use β-NAD as a substrate and catalyze the formation of α-ADP-ribose-L-arginine. In the absence of proteins or guanidino compounds, both toxins catalyze the hydrolysis of NAD (Reaction 2) (Moss *et al.*, 1976c, 1981b; Moss and Richardson, 1978).

$$NAD + H_2O \longrightarrow ADP\text{-}ribose + nicotinamide + H^+ \quad (2)$$

The holotoxins are relatively inactive; in the presence of thiols, such as dithiothreitol, there is a large increase in enzymatic activity (Moss *et al.*, 1976c; Moss and Vaughan, 1979b). Activation appears to result from reduction of the single disulfide bond linking the A1 and A2 peptides in the A subunits of the toxins. It is the free A1 peptide that is catalytically active (Wodnar-Filipowicz and Lai, 1976; Moss *et al.*, 1979c). The A subunits of the toxins are synthesized as a single polypeptide chain which is nicked by proteases to yield the A1 and A2 peptides (Gill and Rappaport, 1979; Mekalanos *et al.*, 1979; Moss *et al.*, 1981b). The intact A subunit is enzymatically inactive, even in the presence of thiol (Mekalanos *et al.*, 1979). Following incubation with proteases, the A1 peptide is generated, linked to A2 by a single disulfide; thiols then can release A1 which exhibits its intrinsic ADP-ribosyltransferase activity (Mekalanos *et al.*, 1979).

Role of ADP-Ribosylation in the Activation of Adenylate Cyclase

Adenylate cyclase is a membrane-bound enzyme through which many hormones and drugs, as well as the bacterial toxins, exert their effects on cells (Rodbell, 1980; Ross and Gilman, 1980). A cyclase system consists of at least three well-defined components: the hormone receptor, a regulatory component known as G/F or guanyl nucleotide-binding protein, and the catalytic unit (C) responsible for the conversion of ATP to cyclic AMP. Hormone-stimulated adenylate cyclase activity requires all three components as well as GTP, which exerts its effect by interacting with G/F (Rodbell, 1980; Ross and Gilman, 1980). Activation of adenylate cyclase by toxin is independent of hormone receptor but requires G/F and C. GTP also appears to be critical to the activation; GTP enhances ADP-ribosylation of G/F, increases stability of the activated enzyme, and results in higher cyclase activity in the assay (Nakaya *et al.*, 1980a,b). In the presence of GTP, the enzyme exists in an active form (Reaction 3A).

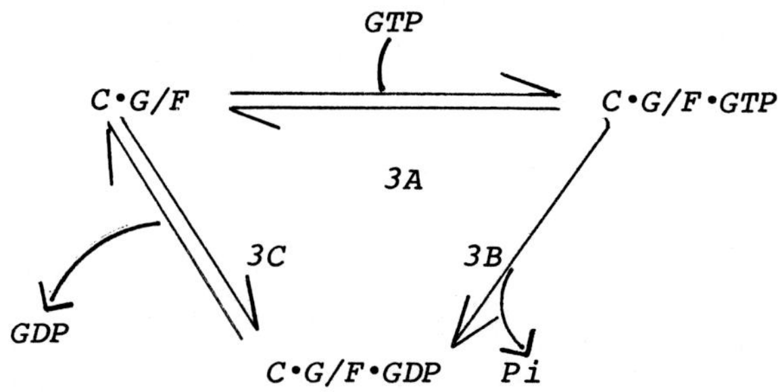

(3)

It is currently believed that the C•G/F•GTP complex is the substrate for a specific GTPase that converts it to C•G/F•GDP (Reaction 3B) (Cassel and Selinger, 1977); this species is inactive. Cassel and Selinger (1977) demonstrated that choleragen inhibits the GTPase and thereby prolongs the existence of C•G/F•GTP. Choleragen also stimulates release of GDP from C•G/F•GDP, leading to the formation of C•G/F (Reaction 3C) (Burns *et al.*, 1982, 1983). With the guanyl nucleotide site on G/F empty, C•G/F can bind GTP to again form an active cyclase complex. Thus, ADP-ribosylation of G/F by toxin appears to have multiple effects on the cyclase system. The net result is to enhance the enzymatic activity of the cyclase complex.

NAD:ARGININE ADP-RIBOSYLTRANSFERASES ENDOGENOUS TO ANIMAL TISSUES

ADP-Ribosyltransferase A: An Enzyme Activated by Histones and Chaotropic Salts

This ADP-ribosyltransferase was first identified in the soluble fraction of turkey erythrocytes by its ability to use arginine and other low molecular weight guanidino compounds as ADP-ribose acceptors (Moss and Vaughan, 1978). The enzyme was purified by successive chromatography on phenyl-Sepharose, carboxymethyl cellulose, NAD-agarose, and concanavalin A-agarose (Moss *et al.*, 1980). The purified enzyme exhibited one major band on sodium dodecyl sulfate-polyacrylamide gel electrophoresis which corresponded to a protomeric molecular weight of 28,300. On Ultrogel AcA 54 columns, ADP-ribosyltransferase activity and protein cochromatographed as a single peak (Moss *et al.*, 1980). The Kav of the transferase was slightly less than that of chymotrypsinogen (molecular weight, 25,300). The enzyme utilized arginine methyl ester > agmatine > arginine > guanidinopropionate as ADP-ribose acceptors (Moss *et al.*, 1980); it appeared that those compounds possessing a positive charge in the vicinity of the guanidino group (arginine methyl ester, agmatine) were more effective substrates than those having a negative charge (guanidinopropionate). The transferase preferred NAD to NADP as donors of ADP-ribose and phosphoribosyl-ADP-ribose, respectively (Moss *et al.*, 1979d). NADP was, however, an effective substrate. In contrast, choleragen did not utilize NADP in the activation of adenylate cyclase (Gill, 1975), in the stimulation of guanyl nucleotide release from the G/F protein (Burns *et al.*, 1982), nor did the toxin catalyze the hydrolysis of NADP (Moss *et al.*, 1977). As noted with the toxin, the transferase-catalyzed reaction was stereospecific (Moss *et al.*, 1979d); the enzyme, in the presence of β-NAD and L-arginine methyl ester, catalyzed the formation of α-ADP-ribose-L-arginine methyl ester.

The transferase ADP-ribosylated a variety of purified proteins as well as proteins from the soluble and particulate fractions of erythrocytes (Moss and Vaughan, 1978, unpublished data). The proteins modified by the transferase differed from those ADP-ribosylated by toxin. The presence of arginine methyl ester significantly inhibited the ADP-ribosylation of proteins (Moss and Vaughan, 1978). Presumably, arginine methyl ester competes with the protein substrate for the catalytic site on the enzyme.

Transferase activity was stimulated by salts with SCN^- > Cl^- > PO_4 (Moss *et al.*, 1981a). The effectiveness of the salts corresponded to their positions in the Hofmeister series; those salts having chaotropic properties were more active in enhancing enzymatic activity. Stimulation of transferase activity by salt was accompanied by dissociation of the enzyme from an inactive aggregated form to an active protomeric form (Moss *et al.*, 1981a). Maximal effective concentrations of NaCl were between 200-300 mM; at 300 mM, the transferase was apparently completely dissociated.

In addition to salt, the enzyme was also stimulated by histone at micromolar concentrations (Moss and Stanley, 1981; Moss *et al.*, 1982). Enzymatic activity was enhanced by both lysine- and arginine-rich histones, which increased the ability of the enzyme to use agmatine and other low molecular weight guanidino substrates as ADP-ribose acceptors. Histone stimulated transferase activity at concentrations significantly lower than those necessary for the histones themselves to serve as ADP-ribose acceptors (Moss and Stanley, 1981; Moss *et al.*, 1982). Activation of the enzyme by histone was also associated with dissociation of the enzyme from an inactive aggregated form to the active protomeric species (Moss *et al.*, 1982). Thus, the effect of histone on both activity and size was analogous to that of chaotropic salts: as shown (Reaction 4)

$$\underset{\substack{\textit{inactive}\\ \textit{high molecular weight}}}{(\textit{Protomer})n} \;\underset{}{\overset{\textit{Histone salt}}{\rightleftharpoons}}\; \underset{\substack{\textit{active}\\ \textit{low molecular weight}}}{n(\textit{Protomer})} \qquad (4)$$

both agents dissociate the enzyme and lead to an active protomeric form.

ADP-Ribosyltransferase B: Physical and Kinetic Properties

A second ADP-ribosyltransferase (transferase B) was purified from the soluble fraction of turkey erythrocytes (Yost and Moss, in press); transferase B was distinguished from transferase A by the fact that its activity was independent of histones and salt. ADP-ribosyltransferase B was separated from transferase A by chromatography on phenyl-Sepharose; transferase B was not retained, whereas transferase A adhered tightly to the resin and required propylene glycol for elution. Transferase B was then further purified by chromatography on concanavalin A-Sepharose, carboxymethyl cellulose, procion Red-agarose, and Ultrogel AcA 54. The purified enzyme exhibited one band by sodium dodecyl sulfate-polyacrylamide gel electrophoresis of $M_r \sim 32{,}000$. A similar molecular weight was obtained by gel permeation chromatography on Ultrogel AcA 54 or TSK 3000 SW. Transferase B thus appears to be slightly larger than the protomeric form of transferase A. In the presence of NAD, both transferases catalyze the ADP-ribosylation of low molecular weight guanidino compounds and proteins. For transferase B, the Km's for NAD and arginine methyl ester were 36 μM and 3 mM, respectively, with a V_{max} of 60 μmol • min-1 • mg-1. Transferase B utilized various purified proteins as ADP-ribose acceptors; the substrate specificity of transferase B clearly differed from that of transferase A. Arginine methyl ester inhibited the ADP-ribosylation of proteins by both enzymes. Transferase B, unlike transferase A, apparently existed as a protomeric

TABLE I. Comparison of the Kinetic and Physical Properties of ADP-Ribosyltransferases A and B

Property	ADP-ribosyl-tranferase A	ADP-ribosyl-transferase B
Km for NAD	15 μM	36 μM
Km arginine methyl ester	1.3 mM	3 mM
Vmax (μmol•min-1•mg-1)[a]	350	60
Effect of histones		
on size	decrease	none
on activity	increase	none
Effect of salts		
on size	decrease	none
on activity	increase	decrease
Molecular weight	28,000	32,000

[a]With arginine methyl ester as ADP-ribose acceptor.

species under experimental conditions. Thus, there was no dissociation induced by salts or histones of aggregated forms. Presumably for the same reason, the activity of transferase B was not stimulated by salts or histones; indeed, salt was inhibitory (Table I).

Summary

The bacterial toxins, choleragen and *E. coli* heat-labile enterotoxin, exert their effects on cells by catalyzing the ADP-ribosylation of a regulatory subunit of adenylate cyclase. This covalent modification results in an increase in enzymatic activity and in an enhanced responsiveness to the physiological effector of the system, GTP. The toxins, in the absence of cellular components, act as ADP-ribosyltransferases, using purified protein and low molecular weight guanidino compounds, such as arginine, as ADP-ribose acceptors. This enzymatic activity is not specific for the toxins; two different transferases were identified in avian erythrocytes that catalyze similar reactions. These enzymes have a different substrate specificity than do the toxins; their biochemical role(s) have not been determined.

REFERENCES

Bennett, V., Craig, S., Hollenberg, M.D., O'Keefe, E., Sahyoun, N., and Cuatrecasas, P. (1976). *J. Supramol. Struct. 4,* 99.

Burns, D.L., Moss, J., and Vaughan, M. (1982). *J. Biol. Chem. 257,* 32.

Burns, D.L., Moss, J., and Vaughan, M. (1983). *J. Biol. Chem, 258,* 1116.

Burzio, L.O., Riquelme, P.T., and Koide, S.S. (1979). *J. Biol. 254,* 3029.

Cassel, D., and Pfeuffer, T. (1978). *Proc. Natl. Acad. Sci. USA 75,* 2669.

Cassel, D., and Selinger, Z. (1977). *Proc. Natl. Acad. Sci. USA 74,* 3307.

Finkelstein, R.A. (1973). *CRC Crit. Rev. Microbiol. 2,* 553.

Fishman, P.H., Moss, J., and Vaughan, M. (1976). *J. Biol. Chem. 251,* 4490.

Fishman, P.H., Moss, J., and Osborne, J.C., Jr. (1978). *Biochemistry 17,* 711.

Gill, D.M. (1975). *Proc. Natl. Acad. Sci. USA 72,* 2064.

Gill, D.M. (1976). *J. Infect. Dis. 133* (suppl), S55.

Gill, D.M. (1977). *In* "Advances in Cyclic Nucleotide Research" (P. Greengard, and G.A. Robison, eds.), Vol. 8, p. 85. Raven Press, New York.

Gill, D.M., and King, C.A. (1975). *J. Biol. Chem. 250,* 6424.

Gill, D.M., and Meren, R. (1978). *Proc. Natl. Acad. Sci. USA 75,* 3050.

Gill, D.M., and Rappaport, R.S. (1979). *J. Infect. Dis. 139,* 674.

Holmgren, J. (1981). *Nature (Lond.) 293,* 413.

Johnson, G.L., Kaslow, H.R., and Bourne, H.R. (1978). *J. Biol. Chem. 253,* 7120.

Mekalanos, J.J., Collier, R.J., and Romig, W.R. (1979). *J. Biol. Chem. 254,* 5855.

Moss, J., and Richardson, S.H. (1978). *J. Clin. Invest. 62,* 281.

Moss, J., and Stanley, S.J. (1981). *Proc. Natl. Acad. Sci. USA 78,* 4809.

Moss, J., and Vaughan, M. (1977). *J. Biol. Chem. 252,* 2455.

Moss, J., and Vaughan, M. (1978). *Proc. Natl. Acad. Sci. USA 75,* 3621.

Moss, J., and Vaughan, M. (1979a). *Annu. Rev. Biochem. 48,* 581.

Moss, J., and Vaughan, M. (1979b). *In* "Receptors and Human Diseases" (A.G. Bearn, and P.W. Choppin, eds.), p. 230.

Josiah Macy Jr. Foundation, New York.
Moss, J., and Vaughan, M. (1980). *In* "Developments in Cell Biology," Vol. 6 (M.E. Smulson, and T. Sugimura, eds.), p. 391. Elsevier/North-Holland, New York.
Moss, J., and Vaughan, M. (1981). *Mol. Cell. Biochem. 37*, 75.
Moss, J., Fishman, P.H., Manganiello, V.C., Vaughan, M., and Brady, R.O. (1976a). *Proc. Natl. Acad. Sci. USA 73*, 1034.
Moss, J., Fishman, P.H., Richards, R.L., Alving, C.R., Vaughan, M., and Brady, R.O. (1976b). *Proc. Natl. Acad. Sci. USA 73*, 3480.
Moss, J., Manganiello, V.C., and Vaughan, M. (1976c). *Proc. Natl. Acad. Sci. USA 73*, 4424.
Moss, J., Osborne, J.C., Jr., Fishman, P.H., Brewer, H.B., Jr., Vaughan, M., and Brady, R.O. (1977). *Proc. Natl. Acad. Sci. USA 74*, 74.
Moss, J., Garrison, S., Oppenheimer, N.J., and Richardson, S. H. (1979a). *J. Biol. Chem. 254*, 6270.
Moss, J., Garrison, S., Oppenheimer, N.J., and Richardson, S. H. (1979b). *In* "Proceedings of the 14th Joint Conference, US-Japan Cooperative Medical Science Program, Cholera Panel, Symposium on Cholera, Karatsu, 1978" (K. Takeya, and Y. Zinnaka, eds.), p. 274. Fuji Printing Co., Ltd., Tokyo.
Moss, J., Stanley, S.J., and Lin, M.C. (1979c). *J. Biol. Chem. 254*, 11993.
Moss, J., Stanley, S.J., and Oppenheimer, N.J. (1979d). *J. Biol. Chem. 254*, 8891.
Moss, J., Stanley, S.J., and Watkins, P.A. (1980). *J. Biol. Chem. 255*, 5838.
Moss, J., Stanley, S.J., and Osborne, J.C., Jr. (1981a). *J. Biol. Chem. 256*, 11452.
Moss, J., Osborne, J.C., Jr., Fishman, P.H., Nakaya, S., and Robertson, D.C. (1981b). *J. Biol. Chem. 256*, 12861.
Moss, J., Stanley, S.J., and Osborne, J.C., Jr. (1982). *J. Biol. Chem. 257*, 1660.
Mullin, B.R., Aloj, S.M., Fishman, P.H., Lee, G., Kohn, L.D., and Brady, R.O. (1976). *Proc. Natl. Acad. Sci. USA 73*, 1679.
Nakaya, S., Moss, J., and Vaughan, M. (1980a). *In* "Proceedings of the Fifteenth Joint Conference on Cholera," p. 299. DHEW Publ. No. (NIH) 80-2003.
Nakaya, S., Moss, J., and Vaughan, M. (1980b). *Biochemistry 19*, 4871.
Ogata, N., Ueda, K., and Hayaishi, O. (1980a). *J. Biol. Chem. 255*, 7610.

Ogata, N., Ueda, K., Kagamiyama, H., and Hayaishi, O. (1980b). *J. Biol. Chem. 255*, 7616
Oppenheimer, N.J. (1978). *J. Biol. Chem. 253*, 4907.
Riquelme, P.T., Burzio, L.O., and Koide, S.S. (1979). *J. Biol. Chem. 254*, 3018.
Rodbell, M. (1980). *Nature (Lond.) 284*, 17.
Ross, E.M., and Gilman, A.G. (1980). *Annu. Rev. Biochem. 49*, 533.
van Heyningen, S. (1977). *Biol. Rev. 52*, 509.
Van Ness, B.G., Howard, J.B., and Bodley, J.W. (1980a). *J. Biol. Chem. 255*, 10710.
Van Ness, B.G., Howard, J.B., and Bodley, J.W. (1980b). *J. Biol. Chem. 3255*, 10717.
Watkins, P.A., Moss, J., and Vaughan, M. (1981). *J. Biol. Chem. 256*, 4895.
Wodnar-Filipowicz, A., and Lai, C.Y. (1976). *Arch. Biochem. Biophys. 176*, 465.
Yost, D.A., and Moss, J. (1983). *J. Biol. Chem.*, in press.

DISCUSSION

Ortez (Oklahoma State U.): What can you tell me about the distribution of the avian enzyme, either within the red cell or among other cell populations?

Moss: In turkey erythrocytes, we found the enzyme both in cytosol and in membrane fractions. We have purified the enzyme approximately 600,000 fold from human erythrocytes where it is present in much, much lower quantities than in turkey erythrocytes; it is still not pure.

Ortez; What about non-erythrocytes?

Moss: We have also found the enzyme in very small quantities in rat liver.

Jagus: Would you like to comment on what you think are the natural substrates for the erythrocyte ADP-ribosyltransferase? Do you have any idea at all? Do you find it a problem that you can ADP-ribosylate turkey erythrocyte membranes with choleragen toxin which means that the endogenous ribosyl—transferase is not using adenylate cyclase as a natural substrate?

Moss: We do not know what the substrates are. We do know that the 42,000 and 47,000 dalton substrates for cholera toxin do not appear to be favored substrates for the avian enzyme.

Graves (Iowa State U.): You said that the ADP-ribosyltransferase B has a different specificity for protein substrates. I wonder if you would elaborate on that?

Moss: What we did with transferase B was similar to what we did with the transferase A and choleragen. We used purified proteins, various membranes and soluble fractions to examine the ADP-ribosylation pattern. What we found was that indeed they differ.

Graves: How about something like poly-arginine and things like that?

Moss: Poly-arginine will serve as an acceptor for both enzymes.

Jacobson: May I presume that there is still unpurified activity in the membrane fraction from these cells?

Moss: We have examined some of the activities that were present in the turkey erythrocyte membranes; it has the characteristics of transferase A. We can elute it and purify it by essentially the same procedure as used for transferase A.

Kun: I think it is very difficult to compare the enzymatic activity of the mono-ADP-transferases with the nuclear poly ADP-transferase for NRS because in your enzyme system, you have appropriate acceptors, second substrates and you are actually measuring V-max. In the nuclear enzyme during purification, you lose all the polymer acceptors and the polymerase is auto-poly-ADP-ribosylated and the enzyme is inhibited from time zero on; therefore the activity of the purified enzyme cannot be assessed at the moment.

Moss: I think it is very difficult to compare the two enzymes. The poly(ADP-ribose) synthetase may be catalyzing two different types of reactions. The problem, I think, with characterizing V-max too, is the fact that we are using monosubstrates, which may be better or worse than the true substrate *in vivo*.

Nishizuka: Many glycohydrolases found in mammalian tissues catalyze the exchange reaction of nicotinamide in NAD. I wonder, in your system, if you have detected any exchange of nicotinamide?

Moss: We have looked for various exchange reactions with this enzyme. These enzymes will not catalyze this reaction. We have taken purified NAD glycohydrolases that will catalyze the exchange reaction and looked at their ability to ADP ribosylate arginine and they do not. We, thus, looked at it both ways. We have looked at it with nicotinamide, looking at formation of radioactive NAD with labeled nicotinamide.

Nishizuka: Many years ago N.O. Kaplan reported that some glycohydrolases in membranes contain polysaccarides. Could you find any sugar in your protein?

Moss: We do not have enough of either of these proteins to do a good carbohydrate analysis. What is interesting in both is that they do stick to Concanavalin A Sepharose, so we think maybe they are glycoproteins.

Cohen: I didn't get any feel from your talk as to how good proteins were as substrates as compared to, say, arginine. Could you comment on that, in terms of Km's and V-max?

Moss: We have not looked at the purified proteins themselves, with a good kinetic analysis. They are fairly good substrates. Normally we have used protein about 1 mg/ml in routine assays.

Cohen: What is the arginine concentration? If you used equimolar amounts of arginine and protein, what would the rates be?

Moss: I don't think I could assess that right now.

Cohen: Would physiological concentrations of free arginine completely block the ADP ribosylation of proteins?

Moss: No, in that case it would not. If we take various membrane and soluble proteins, which are present in much lower quantities, then we are dealing probably with 100 μg/ml or less. Some of these proteins will be ADP-ribose acceptors in the presence of arginine, and this would suggest that their arginine groups are simply more available and conceivably they could be physiological substrates.

Kahn: How do you identify the transferase in membranes?

Moss: We basically purified membranes out of turkey erythrocytes, then extracted the transferase from the membranes.

Kahn: Have you compared either peptide mapping or fingerprinting of transferases A and B?

Moss: No.

Kun: In answer to Dr. Nishizuka's question, not every glycohydrolase can transglycosylate; only some enzymes do.

MONO- AND POLY(ADP-RIBOSYLATION) IN INTACT CELLS[1]

Myron K. Jacobson
D. Michael Payne

Departments of Chemistry and Biochemistry
North Texas State University
Texas College of Osteopathic Medicine
Denton, Texas

Elaine L. Jacobson

Department of Biology
Texas Woman's University
Denton, Texas

INTRODUCTION

Advances in our knowledge of the biological functions of the vitamin niacin may be viewed as having occurred in several distinct stages. The pioneering studies of pellagra by Goldberger in the period 1914 to 1927 (Etheridge, 1972) led to the subsequent identification by Elvehjem and coworkers of nicotinic acid and nicotinamide as anti-pellagra factors (Elvehjam *et al.*, 1937). A second stage led to the finding that nicotinic acid and nicotinamide are converted within cells into NAD^+ and $NADP^+$ and to the realization that these nucleotides play a fundamental role in biological chemistry

[1]The research described was supported in part by Grants CA23994 and CA29357 from the National Institutes of Health, B-633 from the Robert A. Welch Foundation, and institutional grants from North Texas State University and Texas Woman's University. D.M.P. was supported by a predoctoral fellowship from The Samuel Roberts Noble Foundation.

ISBN 0-12-387560-9

TABLE I. Enzymes that Utilize NAD^+ as a Substrate

Enzyme	*Product*
Numerous dehydrogenases	*NADH*
NAD^+ kinase	*$NADP^+$*
DNA ligase, NAD pyrophosphatase	*NMN + AMP*
NAD pyrophosphorylase	*NMN + ATP*
ADP-ribosyl transferases	*$X\text{-}(ADPR)_n + nN$*

as substrates in reactions that generate or consume a hydride ion. As a substrate for numerous dehydrogenases, the oxidized and reduced forms of these nucleotides are rapidly interconverted but the nicotinamide and adenine rings of the molecule remain covalently joined. However, a number of other enzymes utilize NAD^+ as a substrate in reactions in which covalent bonds between the nicotinamide and adenine rings are cleaved. The types of reactions that involve NAD^+ are listed in Table I. The studies of Olivera and coworkers (Rechsteiner *et al.*, 1976a,b) have shown that the half life of the adenine ring of the NAD molecule in mammalian cells is approximately 1 hr. This observation demonstrates that NAD consuming reactions constitute an active metabolism within living cells. The elucidation of the function of these NAD requiring reactions represents a third stage in our understanding of the biological functions of the vitamin niacin. Reactions catalyzed by ADP-ribosyl transferases may play a major role in the metabolism that consumes NAD (for a recent review, see Mandel *et al.*, 1982). The ADP-ribosyl transferases can be subclassified into mono(ADP-ribosyl) transferases and poly(ADP-ribosyl) transferases, more commonly referred to as poly(ADP-ribose) polymerase (Figure 1). *In vitro*, mono(ADP-ribosyl) transferases have been shown to catalyze the transfer of ADP-ribosyl groups to water, a variety of pyridine derivatives, and to protein acceptors (Fukushima *et al.*, 1976; Hilz and Stone, 1976). Poly(ADP-ribose) polymerase is a chromosomal enzyme which catalyzes the transfer of ADP-ribosyl groups *in vitro* to both histone and nonhistone proteins (Hayaishi and Ueda, 1977) and to nucleotides such as diadenosine tetraphosphate (Yoshihara and Tanaka, 1981).

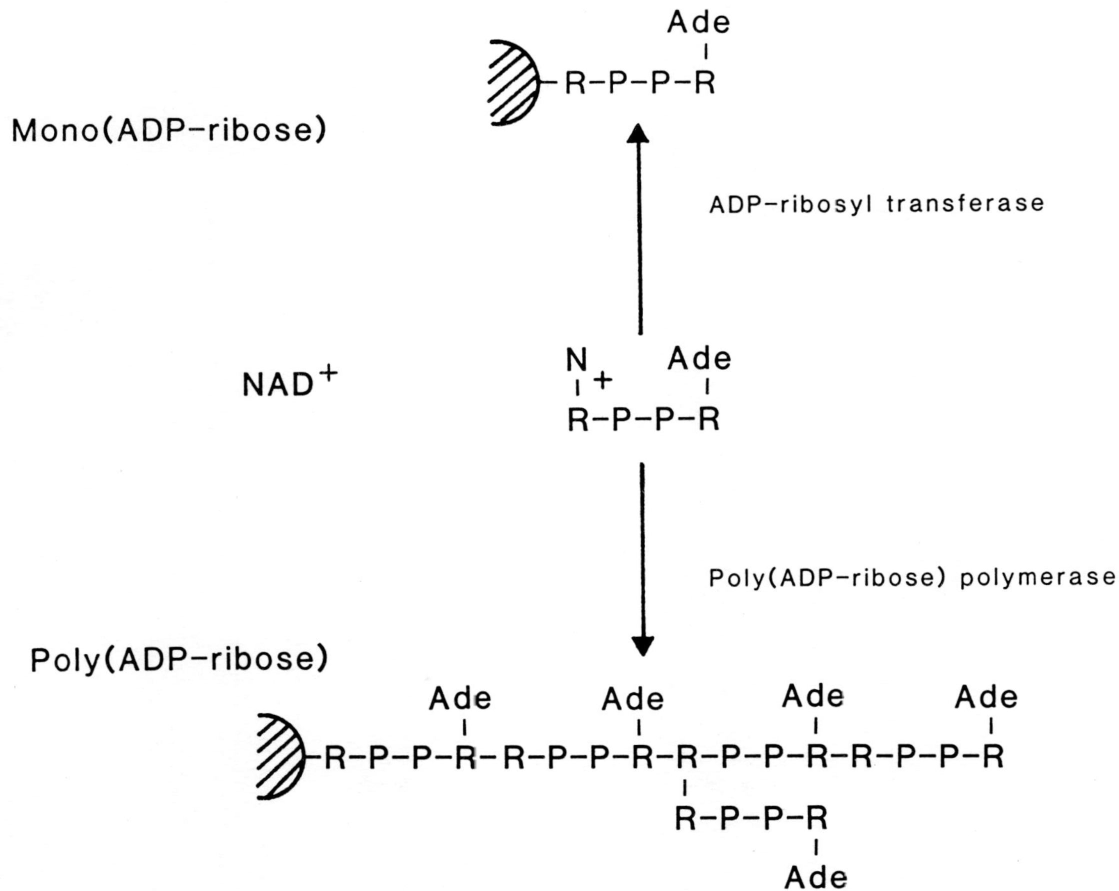

FIGURE 1. Utilization of NAD^+ by enzymes that catalyze ADP-ribosyl transfer to proteins and other acceptors.

While many studies of ADP-ribosyl transferases have been conducted *in vitro*, very few detailed studies of ADP-ribose metabolism in intact cells have been possible due to the lack of suitable experimental methods for such studies. This paper will review recent work in our laboratories concerning the study of ADP-ribose metabolism in intact cells.

STRATEGY FOR MEASUREMENT OF ADP-RIBOSYL RESIDUES FROM INTACT CELLS AND TISSUES

The study of mammalian ADP-ribose metabolism *in vivo* requires methodology capable of measuring the amount of protein-bound monomers and polymers of ADP-ribose present in a limited quantity of cultured cells or animal tissues. The methods for making such measurements must be both highly selective and highly sensitive, since the cellular content of ADP-ribosyl residues is very low relative to chemically similar compounds (Table II). Figure 2 shows the overall approach developed in our laboratory for the measurement of both monomeric and polymeric ADP-ribosyl residues in intact cells. These complementary methods have two common features that provide the needed selectivity and sensitivity. The first is the use of an immobilized boronate resin which allows the quantitative separation of small amounts of ADP-ribosyl residues from the bulk of interfering substances present in crude extracts of acid-insoluble material. The boronate affinity resin functions by the reversible formation of a complex between boronate and molecules that contain cis 1,2-diol groups

TABLE II. Relative Amount of Adenine Ring Compounds in Cultured 3T3 Cells

Compound	*Adenine rings μmol/10^8 cells*
RNA	*3.0*
DNA	*1.0*
ATP	*0.6*
NAD	*0.2*
Poly(ADP-ribose)	*0.000004*

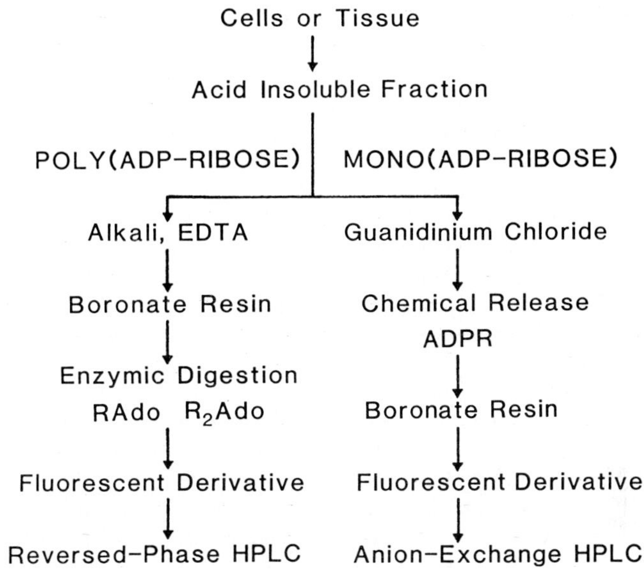

FIGURE 2. Analytical scheme for the measurement of monomeric and polymeric ADP-ribosyl residues from intact cells and tissues.

(Rosenberg *et al.*, 1972). The second key feature of both methods involves the quantitative conversion of adenine-containing compounds to highly fluorescent 1,N^6-etheno derivatives (Barrio *et al.*, 1972), greatly enhancing the sensitivity of detection of the desired compounds.

MEASUREMENT OF POLYMERIC ADP-RIBOSYL RESIDUES

Treatment of the acid insoluble fraction of cells with 1 M KOH, 0.05 M EDTA at 60°C for 1 hr (Figure 2) has several advantages with regard to the quantification of poly(ADP-ribose). First, the acid-insoluble material is conveniently dissolved under conditions in which poly(ADP-ribose) is stable. In addition, this treatment releases the polymer from protein and also results in the degradation of monomeric ADP-ribosyl residues to 5'-AMP and the hydrolysis of RNA to a mixture of 2'-and 3'-ribonucleotides. Following alkali treatment, the sample is mixed with an equal volume of 6 M guanidinium chloride and the pH is adjusted to 9. The poly(ADP-ribose) is then batch adsorbed to a boronate derivative of Bio-Rex 70 (Wielckens *et al.*, 1981). Under these conditions, this resin requires two or more sets of vicinyl hydroxyl groups for effective binding, thus poly(ADP-ribose) is bound to the resin without interference from other adenine-containing compounds or their degradation products. After washing the resin, poly(ADP-ribose) is eluted and enzymatically digested to nucleosides by treatment with venom phosphodiesterase and alkaline phosphomonoesterase (Figure 3). This procedure yields the unique nucleosides ribosyladenosine (RAdo) and diribosyladenosine (R_2Ado) from the internal linear and branched residues respectively. These nucleosides are then converted to fluorescent 1,N^6-etheno derivatives and analyzed by reversed-phase HPLC with fluorescence monitoring. Figure 4 (Panel A) shows results obtained from the analysis of adult rat liver; a peak of fluorescent material, equivalent to approximately 1 pmole was observed at the elution position of ethenoribosyladenosine. Panels B and C show that omission of phosphodiesterase/phosphomonoesterase treatment (Panel B) or chloroacetaldehyde treatment (Panel C) resulted in the absence of a peak eluting at the position of ethenoribosyladenosine. Panel D shows the result obtained when [^{14}C] poly(ADP-ribose) synthesized *in vitro* was added to a liver sample prior to analysis. It can be seen that a large enhancement of the ethenoribosyladenosine peak was observed and that the radioactivity eluted primarily at the position of ethenoribosyladenosine. The radioactivity eluting at the position of ethenoadenosine was derived from the terminal residues of the [^{14}C]poly(ADP-ribose)(Figure 3). Underivatized ribosyladenosine is also separated from the etheno derivatives of adenosine and ribosyladenosine in this chromatographic system. This allows both a control for recovery through all steps in the procedure and a check for quantitative conversion of ribosyladenosine to the etheno derivative. The amount of tissue analyzed in the experiment shown in Figure 3 was inadequate to detect the presence of branched residues in the polymer. Figure 5 shows a chromatogram of an analysis of chicken pancreas in which the presence of a branched residue was clearly detected.

This analytical approach has been applied to the study of poly(ADP-ribose) metabolism in cultured mouse and human fibroblasts following treatments that cause DNA damage (Juarez-Salinas *et al.*, 1979; Sims *et al.*, 1980; Jacobson *et al.*, 1982). We have observed that the normal intracellular levels of poly(ADP-ribose) are very low but that the occurrence of DNA damage leads to a rapid increase in the intracellular

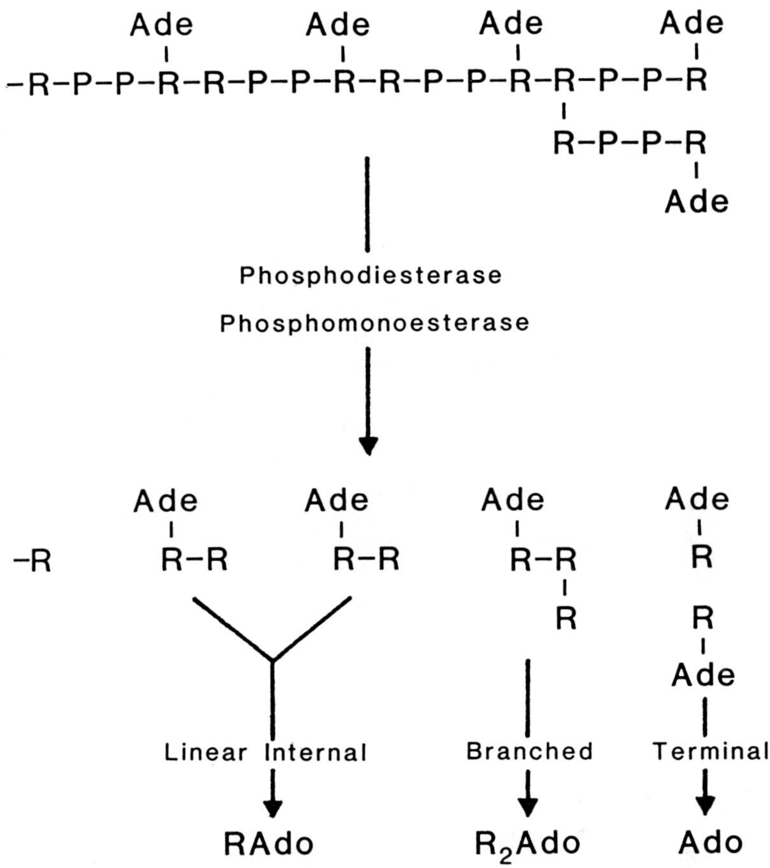

FIGURE 3. Enzymatic degradation of poly(ADP-ribose).

levels. Further, the polymer synthesized in response to DNA damage is turned over rapidly with a half life of 6 min or less (Jacobson *et al.*, 1982). These studies have also concluded that the occurrence of a DNA strand break is the common factor of DNA damage that stimulates polymer synthesis (McCurry and Jacobson, 1981; Jacobson *et al.*, 1982). The application of this approach to the measurement of poly(ADP-ribose) from intact animal tissues should prove useful for studies of poly(ADP-ribose) metabolism in whole animals.

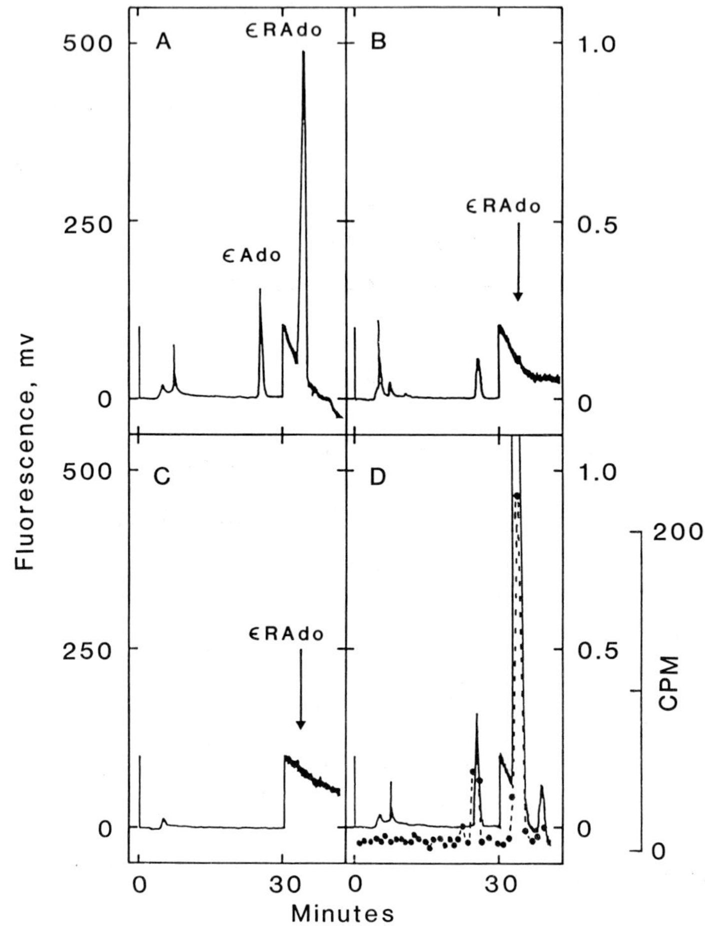

FIGURE 4. Determination of poly(ADP-ribose) from rat liver using reversed-phase HPLC with fluorescence monitoring. Panel A shows an analysis of a sample subjected to the entire procedure. An analysis in which treatment with phosphodiesterase/phosphomonoesterase was omitted is shown in panel B and an analysis in which treatment with chloroacetaldehyde was omitted is shown in panel C. Panel D shows the result when [^{14}C]poly(ADP-ribose) was added to the sample prior to analysis. The sensitivity of the fluorimeter was changed at 30 min following each injection. The fluorescence scale on the left refers to the period 0 to 30 min, the scale on the right from 30 min to the end of the chromatogram.

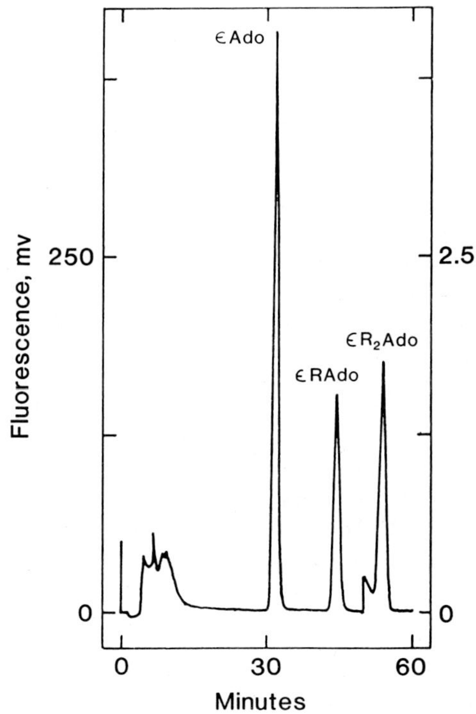

FIGURE 5. Analysis of poly(ADP-ribose) from chicken pancreas.

MEASUREMENT OF MONOMERIC ADP-RIBOSYL RESIDUES

A major analytical problem with regard to the measurement of the levels of protein-bound monomers of ADP-ribose from intact cells is related to the high tissue levels of NAD^+ and NADH relative to the tissue levels of protein-bound ADP-ribose For example, a small fraction of the total tissue levels of NADH trapped in an acid-insoluble cell fraction could lead to a significant overestimation of ADP-ribose, since the acid treatment of NADH would convert it to ADP-ribose (Bredehorst *et al.*, 1981). To rigorously rule out this possibility, it is necessary to remove noncovalently bound material from acid-insoluble pellets prior to release of ADP-ribose from protein. This can be accomplished by dissolving the pellets in a buffer containing 6 M guanidinium chloride followed by gel filtration chromatography or dialysis. Alternatively, the pellet can be dissolved in concentrated formic acid, diluted and reprecipitated with trichloroacetic acid prior to dissolution in buffer containing guanidinium chloride. The cellular mono(ADP-ribosyl) transferases studied thus far catalyze the transfer of ADP-ribose to an arginine residue in the acceptor protein (Moss and Vaughn, 1978). This linkage is very stable at neutral pH in this buffer; however, the ADP-ribosyl residues can be released by treatment with neutral hydroxylamine. Following release, the ADP-ribose is separated from the bulk of material present in the crude extract by batch adsorption to the boronate affinity resin. Following elution from the resin, ADP-ribose is converted to the fluorescent etheno derivative and quantified by an ion exchange HPLC with fluorescence monitoring. Figure 6 (Panel A) shows an analysis of adult rat liver along with a series of controls that demonstrate the selectivity of the method. Panel A shows a peak of fluorescent material corresponding to approximately 50 picomoles at the elution position of etheno-ADP-ribose. Panel B shows that omission of chloroacetaldehyde treatment resulted in the absence of a fluorescent peak at the position of etheno-ADP-ribose. Panel C shows that treatment with venom phosphodiesterase prior to chromatography resulted in the absence of a fluorescent peak at the position of etheno-ADP-ribose, and, as expected, the appearance of a peak at the elution position of etheno-5'-AMP. Panel D shows the result obtained when [^{14}C]ADP-ribosyl-histone H1 synthesized *in vitro* was added to a liver sample prior to analysis. The radioactivity was coincident with fluorescence and eluted at the position of etheno-ADP-ribose.

ADP-RIBOSYL RESIDUES *IN VIVO*

Table III shows the content of polymeric and monomeric residues of ADP-ribose in adult rat liver as determined by the analytical approach described here. The amount of monomeric residues of ADP-ribose in liver exceeds that of the polymeric residues by a factor greater than 250. However, we have previously observed that the levels of poly(ADP-ribose) in cultured cells can increase at least 200 fold following DNA damage (Juarez-Salinas *et al.*, 1979). Further, only a small fraction of the polymeric residues are branched, approximately 0.8% of the total. We have observed a similar frequency of branched residues in other cells and tissues over a 50-fold range of total levels of polymer (Juarez-Salinas *et al.*, 1982). This finding suggests that branching is a general feature of poly(ADP-ribose) as it occurs in intact cells.

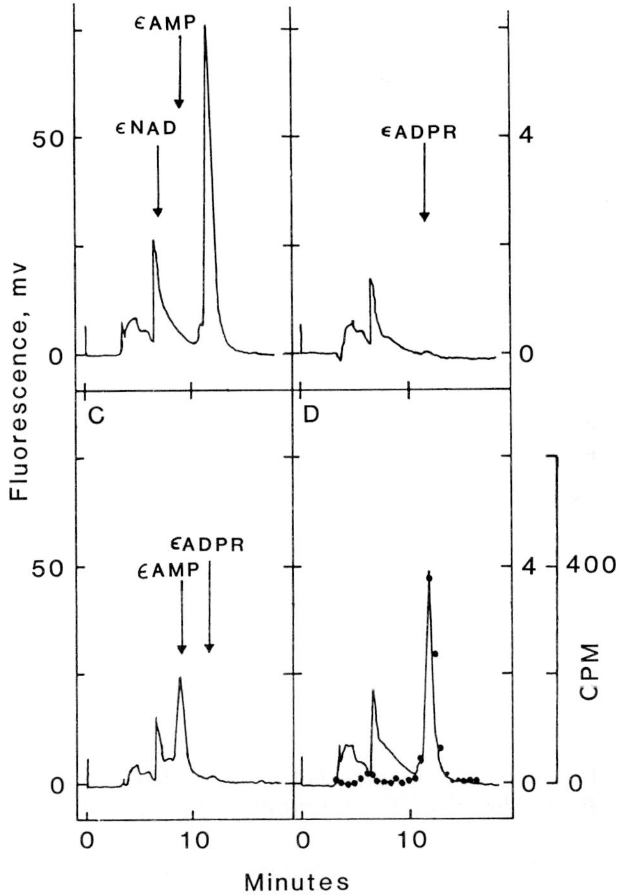

FIGURE 6. Determination of protein-bound mono(ADP-ribose) from rat liver using anion-exchange HPLC with fluorescence monitoring. Panel A shows an analysis of a sample subjected to the entire procedure. An analysis in which treatment with chloroacetaldehyde was omitted is shown in panel B. Treatment of the sample with venom phosphodiesterase prior to chromatography is shown in panel C. Panel D shows the result obtained when [^{14}C]ADP-ribosyl-histone H1 was added to the sample prior to analysis. The sensitivity of the fluorimeter was increased at 7.5 min following each injection. The fluorescence scale on the left refers to the period 0 to 7.5 min and the scale on the right from 7.5 min to the end of the chromatogram.

TABLE III. ADP-Ribosyl Residues in Rat Liver

Type	*ADP-ribose residues pmol/mg DNA*
Monomeric	
minus NH_2OH	*2,200*
released by NH_2OH	*1,600*
total	*3,800 ± 144 (n = 3)*
Polymeric	
branched	*0.15*
linear	*14.1*
total	*14.3 ± 1.5 (n = 6)*

A significant fraction of the total monomeric residues do not require hydroxylamine for release from protein. However, the (ADP-ribosyl) arginine linkage is stable in the absence of hydroxylamine. These data provide evidence that monomers of ADP-ribose are attached to protein by at least two types of chemical linkages. Previously, Hilz and coworkers (Bredehorst *et al.*, 1978) reached a similar conclusion using methodology different from that described here.

At the present time our knowledge of the biological functions of ADP-ribosyl transferase reactions is very limited. The availability of methodology that allows the study of in- mono- and poly(ADP-ribose) metabolism in cultured cells and tact tissues should prove useful in the elucidation of the functions of these novel post-translational modifications of proteins.

ACKNOWLEDGMENTS

We thank Dr. Joel Moss for a gift of [^{14}C]ADP-ribosyl-histone H1 and for a sample of chicken pancreas, Linda Hemingway and Hilda Mendoza-Alvarez for technical assistance, Hector Juarez-Salinas for many helpful discussions and Toni Benjamin and Lynne Gracy for assistance in the preparation of the manuscript.

REFERENCES

Barrio, J., Secrist, J.A., III., and Leonard, N.J. (1972). *Biochem. Biophys. Res. Commun. 46*, 597.

Bredehorst, R., Schluter, M., and Hilz, H. (1981). *Biochim. Biophys. Acta 652*, 16.

Bredehorst, R., Wielckens, K., Gartemann, A., Lengyel, H., Klapproth, K., and Hilz, H. (1978). *Eur. J. Biochem. 92*, 129.

Elvehjem, C., Madden, F., Strong, D., and Wooley, D. (1937). *J. Am. Chem. Soc. 59*, 1767.

Etheridge, E. (1972). "The Butterfly Caste: A Social History of Pellagra in the South." Greenwood Publishing Company, Westport, Connecticut.

Fukushima, M., Okayama, H., Takahashi, Y., and Hayaishi, O. (1976). *J. Biochem. 80*, 167.

Hayaishi, O., and Ueda, K. (1977). *Annu. Rev. Biochem. 46*, 95.

Hilz, H., and Stone, P. (1977). *Rev. Physiol. Pharmacol. 76*, 1.

Jacobson, E.L., Antol, K.M., Juarez-Salinas, H., and Jacobson, M.K. (1983). *J. Biol. Chem. 258*, 103.

Juarez-Salinas, H., Levi, V., Jacobson, E.L., and Jacobson, M.K. (1982). *J. Biol. Chem. 257*, 607.

Juarez-Salinas, H., Sims, J.L., and Jacobson, M.K. (1979). *Nature 282*, 740.

Mandel, P., Okazaki, H., Niedergang, C. (1982). *Prog. Nucleic Acid Res. Mol. Biol. 27*, 1.

McCurry, L.S., and Jacobson, M.K. (1981). *J. Biol. Chem. 256*, 551.

Moss, J., and Vaughan, M. (1978). *Proc. Natl. Acad. Sci. USA 75*, 3621.

Rechsteiner, M., Hillyard, D., and Olivera, B.M. (1976). *J. Cell. Physiol. 88*, 207.

Rechsteiner, M., Hillyard, D., and Olivera, B.M. (1976). *Nature 259*, 695.

Rosenberg, M., Wiebers, J.L., and Gilham, P.T. (1972). *Biochemistry 11*, 3623.

Sims, J.L., Juarez-Salinas, H., and Jacobson, M.K. (1980). *Anal. Biochem. 106*, 296.

Wielckens, K., Bredehorst, R., Adamietz, P., and Hilz, H. (1981). *Eur. J. Biochem. 117*, 69.

Yoshihara, K., and Tanaka, Y. (1981). *J. Biol. Chem. 256*, 6756.

DISCUSSION

Kun: Did you say that 5'-AMP does not bind to any boronic acid resins?

Jacobson: No, I stated only that 5'-AMP does not bind to the boronic acid resin that we have synthesized in our laboratory.

Kun: I don't think this is a very good situation. In J.B.C. of October (this last issue) there is an extensive study of the various nucleotides binding to phenyl boronic acid and there is no exception.[1] All the diol containing - except phosphoribosyl - AMP bind to phenyl boronic acid. So, I wonder whether your boronate chromatography is really effective?

Jacobson: Well, I think it certainly is. The resin you are referring to is probably a polyacrylamide derivative; right?

Kun: It doesn't make any difference.

Jacobson: Well, it certainly does make a difference in our laboratory. Indeed our resins are particularly effective because we can obtain the quantitative binding of NAD, ADP-ribose and poly(-ADP-ribose) from crude extracts with these resins without interference from AMP. While we have been puzzled as to the reason for the specificity, this selectivity is really advantageous.

Kun: I doubt that they are useful if the chemical argument behind them is somewhat nebulous. I think that perhaps one of the problems is that you do not separate proteins, and you hydrolyze everything on your columns, which could interfere with binding.

Jacobson: No, we are not currently hydrolyzing the polymer on the column. We are now using a resin in which we can elute the poly(-ADP-ribose) from the resin.

Kun: Yes, but the binding of 5'-AMP should happen on any phenyl-boronic acid resin. The other point I would like to

---

[1]*Editor's note - Dr. Jacobson reported that a careful reading of the paper shows that the entire point of the paper deals with an exception.*

make here is that the argument for a branching structure is not really convincing. The quantity of this derivative one finds by digesting the poly-ADP-ribose phosphodiesterase is less than 1%. In fact, we see that it varies with the degree of digestion and we think it might be an artifact of the phosphodiesterase digestion and signalling an artificial phosphorylation. I think that it might be a little premature to argue about the poly-ADP-ribose structure as being significantly branched.

Jacobson: We have to face the data, and the best interpretation of the data we have is that branched residues are there. Certainly, phosphorylated artifacts could not be part of our methodology because we are measuring nucleosides and a molecule which has only a single phosphate is going to elute from our chromatographic system very early. That is an advantage of the methodology that we use. We are analyzing nucleosides. We see the same content of branching upon treatment with enzymes or treatment with alkali in the presence of magnesium which will chemically degrade the polymer residues. It is possible that we are underestimating the amount of branching, but the data clearly shows that the branched residues are there. Now, the significance of them is debatable, but I think the best interpretation of our data is that the branched residues are there. I am not saying they are long branches or short branches, but that there are branched residues.

Kun: But you have to say less than 1%?

Jacobson: Yes, well less than 2% of the total residues.

Nishizuka: Due to my experiences many years ago, one has to take care to prevent poly-ADP-ribose synthesis during the preparation of tissue extracts. Otherwise, a small quantity of poly ADP-ribose may be produced during the homogenization, because a large excess of endogenous radioactive NAD is present in the extract. How did you make the tissue extract; how did you treat the cells before measuring poly-ADP-ribose?

Jacobson: With regard to cultured cells, the procedure involved decanting the medium and adding ice-cold 20% trichloroacetic acid to the monolayers directly. With regard to tissues, we excise the tissue and drop it into liquid nitrogen. With regard to measurements of molecules from intact cells, there is always the possibility that something may be happening while you are sampling.

Nishizuka: Could you eliminate the possibility that the observed increase in poly ADP-ribose is simply due to the cell damage by carcinogen rather than the response of the cell?

Jacobson: Well, that is somewhat hard to rule out at this point in the game. I think that clearly one always has to consider that possibility. I think that it is simple to point out the pitfalls that one ought to reduce. I mean I think this information will ultimately come together with *in vitro* information in terms of a consistent idea of what is happening.

Stricker: Have you shown if poly-ADP-ribosylation has some effect on the organization of the chromatin, especially on the nucleosomal organization?

Jacobson: We have no data that speak to that point. Other data have been published that do speak to that point.

Stricker: What do you think about it?

Jacobson: One would expect changes in chromatin conformation. Our kinetics would argue that those changes are localized since we do not observe a large accumulation of poly-ADP-ribose *in vivo*. That is certainly in contrast to what one sees *in vitro*. We don't know the size of the chains in intact cells, so studies of conformational changes done with isolated nuclei which have been done with very long polymers may not be relevant. We really do not know how long they are in intact cells. It is certainly an interesting idea which is likely to be true but I think it is probably premature to conclude that indeed the long polymer chains formed in permeabilized cells are reflective of what happens in intact cells, based on the big differences in kinetics that one sees.

PROTEIN ADP-RIBOSYLATION IN RESPONSE TO DNA DAMAGE: EFFECTS OF Mg^{2+} AND Ca^{2+} ON AUTO ADP-RIBOSYLATION[1]

Nathan A. Berger
Barbara A. Booth
Shirley J. Petzold
James L. Sims

Washington University School of Medicine
Division of Hematology/Oncology
The Jewish Hospital of St. Louis
St. Louis, Missouri

Poly(ADP-ribose) polymerase is a chromosomal enzyme that uses NAD+ as substrate to synthesize the polymer (ADP-ribose), which is covalently attached to chromosomal proteins (Hayaishi and Ueda, 1977). Studies at the cellular level demonstrate that enzyme activity is stimulated by the induction of DNA damage and DNA strand breaks (Berger *et al.*, 1978b, 1979) and the poly(ADP-ribose) synthesized in response to DNA damage may function to open the chromatin structure to provide access for the enzymes of DNA repair (Berger and Sikorski, 1981). Using electron microscopy of isolated pancreatic chromatin, Poirier *et al.* (1982) have recently shown that poly(ADP-ribose) synthesis does result in an unfolding of the chromatin structure.

In order to obtain a better understanding of poly(ADP-ribose) synthesis, we purified the enzyme from lamb thymus and studied the factors affecting its activity *in vitro* (Petzold *et al.*, 1981). The purified enzyme is a single polypeptide chain with a molecular weight of 116,000. DNA is required as an essential enzyme activator and studies at the

[1]*These studies were supported by NIH grants CA24986 and GM26463. NAB is a Leukemia Society of America Scholar. JLS is a recipient of a training fellowship, part of Institutional NSRA Training Grant T32CA09118.*

ISBN 0-12-387560-9

molecular level demonstrate that enzyme activity is stimulated by the presence of DNA strand breaks (Benjamin and Gill, 1980; Ohgushi *et al.*, 1980; Cohen *et al.*, 1982). Enzyme activity is inhibited by substrate analogs such as nicotinamide, and by sulfhydryl binding agents such as PHMB (Petzold *et al.*, 1981).

Figure 1 shows autoradiographs of the proteins that are ADP-ribosylated in reactions conducted in the presence and absence of histone H1 with [^{32}P]NAD as substrate, DNA as enzyme activator, purified poly(ADP-ribose) polymerase and dithiothreitol. Lane 1 shows the proteins that were ADP-ribosylated when the reaction system contained both poly(ADP-ribose) polymerase and histone H1. The band of radioactivity detected at a molecular weight of 116,000 and above corresponds to the Coomassie Blue stained band of poly(ADP-ribose) polymerase and represents ADP-ribosylated enzyme. The ADP-ribosylated material distributed between the stacking gel and the 116,000 molecular weight marker represents enzyme which electrophoresed at higher molecular weight due to the attachment of long chains of poly(ADP-ribose). Lanes 2-4 in Figure 1 are autoradiographs showing proteins that were ADP-ribosylated when the reactions were run for 5 seconds, 20 seconds or 5 minutes with the purified enzyme being the only protein present. The initial labelled product corresponds to the molecular weight of the purified enzyme. As the incubation time increased, the migration of the labelled product indicated that its molecular weight progressively increased. As noted above, this increase in molecular weight is due to elongation of the bound poly(ADP-ribose) chains.

Table 1 shows the effects of Mg^{2+} and Ca^{2+} on the synthesis of poly(ADP-ribose) by purified poly(ADP-ribose) polymerase in the presence and absence of histone H1. The standard reaction contained [^{32}P]NAD as substrate, DNA as enzyme activator, purified poly(ADP-ribose) polymerase and dithiothreitol. In this system, 128 pmols ADPR were incorporated into macromolecular form in 5 minutes. When the same reaction was carried out in the presence of histone H1, there was a 27-fold increase in product so that 3,500 pmols ADPR were incorporated. Addition of Mg^{2+} to the reaction conducted with the enzyme alone resulted in a 1.6-fold increase in the amount of polymer synthesized. Addition of Ca^{2+} resulted in a 4.2-fold increase in poly(ADP-ribose) synthesis. Addition of a combination of Ca^{2+} and Mg^{2+} resulted in no greater stimulation of enzyme activity than occurred with Ca^{2+} alone. In contrast to the effect of Ca^{2+} and Mg^{2+} on the auto ADP-ribosylation reaction, addition of either or both of these cations to the reaction run in the presence of histone H1 resulted in no further increase in the level of poly(ADP-ribose) synthesis.

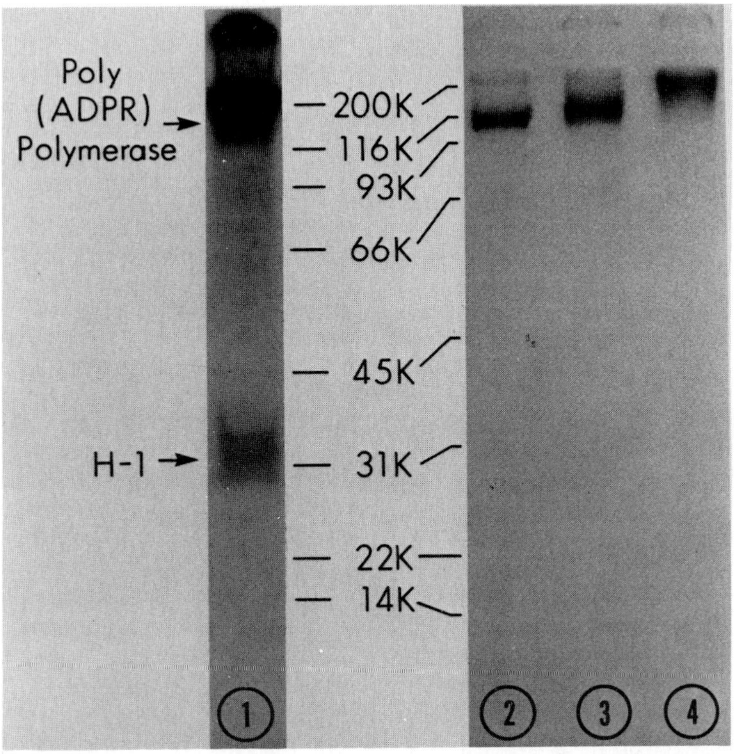

FIGURE 1. Autoradiographs of [^{32}P] ADP-ribosylated proteins. Reactions were run at 37°C with [^{32}P]NAD^+ in the presence or absence of histone H1. Reactions were stopped by precipitation with trichloroacetic acid. Precipitates were solubilized, electrophoresed on SDS, 7.5 to 20% exponential gradient polyacrylamide gels with a 5% polyacrylamide stacking gel and then autoradiographed. Lane 1 shows the results of the reaction run in the presence of 10 μM [^{32}P]NAD^+ (specific activity 680 cpm/pmol) and histone H1 for 20 seconds. Lanes 2, 3 and 4 show the results obtained with the enzyme as the only protein present and 5 μM [^{32}P]NAD^+ (specific activity 4,700 cpm/pmol) for 5 seconds, 20 seconds and 5 minutes respectively. Autoradiographs were exposed for 24 hours. Molecular weights, indicated between lanes 1 and 2, were determined by Coomassie Brilliant Blue staining of standards included in each slab gel.

TABLE 1. Effect of Mg^{2+} and Ca^{2+} on Poly(ADP-Ribose) Synthesis in vitro

	Poly(ADP-ribose) synthesis pmol [^{32}P] ADPR incorporated	
	Proteins present	
Reaction components	*Enzyme*	*Enzyme + H1*
Standard reaction	*128*	*3,500*
Plus 10 mM Mg^{2+}	*212*	*3,520*
Plus 10 mM Ca^{2+}	*545*	*3,250*
Plus 10 mM Mg^{2+} + 10 mM Ca^{2+}	*514*	*2,900*

Poly(ADP-ribose) polymerase activity was measured with purified lamb thymus poly(ADP-ribose) polymerase and other components described in text. Reactions contained 500 μM [^{32}P] NAD^+ (specific activity 66 cpm/pmol). Where indicated, reactions contained 10 μg histone H1. Incubations were for 5 minutes at 37°C.

To further characterize the ability of Ca^{2+} and Mg^{2+} to stimulate the auto ADP-ribosylation reaction, we examined the time course of their effects on reactions conducted with the purified enzyme. Figure 2 shows that in the absence of added cations, poly(ADP-ribose) synthesis leveled off after approximately 5 to 10 minutes. Addition of Mg^{2+} increased both the reaction rate and the duration of synthesis. Addition of Ca^{2+} produced an even greater increase in reaction rate and duration of synthesis. By 60 minutes the amount of poly(ADP-ribose) synthesized in the Mg^{2+}-stimulated reaction was 2.8-fold greater than that in the absence of divalent cations and the amount synthesized in the Ca^{2+}-stimulated reaction was 8-fold greater.

The auto ADP-ribosylation of poly(ADP-ribose) polymerase, as illustrated in Figure 1, has also been shown to occur at the cellular level (Ogata *et al.*, 1981; Surowy and Berger, 1983). Since our studies show that Mg^{2+} and Ca^{2+} stimulate the auto (ADP-ribosylation) reaction, it was of interest to use these cations to probe the mechanisms involved in this auto modification. Thus, we conducted experiments to determine whether the auto ADP-ribosylation reaction consisted of mono- or poly(ADP-ribosylation) and also to determine the

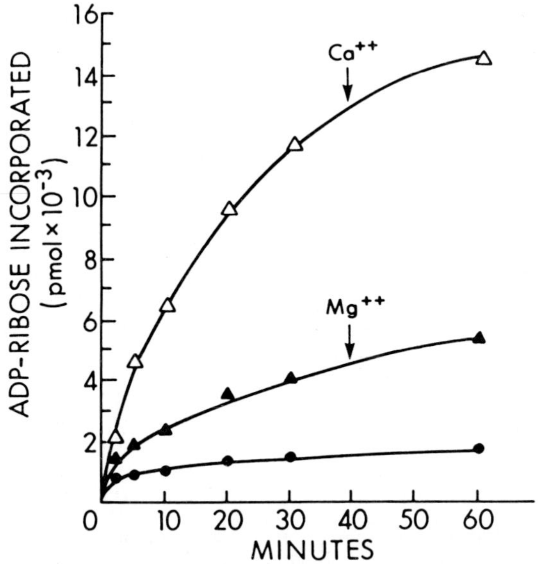

FIGURE 2. Effect of Mg^{2+} and Ca^{2+} on the time course of poly-(ADP-ribose) synthesis using purified components as described in the text. Reactions contained poly(ADP-ribose) polymerase as the only protein in the reaction and 500 μM [^{32}P]NAD^+ (specific activity 60 cpm/pmol). Incubations were at 37°C for the indicated times. (●), control-no divalent cations; (▲), 20 mM Mg^{2+}; (△), 20 mM Ca^{2+}.

number and length of chains attached to each enzyme molecule. Studies were also conducted to determine whether the auto ADP-ribosylation reaction is an intermolecular or intramolecular reaction with respect to enzyme protein.

Table 2 shows the results obtained when the auto ADP-ribosylation reaction was carried out without added cations or with 20mM Mg^{2+} or Ca^{2+}. The total incorporation, presented in the first line of Table 2, shows that Mg^{2+} and Ca^{2+} stimulated the level of poly(ADP-ribose) synthesis. The products of each reaction were subjected to complete digestion with proteinase K and snake venom phosphodiesterase and then separated and analyzed by HPLC using a Varian Micropak AX-10 ion exchange column. All components were normalized to the AMP content which in each case was set equal to 1. AMP residues are derived from chain ends, phosphoribosyl AMP residues represent internal chain units and di(phosphoribosyl) AMP residues represent branch points in the poly(ADP-ribose) chain (Miwa *et al.*, 1979; Juarez-Salinas *et al.*,

TABLE 2. Effect of Mg^{2+} and Ca^{2+} on Number and Length of Poly(ADP-ribose) Chains Synthesized in the Auto ADP-Ribosylation Reaction

	Added cation		
	None	20 mM Mg^{2+}	20 mM Ca^{2+}
[^{14}C] ADPR incorp. (nmol)	49	60	95
Distribution of [^{14}C] labelled Residues from polymer digest			
AMP	1	1	1
PR-AMP	15.2	15.4	12.5
Di(PR)-AMP	.28	.23	.23
Polymer chains/enzyme molecule	6.9	9.1	17.4
Average chain length	23.0	21.6	17.9

Each reaction contained 37 μg purified lamb thymus poly(ADP-ribose) polymerase and 200 μM [^{14}C] NAD^+ and other components indicated in text. Incubations were for 10 minutes at 37°C, following which the products were collected, digested and analyzed by HPLC. For each condition the distribution of residues is expressed relative to AMP which is taken as 1. Average chain length was calculated using the formula described by Kawaichi et al. *(1981):*

$$\frac{(AMP)+(PR\text{-}AMP)+[Di(PR)\text{-}AMP]}{(AMP)-[Di(PR)\text{-}AMP]}$$

1982). The relative numbers of these residues are used to determine average chain lengths of poly (ADP-ribose) using the formula developed by Kawaichi *et al.* (1981). The number of polymer chains per enzyme molecule is determined by relating the total amount of polymer synthesized and the average chain length to the amount of enzyme present in the reaction.

The data in Table 2 indicate that the auto ADP-ribosylation reaction consists of poly(ADP-ribosyl)ation rather than mono ADP-ribosylation. In all cases, the proportion of branch residues was relatively constant. The increase in poly(ADP-ribose) synthesis stimulated by Mg^{2+} and Ca^{2+} was associated with an increase in the number of polymer chains attached to each enzyme molecule. Stimulation was not associated with an increase in chain length. In fact, just the opposite occurred. Ca^{2+}, which caused the greatest increase in polymer synthesis, led to the accumulation of more chains of shorter average chain length.

Poly(ADP-ribose) polymerase could be auto (ADP-ribosylated) by at least two mechanisms. One possibility is an intramolecular reaction in which each enzyme molecule attaches ADP-ribose residues to itself. Another possibility is an intermolecular reaction in which one enzyme molecule ADP-ribosylates another. In order to distinguish between these possibilities, we added inactivated enzyme to determine if the inactive protein could act as acceptor for ADP-ribosylation. As shown in Table 3, in the absence of added divalent cation, doubling the amount of enzyme protein resulted in a doubling of the amount of polymer synthesized. This is the expected increase for an intramolecular reaction where there is no interaction of the individual enzyme molecules. If the reaction was intermolecular, then the activity would be expected to increase in a second order fashion with the square of the enzyme concentration.

As shown in Table 3, enzyme inactivated by treatment at 45°C for 1 hour or at 60°C for 5 minutes showed essentially no activity when assayed in the absence of divalent cations. When the inactivated enzyme was added to active enzyme there was a slight increase in the amount of reaction product formed, suggesting the possibility of a small amount of an intermolecular reaction. Thus the auto ADP-ribosylation reaction, conducted in the absence of divalent cations, appears to be mostly an intramolecular reaction.

Reactions conducted in the presence of Mg^{2+} were intermediate between those conducted in the absence of divalent cation and those conducted in the presence of Ca^{2+}. In the presence of Ca^{2+}, the reaction showed more than a 2-fold increase in product formation associated with doubling of the enzyme content. These results indicate that the reactions conducted in the presence of Ca^{2+} have an intermolecular component. This is further supported by the demonstration that, when the reaction was conducted in the presence of Ca^{2+}, addition of inactive enzyme to active enzyme resulted in a doubling of product formation. The latter experiment suggests that the inactive enzyme was ADP-ribosylated in an intermolecular reaction. Thus it appears that poly(ADP-ribose) polymerase can auto-modify itself in an intramolecular reaction and in the presence of Mg^{2+} or Ca^{2+} it can also perform the auto modification via an intermolecular mechanism.

TABLE 3. Effect of Addition of Inactivated Enzyme to Activated Enzyme on the Level of Poly(ADP-ribose) Synthesis in the Presence or Absence of Ca^{2+} and Mg^{2+}

		Added cation		
		None	Mg^{2+}	Ca^{2+}
Native enzyme	Treated enzyme	Incorporation ADP-ribose % standard reaction		
1 μg	----	100	100	100
2 μg	----	206	212	309
----	1 μg (45°C, 1 hr.)	4	11	34
1 μg	1 μg (45°C, 1 hr.)	127	143	220
----	1 μg (60°C, 5 min.)	0	0	0
1 μg	1 μg (60°C, 5 min.)	119	113	226

The auto ADP-ribosylation reaction contained 500 μM[^{32}P] NAD^+ and other components described in text. The amount of polymer synthesized in the standard reaction containing 1 μg of purified enzyme was 263 pmol in the absence of divalent cations, 320 pmol in the presence of Mg^{2+} and 703 pmol in the presence of Ca^{2+}. In each case, these levels were set equal to 100% and changes in the level of incorporation due to the presence of additional proteins are expressed relative to the 100% values.

To explore the potential significance of the Ca^{2+} and Mg^{2+} stimulation of poly(ADP-ribose) polymerase activity at the cellular level, we examined the effect of these cations on poly(ADP-ribose) synthesis in resting human lymphocytes. These cells have previously been shown to have very low levels of poly(ADP-ribose) polymerase activity (Berger *et al.*, 1979) and were therefore useful for examining the cation effects. For these studies, freshly prepared human lymphocytes were permeabilized (Berger *et al.*, 1978a) in a buffer containing no divalent cations and then incubated with [^{32}P]-NAD in the absence of divalent cations or in the presence of either Mg^{2+} or Ca^{2+}. Figure 3 shows that both Mg^{2+} or Ca^{2+} caused a marked increase in the level of poly(ADP-ribose) synthesis such that at 15 minutes the level of poly(ADP-

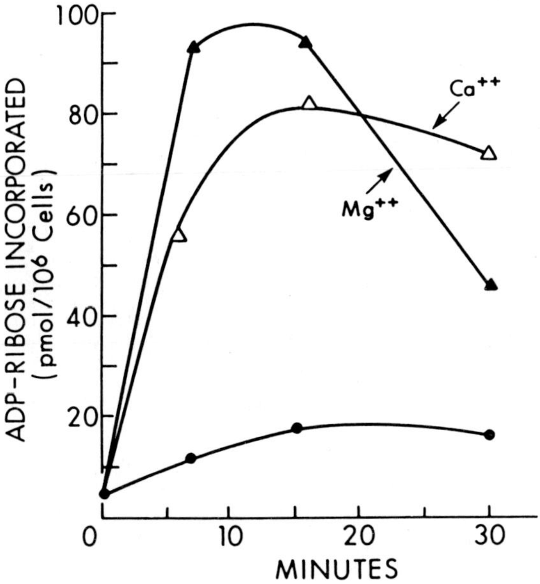

Figure 3. Effect of Mg^{2+} and Ca^{2+} on the time course of poly-(ADP-ribose) synthesis in permeabilized human lymphocytes. Reactions contained 2 x 10^6 permeable cells, and 300 μM [^{32}P]NAD^+ (specific activity 83 cpm/pmol). Incubations were at 37°C for the indicated times. (●), control-no divalent cations; (▲), 10 mM Mg^{2+}; (△), 10 mM Ca^{2+}.

ribose) synthesized in the Mg^{2+}-stimulated reactions was 5 times greater than controls and 4.4 times greater than controls in the Ca^{2+}-stimulated reactions. After 30 minutes incubation there was a decrease in the amount of poly(ADP-ribose) in both Mg^{2+}- and Ca^{2+}-treated cells. This decrease in polymer is probably due to the activities of endogenous glycohydrolases and/or phosphodiesterases which may also be affected by the presence or absence of Mg^{++} and Ca^{2+}. Thus, while the Mg^{2+} and Ca^{2+} effect in permeable cells may be much more complicated than that in the purified enzyme system, both cations appear to cause a marked stimulation of poly-(ADP-ribose) synthesis at the cellular level. It will be interesting to determine how poly(ADP-ribose) polymerase activity is affected by cellular fluxes of Mg^{2+} and Ca^{2+} and how these alterations affect the processes of DNA repair and cellular differentiation.

REFERENCES

Benjamin, R.C., and Gill, D.M. (1980). *J. Biol. Chem. 255,* 10502.

Berger, N.A., Adams, J.W., Sikorski, G.W., Petzold, S.J., and Shearer, W.T. (1978a). *J. Clin. Invest. 62,* 111.

Berger, N.A., Weber, G., and Kaichi, A.S. (1978b). *Biochim. Biophys. Acta. 519,* 87.

Berger, N.A., Sikorski, G.W., Petzold, S.J., and Kurohara, K. K. (1979). *J. Clin. Invest. 63,* 1164.

Berger, N.A., and Sikorski, G.W. (1981). *Biochemistry 20,* 3610.

Cohen, J.J., Catino, D.M., Petzold, S.J., and Berger, N.A. (1982). *Biochemistry 21,* 4931.

Hayaishi, O., and Ueda, K. (1977). *Annu. Rev. Biochem. 46,* 95.

Juarez-Salinas, H., Levi, V., Jacobson, E.L., and Jacobson, M.K. (1982). *J. Biol. Chem. 257,* 607.

Kawaichi, M., Ueda, K., and Hayaishi, O. (1981). *J. Biol. Chem. 256,* 9483.

Miwa, M., Saikawa, N., Yamaizumi, Z., Nishimura, S., and Sugimura, T. (1979). *Proc. Natl. Acad. Sci. USA 76,* 595.

Ogata, N., Ueda, K., Kawaichi, M., and Hayaishi, O. (1981). *J. Biol. Chem 256,* 4135.

Ohgushi, H., Yoshihara, K., and Kamiya, T. (1980). *J. Biol. Chem. 255,* 6205.

Poirier, G.G., DeMurcia, G., Jongstra-Bilen, J., Niedergang, C., and Mandel, P. (1982). *Proc. Natl. Acad. Sci. USA 79,* 3423.

Petzold, S.J., Booth, B.A., Leimbach, G.A., and Berger, N.A. (1981). *Biochemistry 20,* 7075.

Surowy, C.S., and Berger, N.A. (1983). *J. Biol. Chem. 258,* 579.

DISCUSSION

Kun: The mechanism of action of DNAase I has been extensively studied by Price, Shachman, Dekker and many others; and the ionic requirements have been well established. If you take DNAase and supply it only with calcium, the enzyme is inactive. If you add magnesium alone, it only produces a single-stranded cut in a double-stranded DNA and it does not go beyond that. It requires both calcium and magnesium, so your experiments would suggest, as we also confirm, that in the presence of calcium, DNA strand breaks really do not occur and yet you have an increase in the activity of the polymerase enzyme. We also see here that calcium is a potent activator of the enzyme itself, so the correlation with DNA strand breaks and the number of strand breaks here, does not seem to be substantiated.

Berger: Poly(ADP-ribose) polymerase requires DNA as an essential enzyme activator. Using well-defined DNA templates cut with restriction endonucleases to produce a precise number of strand breaks, others have shown that enzyme activity is proportional to the number of DNA strand breaks. Calcium and magnesium are not enzyme activators; rather as I've indicated today, they modulate the level of activity of enzyme already activated by DNA. In the study that we reported using DNAase-treated SV40 minichromosomes, magnesium was the only divalent cation present and it was present at a constant concentration. The variable factor was the concentration of DNAase. We followed the effect of DNAase with gel electrophoresis and these studies showed that at low concentrations of DNAase SV40 form II was gradually converted to Form I DNA, indicating the formation of single strand breaks. At higher concentrations of DNAase, form I DNA was converted to low molecular weight fragments indicating that the DNA had sustained so many single strand breaks that pieces were falling apart. The increase in Poly(ADPR) polymerase activity correlated with the change in the number of DNA strand breaks and not with the magnesium concentration which was constant.

Kun: We actually determined the number of strand breaks, and there is only one (if you use magnesium), and the number of strand breaks by various endo nucleases does not really mean that the number is the only significant factor. With various restriction enzymes, you obtain qualitatively different subunits of DNA. There is a possibility that the activation of the poly(ADP-ribose) polymerase is by various DNA restriction fragments, reflecting the qualitative composition of DNA fragments, which is quite likely, because there is a specific DNA isolated with the enzyme which is the most active DNA as an enzyme activator.

Berger: I agree. Gill's laboratory has shown that fragments with flush ends are more effective than those with staggered ends as enzyme activators. The activity of the small DNA isolated with the enzyme appears to be due to its small size, thereby providing a higher ratio of ends per unit DNA.

Jacobson: Do you have any evidence as to whether or not DNA would actually serve as an acceptor for ADP-ribosyl chains in your system?

Berger: No, we have no specific evidence. However, this has been demonstrated by Yoshihara's group in a system with purified enzyme.

Stricker: Do you have any idea as to which region of the histone H1 can be poly ADP-ribosylated?

Berger: The group in Kyoto has shown that H1 is ADP-ribosylated at 3 glutamate residues and also at the C terminal lysine. All of the ADP-ribosylated glutamate residues appear to be in polar regions of the molecule.

Stricker: Have you tried to get poly ADP-ribosylation after phosphorylation of the histone H1?

Berger: No.

Wold: The last step in the repair is the ligation, and there are ligases that require NAD; is that correct?

Berger: Eukaryotic ligases do not require NAD, whereas prokaryotic ligases do. Shall's group (C.B. Shall, Sussex) recently suggested that ADP-ribosylation activates eukaryotic ligase II; however, the mechanism of the activation has not yet been defined.

Ortez (Oklahoma State U.): What is known about the distribution or the activity of this enzyme in embryonic tissue relative to somatic tissue? What has been shown here is activation of the enzyme as a result of DNA damage. My question is, is this enzyme perhaps also active in gene rearrangement?

Berger: There is a considerable amount of work indicating that the enzyme is involved in differentiation. There are a number of studies using nicotinamide analogues, which inhibit poly(ADP-ribose) polymerase and produce developmental abnormalities in chick embryos. Kaplan (Northwestern) has shown that NAD levels are important in determining whether mesenchymal cells will differentiate into muscle or cartilage; and Shall's group has shown that conversion of myoblasts to myotubes involved DNA strand breakage and that activation of poly(ADP-ribose) polymerase occurs in this process and if one can block poly(ADP-ribose) one blocks muscle differentiation. Finally, Koide's group has evidence that ADP-ribosylation is involved in early stages of development of sea urchin eggs.

So, there certainly seems to be a requirement for poly(ADP-ribose) synthesis in differentiation and development.

Olson: You mentioned that ADP-ribosylation of proteins, presumably histones, alters chromosome structure. What happens? Is the nucleosome disturbed; is the superhelix exchanged? Do you have any ideas?

Berger: At the moment, the best indication of what happens comes from the electron micrographs from the Strasbourg group. They have shown that ADP-ribosylated nucleosomes are more spread out and resistant to salt-induced condensation.

Floyd (OMRF): In the ribosylation, I would expect the protein to be modified conformationally in the same way that hydrophilic surfaces are on the outside of proteins. Possibly ribosylation may be one of the ways it is possible to get conformational change in the chromatin.

Berger: Are you talking about the enzyme itself?

Floyd: Or whatever is ribosylated. If it is H1 or whatever it is, you would *a priori* assume that there would be more hyrophilic surfaces on the surface of the protein which may be compelling the protein to unfold in some manner.

Berger: I agree, it is possible that ADP-ribosylation of the proteins alters their hydrophobicity. In addition, ADP-ribosylation of the proteins alters the charge on the protein which should affect its binding to DNA as well as the structure of the chromatin.

ADP-RIBOSYLATION OF THE REGULATORY COMPONENTS OF ADENYLATE CYCLASE[1,2]

Richard A. Kahn
Toshiaki Katada
Gary M. Bokoch
John K. Northup
Alfred G. Gilman

Department of Pharmacology
University of Texas Health Science Center at Dallas
Dallas, Texas

Adenylate cyclase, first described in 1962 by Sutherland and Rall (Sutherland, Rall, and Menon, 1962; Rall and Sutherland, 1962; Murad *et al.*, 1962; Kleiner *et al.*, 1962), is the membrane-bound enzyme that catalyzes the conversion of ATP to cyclic AMP in response to hormones or other activating ligands. Ten years after the initial description of the enzyme, Rodbell and coworkers (1971) demonstrated an absolute requirement for guanine nucleotides (in particular, GTP) for hormonal stimulation of cyclic AMP synthesis and later showed that nonhydrolyzable analogs of GTP (e.g. GTPγS or Gpp(NH)p) were powerful stimulators of adenylate cyclase activity in the absence of hormones (Londos *et al.*, 1974). Since little was known about the structure of adenylate cyclase at that time, it seemed possible that the enzyme consisted of a single polypeptide with binding sites for hormone, guanine nucleotide, and substrate. This simple, one component model became untenable, however, when the regulatory properties of the system were resolved from its catalytic function.

[1]*This work was supported by United States Public Health Service Grant NS18153 and by a grant (BC240D) from the American Cancer Society.*
[2]*A list of abbreviations used, is included at the end of this chapter.*

ISBN 0-12-387560-9

Bourne, Coffino, and Tomkins (1975) exploited the fact that cyclic AMP is cytocidal to the S49 mouse lymphoma cell line to select mutants with alterations of cyclic AMP metabolism or action. One important mutant, called AC^- or cyc^-, retained its complement of hormone receptors but appeared to be deficient in the ability to synthesize cyclic AMP (Insel *et al.*, 1976). These results, which were soon coupled with the ability to separate hormone binding and catalytic activities physically (Limbird and Lefkowitz, 1977; Haga, Haga, and Gilman, 1977), clearly indicated that hormone-sensitive adenylate cyclase consisted of at least two types of components -- hormone receptors and a catalytic unit. Pfeuffer (1977) then demonstrated that a detergent-solubilized preparation of adenylate cyclase could be further resolved into two components. One of these, which was devoid of catalytic activity, bound to GTP-Sepharose and could then be released from the affinity support with GTP or Gpp(NH)p; the other fraction did not bind to GTP-Sepharose and contained a modest level of catalytic activity which was reduced in its ability to be stimulated by Gpp(NH)p. Recombination of these two fractions restored Gpp(NH)p-stimulated adenylate cyclase activity to its original level.

The membrane solubilization and reconstitution experiments of Ross and Gilman (Ross and Gilman, 1977a; Ross and Gilman, 1977b; Ross *et al.*, 1978; Ross and Gilman, 1980) extended these results and clearly demonstrated that hormone-sensitive adenylate cyclase consists of a minimum of three components -- hormone receptors, a regulatory subunit which binds guanine nucleotides, and a catalytic subunit. The catalytic moiety is essentially devoid of activity (with MgATP as substrate) in the absence of the regulatory protein.

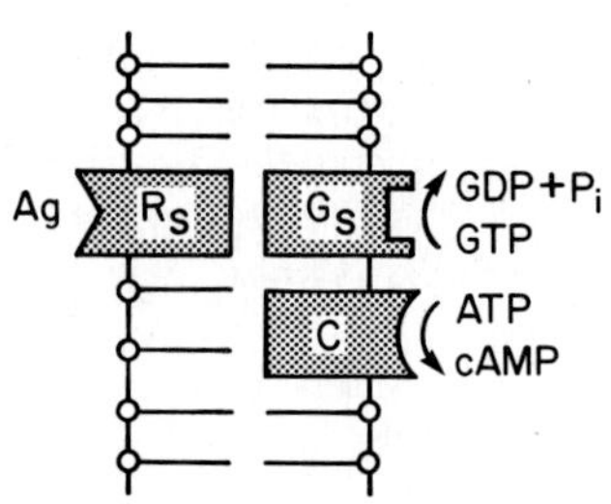

FIGURE 1. Three-Component Model for Hormone-Stimulated Adenylate Cyclase. Abbreviations are: R_S, stimulatory hormone receptor; G_S, guanine nucleotide-binding regulatory component modulating stimulation of adenylate cyclase; C, catalytic moiety of adenylate cyclase.

This three-component model is shown in Figure 1, in which a stimulatory hormone receptor (e.g. β-adrenergic) is abbreviated R_s, the regulatory component G_s, and the catalyst, C. The cyc^- cell line mentioned above was found to lack the regulatory subunit G_s, which explains the lack of hormone- or guanine nucleotide-sensitive adenylate cyclase and the very low level of basal enzymatic activity. In addition to providing an explanation for the cyc^- phenotype, these findings provided the basis for a reconstitutive assay for G_s, which greatly facilitated the purification of this regulatory protein to homogeneity (Northup *et al.*, 1980; Sternweis *et al.*, 1981b; Hanski *et al.*, 1981). Lane 1 of Figure 2 is a sodium dodecyl sulfate-polyacrylamide gel (Coomassie blue stain) of pure G_s from rabbit liver membranes; 3 bands are evident, corresponding to molecular weights of 35,000, 45,000 and 52,000. Pure G_s from human or turkey erythrocyte mem-

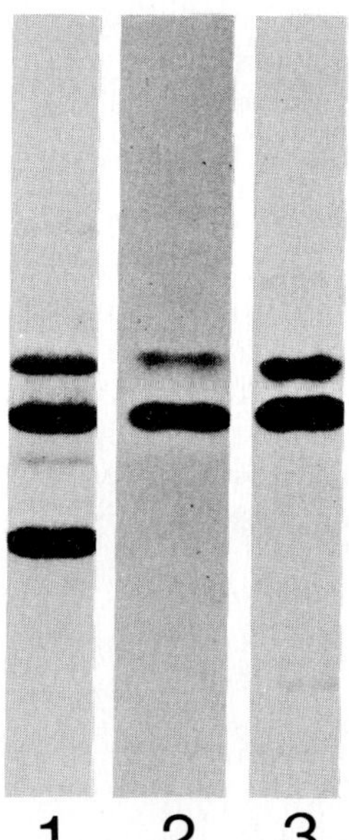

FIGURE 2. Purified Regulatory Component from Rabbit Liver Membranes. Lane 1 shows the Coomassie Blue staining pattern of purified G_s (8 µg) from rabbit liver membranes after electrophoresis in a sodium dodecyl sulfate-polyacrylamide slab gel. Lanes 2 and 3 are autoradiograms showing the cholera toxin-catalyzed labeling pattern of G_s after reconstitution into cyc^- membranes as described by Northup et al. *(1980). Autoradiograms of a single sample were exposed for either 16 hrs (lane 2) or 48 hrs (lane 3).*

branes has the 35,000- and 45,000-dalton subunits but lacks the 52,000-dalton polypeptide (Hanski *et al.*, 1981; Hanski and Gilman, 1983). Hydrodynamic studies (Sternweis *et al.*, 1981b; Hanski *et al.*, 1981) and other data (Northup, Smigel, and Gilman, 1982) suggest that G_s is a heterodimer of approximately 80,000 daltons, consisting of a 35,000-dalton peptide and one 45,000- or 52,000-dalton peptide.

Workers in other laboratories, in particular those of Vaughan (Moss, Manganiello, and Vaughan, 1976; Moss and Vaughan, 1979; Moss and Vaughan, 1981) and Gill (Gill, 1975; Gill and Meren, 1978; Enomoto and Gill, 1979), made notable contributions to the field by their studies of the mechanism of action of cholera toxin. It was demonstrated that the persistent activation of adenylate cyclase caused by the toxin was the result of an inherent ADP-ribosyltransferase activity. Using NAD as donor, cholera toxin was found to catalyze the specific ADP-ribosylation of a membrane protein with a molecular weight of approximately 45,000. A good correlation was seen between the extent of labeling of this band (when [^{32}P]NAD was used) and the extent of activation of adenylate cyclase by toxin. In addition, Pfeuffer's work suggested that the protein that bound to GTP-Sepharose was likely to be the same as the 45,000-dalton polypeptide that could be radiolabeled with cholera toxin and [^{32}P]NAD (Cassel and Pfeuffer, 1978). We were thus not surprised to find that pure G_s is an excellent substrate for ADP-ribosylation by cholera toxin when reconstituted into cyc^- membranes, as seen in lanes 2 and 3 of Figure 2 (Northup *et al.*, 1980). The 45,000- and 52,000-dalton subunits of G_s are substrates for the toxin, while the 35,000-dalton polypeptide is not. The larger subunits of G_s also contain the site for binding of guanine nucleotides (Northup, Smigel, and Gilman, 1982). The next section details the requirements for cholera toxin-dependent ADP-ribosylation of the purified guanine nucleotide-binding regulatory component, G_s.

CHOLERA TOXIN-DEPENDENT ADP-RIBOSYLATION OF G_s

We have utilized two methods to assay cholera toxin-dependent ADP-ribosylation of pure G_s, as illustrated in Figure 3. The first is to use [α^{32}P]NAD as the ADP-ribose donor, thereby tagging the peptide substrate. The labeled products are then separated on sodium dodecyl sulfate-polyacrylamide gels and visualized by autoradiography. For quantitation, the gel bands can be excised and counted. Alternatively, one can monitor the effect of ADP-ribosylation on the

Method 1: Autoradiography

$$G/F + [^{32}P]NAD^{+} \xrightarrow[\text{Factor}]{\text{Activated cholera toxin, GTP}} [^{32}P]G/F$$

$\downarrow$

SDS polyacrylamide gel electrophoresis and autoradiography

Method 2: Two-Stage Adenylate Cyclase Assay

Stage 1:

$$G/F + NAD^{+} \xrightarrow[\text{Factor}]{\text{Activated cholera toxin, GTP}} \text{Activated } G/F$$

Stage 2:

$$\text{Activated } G/F + [^{32}P]ATP \xrightarrow[\text{cyc}^{-}\text{ membranes}]{\text{GTP}} [^{32}P]\text{cyclic AMP}$$

FIGURE 3. Methods for Assay of Cholera Toxin-Catalyzed ADP-Ribosylation of Purified G_s.

activity of G_s. When the regulatory protein is not covalently modified, the adenylate cyclase activity of a mixture of GTP, G_s, and the catalyst is low but can be stimulated markedly by nonhydrolyzable guanine nucleotide analogs such as Gpp(NH)p or GTPγS. When G_s is ADP-ribosylated, however, GTP is essentially as effective as the non-hydrolyzable analogs, presumably because of inhibition of a GTPase (Cassel and Selinger, 1977; Cassel, Levkovitz, and Selinger, 1977). Thus, assay of GTP-dependent adenylate cyclase activity serves as a sensitive index of the extent of ADP-ribosylation of G_s (Schleifer *et al.*, 1982). In practice, this entails a two-step procedure, shown as Method 2. Pure G_s is treated in detergent-containing solution with cholera toxin, NAD, and other required factors. This modified and activated G_s is then reconstituted into cyc^- membranes, which lack G_s but contain the catalyst, and GTP-dependent adenylate cyclase activity is measured as the rate of production of [^{32}P]cyclic AMP from [α^{32}P]ATP. These two methods have always given the same results in our laboratory. It is clear from Figure 4 that the extent of activation of GTP-dependent adenylate cyclase activity correlates very well with the incorporation of [^{32}P]ADP ribose into G_s.

Surprisingly, the ability of activated cholera toxin to catalyze the ADP-ribosylation of purified rabbit liver G_s in detergent-containing solutions was found to be poor. The reaction occurred extremely slowly, as judged by either the incorporation of radioactivity from [^{32}P]NAD into G_s or the

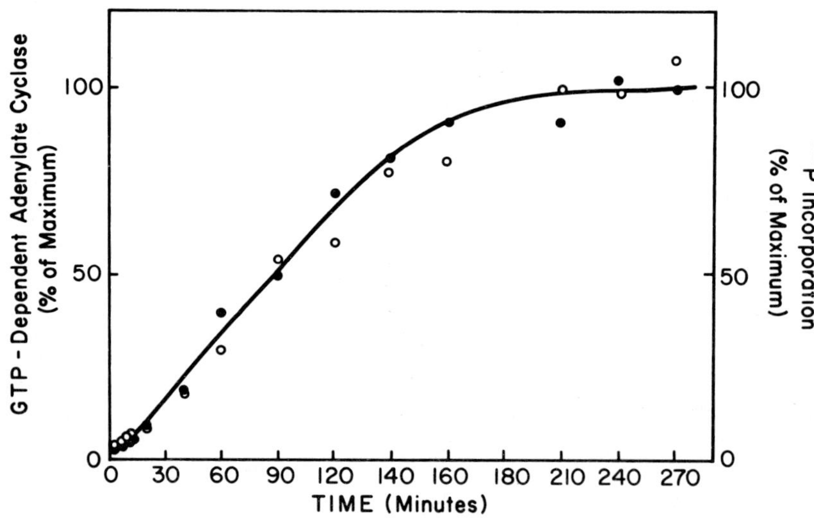

FIGURE 4. Activation and ADP-Ribosylation of G_s by Cholera Toxin. Purified G_s was treated with cholera toxin (80 µg/ml) and [^{32}P]NAD (10 µM, 10,000 cpm/pmol) at 30°C. Aliquots were removed and processed either for quantitative sodium dodecyl sulfate-polyacrylamide gel electrophoresis or for assay of GTP-dependent adenylate cyclase activity. (From Sternweis et al., *1981a.)*

enhancement of the GTP-dependent reconstitutive activity of G_s. However, as noted above, if purified G_s is first reconstituted into cyc⁻ membranes, which lack G_s, and is subsequently treated with activated cholera toxin and NAD, covalent modification is markedly enhanced. Initial experiments indicated that a factor present in detergent extracts of these membranes would facilitate the toxin-catalyzed modification of pure G_s (Schleifer *et al.*, 1982). It thus became desirable to define the requirements for ADP-ribosylation of pure G_s by cholera toxin and the nature of the component(s) supplied by cyc⁻ membranes.

After the second step in the purification of G_s (gel filtration chromatography), it was observed that the ability to label G_s with [^{32}P]NAD and cholera toxin was lost. We looked for and found fractions from this column that have no assayable G_s activity but that restore the ability of G_s to be ADP-ribosylated by cholera toxin (Schleifer *et al.*, 1982). This factor has been named ADP-ribosylation factor, or ARF, and its sole function discovered to date is to confer upon G_s the ability to be covalently modified by cholera toxin. ARF is heat labile, protease-sensitive, and N-ethylmaleimide-insensitive. Figure 5 shows the requirements for toxin-dependent modification of pure G_s from two sources -- rabbit liver and turkey erythrocytes. The striking effect of ARF is obvious. The results shown in Figure 5 were obtained with partially purified ARF from turkey erythrocyte membranes. Identical results are obtained with ARF from rabbit liver membranes. Also shown is the effect of omission of GTP. Gill and others (Enomoto and Gill, 1979; Nakaya, Moss, and Vaughan, 1980) have shown a GTP requirement for cholera toxin-catalyzed ADP-ribosylation of G_s in membranes. We have found that GTP stimulates the covalent modification of pure G_s by only about 2-3 fold. It is not known why GTP enhances ADP-ribosylation of purified G_s, but GTP does appear to be the preferred nucleotide. Also seen in this figure is the effect of dimyristoyl phosphatidylcholine. The presence of detergents in the activation reaction was found to inhibit the toxin-dependent ADP-ribosylation (Lubrol PX and cholate are added with the G_s and ARF, respectively). Lipids were added in an attempt to lower the effective concentration of detergent. Although factor activity cannot be supplied by the addition of lipid to purified G_s, the addition of certain phospholipids or cholesterol to reaction mixtures that contain ARF increases the rate of the reaction. This effect of added lipid can be met equally well by dimyristoyl phosphatidylcholine, dimyristoyl phosphatidylethanolamine, dipalmitoyl phosphatidylethanolamine, or cholesterol. Although originally added to lower the effective concentration of de-

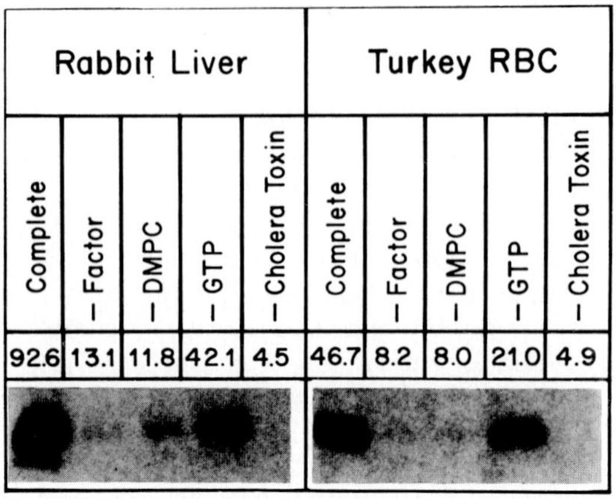

FIGURE 5. Requirements for Cholera Toxin-Dependent Labeling of Purified Rabbit Liver or Turkey Erythrocyte G_s. Purified rabbit liver or turkey erythrocyte G_s (3.4 μg/ml) was incubated at 30°C for 60 min in the presence of [^{32}P]NAD (10 μM, 10,000 cpm/pmol), activated cholera toxin (40 μg/ml), GTP (100 μM), dimyristoyl phosphatidylcholine (DMPC, 3 mM), and partially purified ARF (factor) from turkey erythrocytes (2 μg/ml) in the complete reaction. Effects of omission of each of these components are shown. The autoradiogram of the 45,000-dalton region is shown at the bottom and the numbers in each lane represent the GTP-stimulated adenylate cyclase activity (pmol/15 min/ng G_s) that can be reconstituted with G_s treated similarly. These reactions were the same as those for labeling of G_s, with the exception of omission of labeled NAD and elevation of the concentration of NAD to 100 μM. (From Schleifer et al., *1982.)*

tergent in solution, we do not as yet know the mechanism of this effect of added lipids. Thus, the complete reaction includes phosphate buffer, G_s, NAD, GTP, dimyristoyl phosphatidylcholine, ARF, and cholera toxin.

The activation of G_s as a function of time is shown in Figure 6A. It is clear that the activation is absolutely dependent on cholera toxin and is markedly accelerated by ARF. Under these conditions the rate of activation of G_s in the presence of ARF is constant for at least 30 minutes. When initial rates are plotted as a function of the concentration of ARF in the activation reaction, a linear relationship is obtained (Figure 6B). This can be used as a quantitative assay for ARF.

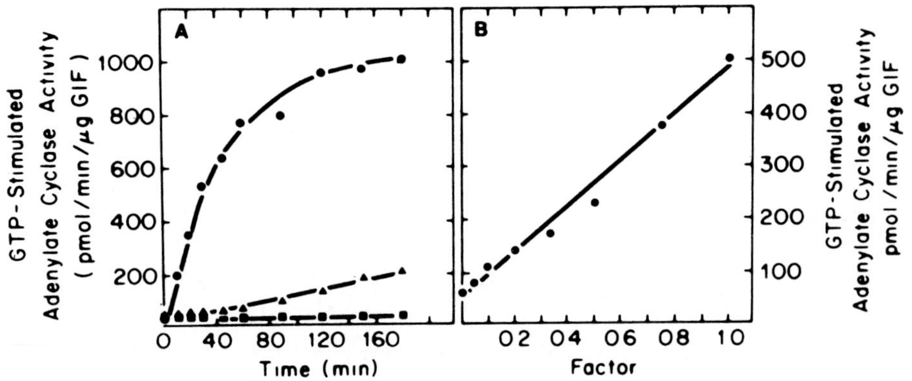

FIGURE 6. Time Course of Activation of G_S by Cholera Toxin and the Effect of ARF on the Initial Rate of Activation. A) Purified rabbit liver G_S (10 µg/ml) was incubated with cholera toxin (40 µg/ml) in a complete reaction mixture (●) or when ARF (▲) or cholera toxin (□) were omitted. At various times aliquots were stopped by dilution into cold buffer and assayed for the ability to reconstitute GTP-stimulated adenylate cylase activity. B) The amount of ARF (factor) in the complete reaction mixture was varied and the reaction was stopped after 20 min at 30°C and reconstituted GTP-stimulated adenylate cyclase activity was determined. A factor concentration of 1.0 represents 5.5 µg/ml of protein, and dilutions of this were made with buffer containing cholate (0.85%) such that the amount of detergent in the reaction was constant. (From Schleifer et al., *1982.)*

ARF has been found in the membrane fraction of every tissue or cell tested so far, including the S49 mouse lymphoma line, rabbit liver, and turkey and human erythrocytes. The factor from the cyc⁻ cell is firmly associated with the membrane, since complete solubilization requires at least 0.8% cholate. ARF activity is not extracted from these membranes with either 0.5 M NaCl or 3 mM EDTA in the absence of detergent. Thus, ARF is distinct from the cytoplasmic factor described by Enomoto and Gill (1979), which lowers the concentration of cholera toxin required to activate adenylate cyclase in pigeon erythrocyte membranes.

A considerable amount of effort has been spent trying to purify ARF as a prelude to attempts to determine its mechanism of action and to test whether or not ARF serves any other role in the regulation of adenylate cyclase activity. A partial purification scheme for ARF is shown in Table I.

TABLE I. Partial Purification of ARF from Rabbit Liver Plasma Membranes[a]

	Vol.	Activity Units/ml	Specific Activity Units/mg	Protein mg/ml	Recovery %	Fold Purification
Cholate Extract	1720	393	83	4.72	100	1
Pool I	234	2540	2230	1.14	88	27
AcA44	102	2480	6050	0.41	37	73
Heptylamine Sepharose	86	970	32,300	0.03	12	390

[a]*A cholate extract of crude rabbit liver membranes was prepared as described by Sternweis* et al., *1981b. Fractions containing ARF activity from the first two steps in the purification of* G_s *(DEAE-Sephacel and Ultrogel AcA34; see Sternweis* et al., *1981b) were pooled, concentrated by ultrafiltration, and chromatographed on Ultrogel AcA44. The AcA44 pool was loaded onto the heptylamine-Sepharose column and eluted with a cholate gradient from 0.4 to 1.8%.*

Starting with a cholate extract of crude rabbit liver membranes, we were able to purify ARF about 400-fold after ion exchange, gel filtration, and hydrophobic chromatography. We have recently utilized one additional step which appears to yield a homogenous peptide. After noting that ARF activity is markedly stabilized by sodium dodecyl sulfate, preparative sodium dodecyl sulfate-polyacrylamide gel electrophoresis has been employed for purification. The pool of ARF activity from the heptylamine-Sepharose column is concentrated on hydroxyapatite and applied to a preparative polyacrylamide slab gel in the presence of 0.1% sodium dodecyl sulfate. After electrophoresis, the gel is sliced horizontally and proteins are allowed to diffuse out of the polyacrylamide. Recovery of activity is excellent and all the ARF activity runs in a single band on the preparative gel. The preparative gel system described yields roughly a 20-fold increase in specific activity, making it likely that pure ARF is obtainable with a yield of 5 to 10% after a 6,000- to 8,000-fold purification from cholate extracts of rabbit liver membranes. The availability of this preparation should greatly facilitate analysis of the mechanism of action of this intriguing protein.

ISLET ACTIVATING PROTEIN AND G_i

Hormone-dependent inhibition of adenylate cyclase activity was described early in the course of investigation of the enzyme. Jakobs, Saur, and Schultz (1978) subsequently demonstrated a requirement for GTP for hormone-dependent inhibition of adenylate cyclase, similar to that for activation of the enzyme. The obvious question is whether G_s somehow mediates both stimulation and inhibition of the catalyst or if there exists yet another component of the system -- an inhibitory guanine nucleotide-binding regulatory protein. The latter possibility is illustrated in Figure 7. This five-component model has recently received considerable support from Ui and coworkers, as a result of their studies of the mechanism of action of islet-activating protein (IAP), one of the toxins of *Bordetella pertussis*.

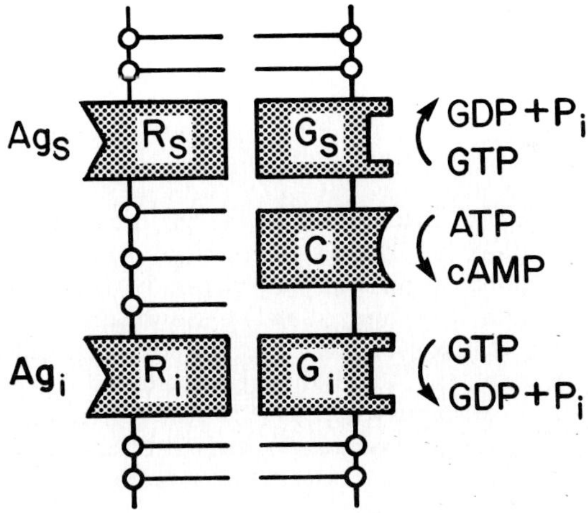

FIGURE 7. Five-Component Model for Adenylate Cyclase Showing Both Stimulatory and Inhibitory Pathways. R_s, stimulatory hormone receptor; R_i, inhibitory hormone receptor; G_s, regulatory component mediating stimulation of adenylate cyclase; G_i, regulatory component mediating inhibition of adenylate cyclase; C, catalytic moiety.

IAP was named as a result of initial findings that injection of the toxin into rats led to sustained potentiation of insulin secretion in response to a number of secretagogues (Sumi and Ui, 1975; Katada and Ui, 1977). In a beautiful series of papers Katada and Ui (Katada and Ui, 1981; Katada, Amano, and Ui, 1982; Katada and Ui, 1982a; Katada and Ui, 1982b) have made the following observations about the mechanism of action of this interesting toxin: 1) Treatment of several types of cells with IAP blocks hormone-induced inhibition of adenylate cyclase. This is true for a variety of inhibitory systems (e.g. α-adrenergic, muscarinic, opioid). IAP does not interfere with the binding of agonists to these receptors. 2) IAP potentiates the increases in cyclic AMP concentrations in C6 glioma cells caused by β-adrenergic agonists in a manner distinct from that of cholera toxin. 3) IAP utilizes NAD to ADP-ribosylate a membrane protein with a molecular weight of 41,000; this protein is distinct from the substrate for cholera toxin. 4) A good correlation exists between the extent of ADP-ribosylation of this 41,000-dalton protein and the effects of IAP on cyclic AMP concentrations and adenylate cyclase activities. Katada and Ui have hypothesized that G_i and the substrate for IAP may be the same protein. We have recently purified the substrate for IAP from rabbit liver membranes and have begun its characterization (Bokoch *et al.*, 1983).

IAP-DEPENDENT ADP-RIBOSYLATION

During the purification of G_s from rabbit liver it was observed that a peak of protein consisting predominantly of 41,000- and 35,000-dalton polypeptides elutes just prior to the peak of G_s from the heptylamine-Sepharose column. With the knowledge of the molecular weight of the IAP substrate described by Katada and Ui, we attempted o label this partially purified fraction with IAP. The result is shown in Figure 8. When this preparation was incubated with IAP and [^{32}P]NAD, there was rapid, covalent incorporation of [^{32}P] ADP-ribose, and labeling was confined to the 41,000-dalton polypeptide (lane 7). The labeling is totally dependent on toxin (lane 5) and is not affected by the presence of ARF (lane 6). Cholera toxin does not label the 41,000-dalton polypeptide either in the presence or the absence of ARF (lanes 4 and 3, respectively). In contrast, the 45,000- and 52,000-dalton subunits of G_s are labeled by cholera toxin when ARF is included (lane 10). This labeling is dependent both on cholera toxin and ARF (lanes 11 and 12), as shown

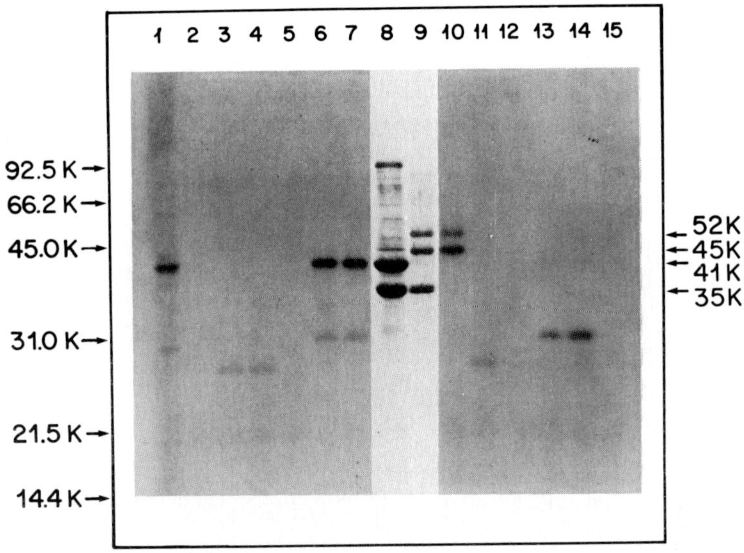

FIGURE 8. Autoradiographic Analysis of Radioactive Products Resulting from IAP- or Cholera Toxin-Catalyzed ADP Ribosylation of Rabbit Liver Membranes and Purified Protein Fractions. Sample proteins were incubated with [^{32}P]NAD in the presence of IAP or activated cholera toxin as indicated below. Lane 1, rabbit liver membranes plus 20 µg/ml IAP; Lane 2, 41,000/35,000-dalton protein fraction; lane 3, 41,000/35,000-dalton protein plus 50 µg/ml cholera toxin; lane 4, 41,000/35,000-dalton protein plus ARF plus 50 µg/ml cholera toxin; lane 5, 41,000/35,000-dalton protein plus ARF; lane 6, 41,000/35,000-dalton protein + ARF + 20 µg/ml IAP; lane 7, 41,000/35,000-dalton protein + 20 µg/ml IAP; lane 8, Coomassie blue stain of 41,000/35,000-dalton protein fraction; lane 9, Coomassie blue stain of G_s; lane 10, G_s + ARF + 50 µg/ml cholera toxin; lane 11, G_s + 50 µg/ml cholera toxin; lane 12, G_s + ARF; lane 13, G_s + ARF + 20 µg/ml IAP; lane 14, G_s + 20 µg/ml IAP; lane 15, G_s. Labeled bands with lower molecular weights correspond to the A_1 subunit of cholera toxin (lanes 3, 4, and 11) and to the S_1 subunit of IAP (lanes 6, 7, 13, and 14). (From Bokoch et al., *1983).*

previously. There is no labeling of G_s by IAP either in the absence (lane 13) or presence (lane 14) of ARF. Thus, the two toxins utilize distinct protein substrates and only the cholera toxin-catalyzed labeling of G_s is dependent on ARF.

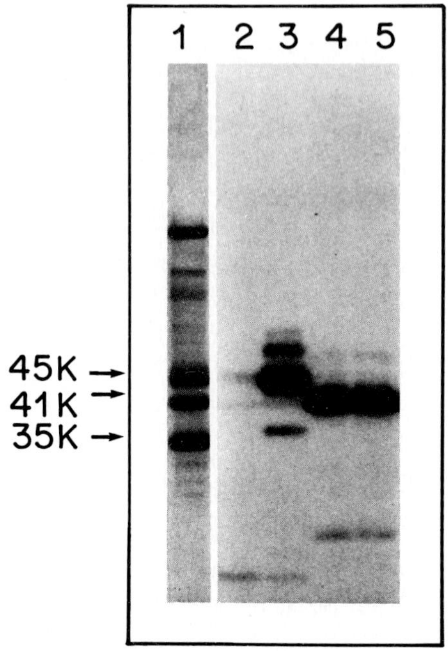

FIGURE 9. Autoradiographic Analysis of Labeling of the 41,000-Dalton Protein by IAP. Proteins were labeled with [^{32}P]NAD and toxins as described below. Lane 1, Coomassie blue stain of a partially purified protein sample containing a mixture of G_s and the 41,000/35,000-dalton substrate for IAP; lane 2, plus 50 μg/ml cholera toxin; lane 3, plus ARF and 50 μg/ml cholera toxin; lane 4, plus 20 μg/ml S_1 subunit of IAP; lane 5, plus ARF and 20 μg/ml S_1 subunit of IAP. (From Bokoch et al., *1983.)*

The specificity of these two toxins is further demonstrated in Figure 9, which shows ADP-ribosylation of a fraction containing both G_s and the IAP substrate. It is clear that each toxin recognizes only the one substrate and that the effect of ARF is specific for the cholera toxin-dependent ADP-ribosylation of G_s.

We have good evidence that the 41,000-dalton peptide that is labeled by IAP contains a guanine nucleotide binding site, similar to that found on the 45,000-dalton subunit of G_s. The photoaffinity probe 8-azido[γ^{32}P]GTP, first synthesized

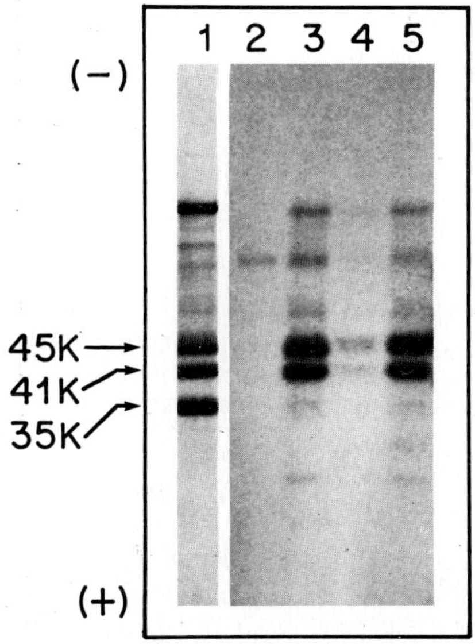

FIGURE 10. Photoaffinity Labeling of G_s and the 41,000/35,000-Dalton Protein with 8-azido [^{32}P]GTP. Labeling of soluble proteins with 8-azido[^{32}P]GTP was carried out as described (Northup, Smigel, and Gilman, 1982). Lane 1, Coomassie blue stain of the protein sample containing a mixture of G_s and the 41,000/35,000-dalton protein; lane 2, sample plus 8-azido[^{32}P]GTP with no ultraviolet exposure; lane 3, with ultraviolet exposure; lane 4, photolyzed in the presence of 1 mM GTP S; lane 5, photolyzed in the presence of 1 mM ATP. After incubation with 8-azido[^{32}P]GTP, samples were run on a standard 11% Laemmli gel and prepared for autoradiographic analysis. (From Bokoch et al., *1983.)*

and described by Czarnecki, Geahlen, and Haley (1979), can be utilized to label GTP-binding proteins covalently. As seen in Figure 10, when a fraction containing both the 45,000-dalton subunit of G_s and the 41,000-dalton polypeptide was photolyzed in the presence of 8-azido[γ^{32}P]GTP, there is prominent labeling of both of these proteins. Labeling was blocked by the addition of GTPγS (lane 4) but was unaffected by ATP (lane 5). After removal of G_s from this preparation we were able to examine this guanine nucleotide binding site more quantitatively with [^{35}S]GTPγS; this ligand binds to the purified substrate for IAP with a K_D of approximately 30 nM. Binding was not affected by inclusion of 1 mM App(NH)p but

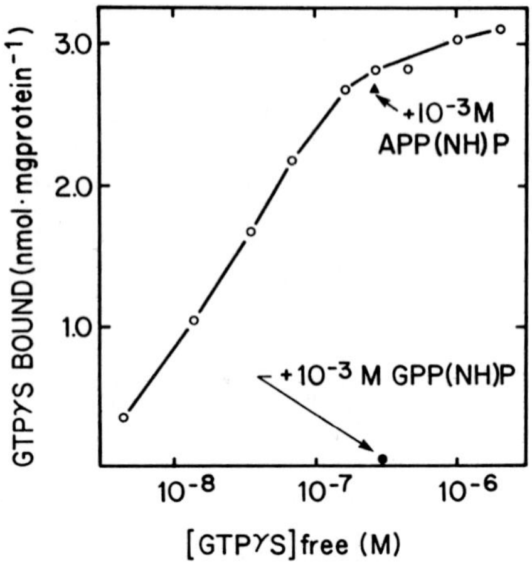

FIGURE 11. Binding of [^{35}S]GTPγS to the 41,000/35,000-Dalton Protein. Binding of [^{35}S]GTPγS was measured as described by Northup, Smigel, and Gilman (1982). The protein fraction utilized was purified to an extent comparable to that seen in Figure 8 (lane 8). The open circles *indicate binding of [^{35}S]GTPγS; the* solid circle *indicates binding of [^{35}S]GTP S in the presence of 1 mM Gpp(NH)p; the* solid triangle *indicates binding of [^{35}S]GTPγS in the presence of 1 mM App (NH)p.*

was abolished by the same amount of Gpp(NH)p (Figure 11). The stoichiometry of labeling of the IAP substrate with either [^{32}P]ADP-ribose or with [^{35}S]GTPγS suggests that there is one site for ADP-ribosylation per GTPγS binding site.

Hydrodynamic characterization of the IAP substrate indicates that it behaves as an 80,000-dalton species in the absence of activating ligands. However the apparent molecular weight is reduced to 50,000 following incubation with either GTPγS or fluoride (Table II). This behavior is essentially identical to that seen with G_s. In addition, the 41,000- and the 35,000-dalton polypeptides in preparations of the IAP substrate cofractionate under a variety of conditions. It seems very likely that the substrate for IAP exists in solution as a dimer of 41,000- and 35,000-dalton subunits. The relationship between the 35,000-dalton subunit of G_s and that of the IAP substrate is under investigation. Studies performed to date indicate that these proteins are very similar and may, in fact, be identical. Evidence thus favors the hypothesis that the IAP substrate is the inhibi-

TABLE II. Hydrodynamic Parameters of the Substrate for IAP[a]

Parameter	Incubation None	Prior to AMF[b]	Analysis GTPγS
Stokes radius, a (nm)	4.43±0.02	3.61±0.05	3.73±0.06
Sedimentation coefficient $s_{20,w}$ (S)	4.31±0.12	3.29±0.17	3.21±0.18
Molecular weight (M_r)	82,000	51,000	51,000
Frictional coefficient, f/fo[c]	1.42	1.35	1.40

[a]Values shown are the means ± standard deviations of 3 separate experiments. Stokes radii (assessed by gel filtration) and sedimentation coefficients were determined by interpolation from graphs of the migration of standard proteins. Assays for IAP substrate activity and for [^{35}S]GTPγS binding activity were performed as described in reference. A value of 0.735 ml/g for the partial specific volume (v) was assumed. (From Bokoch et al., 1983.)

[b]ATP, Mg^{2+}, and F^-

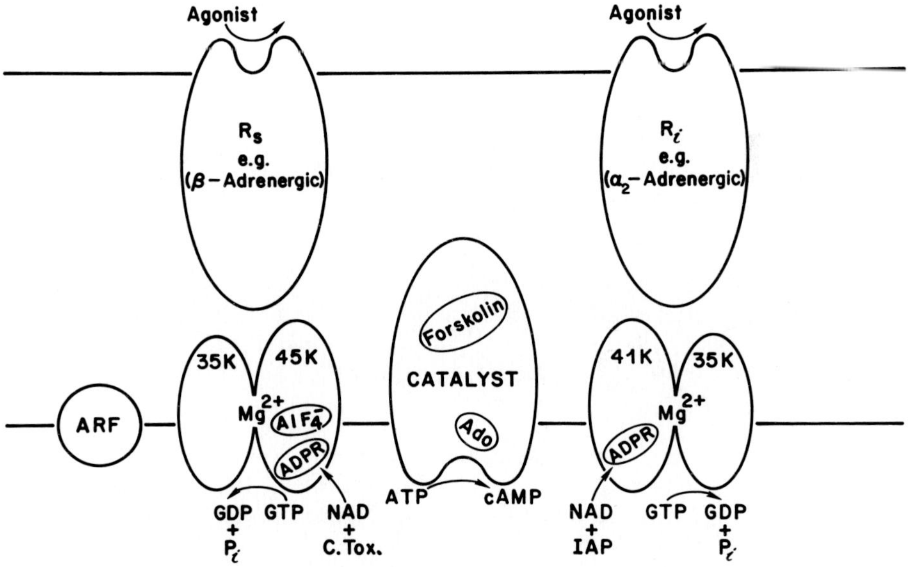

FIGURE 12. Proposed Model Depicting Components of Adenylate Cyclase.

tory guanine nucleotide-binding regulatory component of adenylate cyclase. However, this concept cannot be verified until we develop a reconstitutive assay for this protein and demonstrate an effect of IAP-dependent ADP-ribosylation on its activity. Though not yet complete, we believe that the concept of a five component model of hormone-sensitive adenylate cyclase is a tenable one. These components appear to represent the minimal number required to explain the stimulatory and inhibitory pathways that regulate the rate of synthesis of cyclic AMP. The symmetry of this model is evident in Figure 12. The regulatory components G_s and G_i show remarkable similarities (guanine nucleotide binding, heterodimer composition, dissociation of dimer by activating ligands), even to the extent that they appear to share a very similar, if not identical, 35,000-dalton subunit. In addition, both are highly specific substrates for individual bacterial ADP-ribosyltransferases. These toxins have obviously proven to be very useful in the identification, purification, and characterization of G_s and G_i.

ACKNOWLEDGMENTS

We wish to thank Frank Roganti for technical assistance and Wendy Deaner for excellent editorial assistance.

ABBREVIATIONS USED

App(NH)p -	5'-adenylimidodiphosphate
ATP -	adenosine-5'triphosphate
AMP -	adenosine-5'-monophosphate
GTP -	adenosine-5'-triphosphate
cAMP -	cyclic AMP - adenosine-3':5'-monophosphate
GTPγS -	guanosine-5'-0-(3-thiotriphosphate)
Gpp(NH)p -	5'-Guanylylimidodiphosphate
S49 -	mouse-lymphoma cell line
AC^- or cyc^- -	a variant of the S49 lymphoma mouse cell line that is devoid of G_s, (Insel et al., 1976). This mutant is unable to synthesize cyclic AMP
R_s -	stimulatory hormone receptor (e.g. β-adrenergic)
R_i -	inhibitory hormone receptor (e.g. α-adrenergic)
G_s -	regulatory component of adenylate cyclase = G/F as frequently used in the literature
G_i -	adenylate cyclase inhibitor

C - catalytic subunit of adenylate cyclase
^{32}P - radioactive phosphorus mass 32
NAD - nicotinamide-adenine dinucleotide
ARF - ADP - ribosylation factor
IAP - islet-activating protein

REFERENCES

Bokoch, G.M., Katada, T., Northup, J.K., Hewlett, E.L., and Gilman, A.G. (1983). *J. Biol. Chem. 258,* 2072.
Bourne, H.R., Coffino, P., and Tomkins, G.M. (1975). *Science 187,* 750.
Cassel, D., and Selinger, Z. (1977). *Proc. Natl. Acad. Sci. USA 74,* 3307.
Cassel, D., Levkovitz, H., and Selinger, Z. (1977). *J. Cyclic Nucl. Res. 3,* 393.
Cassel, D., and Pfeuffer, T. (1978). *Proc. Natl. Acad. Sci. USA 75,* 2669.
Czarnecki, J., Geahlen, R.L., and Haley, B.E. (1979). *Methods Enzymol.56,* 642.
Enomoto, K., and Gill, D.M. (1979). *J. Supramol. Struct. 10,* 51.
Gill, D.M. (1975). *Proc. Natl. Acad. Sci. USA 72,* 2064.
Gill, D.M., and Meren, R. (1978). *Proc. Natl. Acad. Sci. USA 75,* 3050.
Haga, T., Haga, K., and Gilman, A.G. (1977). *J. Biol. Chem. 252,* 5776.
Hanski, E., and Gilman, A.G. (1983). *J. Cyclic Nucleotide Res. 8,* 323.
Hanski, E., Sternweis, P.C., Northup, J.K., Dromerick, A.W., and Gilman, A.G. (1981). *J. Biol. Chem. 256,* 12911.
Insel, P.A., Maguire, M.E., Gilman, A.G., Bourne, H.R., Coffino, P., and Melmon, K.L. (1976). *Mol. Pharmacol. 12,* 1062.
Jakobs, K.H., Saur, W., and Schultz, G. (1978). *FEBS Lett. 85,* 167.
Katada, T., and Ui, M. (1977). *Diabetologia 13,* 521.
Katada, T., and Ui, M. (1981). *J. Biol. Chem. 256,* 8310.
Katada, T., and Ui, M. (1982a). *Proc. Natl. Acad. Sci. USA 79,* 3129.
Katada, T., and Ui, M. (1982b). *J. Biol. Chem. 257,* 7210.
Katada, T., Amano, T., and Ui, M. (1982). *J. Biol. Chem. 257,* 3739.
Klainer, L.M., Chi, Y.-M., Freidberg, S.L., Rall, T.W., and Sutherland, E.W. (1962). *J. Biol. Chem. 237,* 1239.
Limbird, L.E., and Lefkowitz, R.J. (1977). *J. Biol. Chem.*

252, 779.
Londos, C., Salomon, Y., Lin, M.D., Harwood, J.P., Schramm, M., Wolff, J., and Rodbell, M. (1974). *Proc. Natl. Acad. Sci. USA 71,* 3087.
Moss, J., and Vaughan, M. (1979). *Ann. Rev. Biochem. 48,* 581.
Moss, J., and Vaughan, M. (1981). *Mol. and Cell. Biochem. 37,* 75.
Moss, J., Manganiello, V.C., and Vaughan, M. (1976). *Proc. Natl. Acad. Sci. USA 73,* 4424.
Murad, F., Chi, Y.-M., Rall, T.W., and Sutherland, E.W. (1962). *J. Biol. Chem. 237,* 1233.
Nakaya, S., Moss, J., and Vaughan, M. (1980). *Biochem. 19,* 4874.
Northup, J.K., Sternweis, P.C., Smigel, M.D., Schleifer, L.S., Ross, E.M., and Gilman, A.G. (1980). *Proc. Natl. Acad. Sci. USA 77,* 6516.
Northup, J.K., Smigel, M.D., and Gilman, A.G. (1982). *J. Biol. Chem. 257,* 11416.
Pfeuffer, T. (1977). *J. Biol. Chem. 252,* 7224.
Rall, T.W., and Sutherland, E.W. (1962). *J. Biol. Chem. 237,* 1228.
Rodbell, M., Birnbaumer, L., Pohl, S.L., and Krans, M.J. (1971). *J. Biol. Chem. 246,* 1877.
Ross, E.M., and Gilman, A.G. (1977a). *Proc. Natl. Acad. Sci. USA 74,* 3715.
Ross, E.M., and Gilman, A.G. (1977b). *J. Biol. Chem. 252,* 6966.
Ross, E.M., and Gilman, A.G. (1980). *Annu. Rev. Biochem. 49,* 533.
Ross, E.M., Howlett, A.C., Ferguson, K.M., and Gilman, A.G. (1978). *J. Biol. Chem. 253,* 6401.
Schleifer, L.S., Kahn, R.A., Hanski, E., Northup, J.K., Sternweis, P.C., and Gilman, A.G. (1982). *J. Biol. Chem. 257,* 20.
Sternweis, P.C., Northup, J.K., Hanski, E., Schleifer, L.S., Smigel, M.D., and Gilman, A.G. (1981a). *Adv. Cyclic Nucleotide Res. 14,* 23.
Sternweis, P.C., Northup, J.K., Smigel, M.D., and Gilman, A.G. (1981b). *J. Biol. Chem. 256,* 11517.
Sumi, T., and Ui, M. (1979). *Endocrinology 97,* 352.
Sutherland, E.W., Rall, T.W., and Menon, T. (1962). *J. Biol. Chem. 237,* 1220.

DISCUSSION

Moss: You mentioned that aluminum fluoride also dissociates the 41-35K protein. Does that indicate that fluoride could, in some systems, cause inhibition of cyclase, and would that explain the inhibition observed in some cholera toxin treated tissues?

Kahn: As far as I know, there have been no reports of fluoride causing inhibition. We are proposing that this dissociation model is quite a complex equilibrium in the membrane, and again, we are hypothesizing from pure solutions to a membrane and thinking the same things go on there. It is clearly going to be a complex kinetic argument to be formulated as these things become tested more. I believe that the fluoride acts on the stimulatory pathway, thereby swamping out the inhibitory pathway and how those two pathways are going to interact when you stimulate both is not known at all now.

Rubin: I am wondering, in the 35K protein in G_i, have you demonstrated that it is identical to the 35K in the activating G protein?

Kahn: What has been done so far is to demonstrate that the protease digests mapping of these 35 proteins from a number of different sources - G_i and G_s are more prevalent -and peptide maps to date have been identical. We can do the dissociation experiments; we can separate 35-45, 35-41; and then try to exchange 35 subunits; preliminary evidence suggests that they both have activity with the other subunit. The actual efficiency of those is not quite known, but it looks as if it will be close to (and they probably are) identical.

Rubin: Have you also carried out peptide mapping experiments to compare the 41K inhibitory and the 45K activating G proteins?

Kahn: Yes. I mentioned that the 45 and 52 are virtually identical, with 52 having an additional spot. The IAP substrate and the cholera-toxin substrate, the 45,000 and the 41,000 dalton bands, do not look alike in protease mapping.

Cohen: You always show in your slides the catalytic subunit as being present in equimolar amounts with the GS and the GI. Could you bring us up-to-date on that? Why is it so difficult to purify the catalytic (C) subunit of adenylate cyclase?

Kahn: Well, first as to the stoichiometry in the membrane, I did not mean to imply by this model that everything was there in a 1:1 stoichiometry. Rather, that is clearly not the case and it looks like, G_i will be there in several fold excess over that of G_s. These numbers are very loose, I hope you understand, and so the actual stoichiometry of these subunits is clearly going to be an important question even as it relates to Dr. Moss's question. Until quite recently C has been very labile. Any of these purifications, because they are membrane proteins, have to be done in detergent solutions and C died at a half-life of less than a day in any combination of detergents tried. We are working on purification of C. Murray Smigle in our lab has made some progress so that we can now separate C from the other adenylate cyclase components that I have described so far, and we hope to be able to utilize this partially purified C to replace the cyclysed-membrane in the reconstitution assay.

Cohen: You mentioned that the ADP ribosylation of the G_i causes a dissociation of the 41 from the 35. Does that also occur with the G_s; that you get a dissociation of the 45 from the 35 on ADP ribosylation of the 45?

Kahn: I realized in passing that I might have misled you on that point. ADP ribosylation is unlike aluminum fluoride, GTPγS or GppNHp. Those activators can completely dissociate dimer into monomer. ADP ribosylation causes an apparent size shift on sucrose gradient but it is significantly smaller than the shift seen with full activation with fluoride or GTPγS. This suggests that ADP ribosylation really just causes a loosening up of the 45-35 dimer and in the case of the 41, it is just the opposite. We believe (from a working model in our lab) that ADP ribosylation of 41 is going to be exactly opposite to that of G_s. That is, ADP ribosylation of 41 is going to increase the association of the heterodimer in 41-35.

Klee: Have you looked at the effect of the pertussis toxin on the G protein of the retina - the GTP-riboprotein of the retina?

Kahn: These are incomplete studies. The rod outer segments from bovine retinal discs have a guanine nucleotide binding component called transducin, which has been purified and is involved in the light activated stimulation of phosphodiesterase. The transducin has been purified in our lab according to methods previously published by Fung, *et al*. The transducin in the membranes appears to be a substrate for both toxins. In purified form it is not a substrate for cholera toxin. As Abood *et al*., published recently in J.B.C. purified transducin is not a substrate for cholera toxin. We have labeled pure transducin with IAP and it is a very good substrate for labeling. We are now in the process of going back and seeing if these two toxins can be active in solution and if there is a single site. I think the transducin, at this point, looks like it is going to be more of an IAP-type substrate, but we may be able to force it to serve as a substrate for either toxin, and structural studies using the site of ADP ribosylation could be very interesting.

Klee: Have you looked at the effect GTP-binding protein from retina has on the cyclase activity?

Kahn: Yes, we have been unable to reconstitute a hormone sensitive adenylate cyclase in cyc- cells with transducin and I have not seen any data which suggests that anybody else has.

Linden (OMRF): Does IAP inhibit a GTPase, and also do you envision the effects of GTP-NADP as a combination of stimulatory effects of G_s and inhibitory effects by G_i?

Kahn: To answer the second part of your question, I think it is too early to say what happens when you turn on these activators in the whole membrane. With the components of cyclase being purified we can begin to ask some of the questions that have confused the literature for so many years; some phenomenology will hopefully begin to be worked out and put on a more rational basis. Regarding GTPase activity, Zelinger proposed that adenylate cyclase has a GTPase activity and that the activation of the catalyst depended on the tight binding of GTP somewhere in the adenylate cyclase and that the turn-off mechanism for hormone stimulated adenylate cyclase is a GTPase reaction. In our hands, we have had G_s for some time. We have never been able to find any GTPase activity inherent in G_s. That does not mean it is not there. It is very likely to require coupling of G_s either to C or to receptor. Neither are presently available in purified enough form that those experiments can be done cleanly. Cassel and Zelinger and others who do this kind of study use a number of inhibitors of less specific GTPases to reveal the adenylate cyclase dependent GTPase. We have not done those types of experiments. As to the 41,000 dalton band, IAP labeling in GTPase, we have not done them, and I really can not comment on it, but I suspect again you will need coupling of these regulatory subunits to other components to get GTPase activity.

MACROMOLECULAR PROPERTIES OF POLY(ADP-RIBOSE) AND THE REGULATION OF ITS BIOSYNTHESIS AND DEGRADATION IN PERMEABILIZED RAT BRAIN 9L-GLIOSARCOMA CELLS[1]

Ernest Kun

The University of California
Department of Pharmacology
and
The Cardiovascular Research Institute
San Francisco, California

Post-transcriptional covalent modification of the chromatin proteins by poly ADP-ribose occupies a unique position among enzymatic protein modifying reactions. Phosphorylation, acetylation, etc., results in the addition of relatively small modifying groups to protein molecules. It would be anticipated that the physicochemical consequences of these types of protein modifications would depend on the extent of modification and on the selectivity of the reactions with respect to modifiable end groups. The same argument holds when predictions related to the effect of protein modifications on the catalytic activity of modified proteins is to be assessed. Macromolecular interactions between protein molecules and between proteins and nucleic acids is likely to be altered by covalent protein modifiers of small size through induced changes in charges. In contrast, poly ADP-ribosylation of proteins not only can alter the charge of acceptor

[1]*Ernest Kun is a recipient of a Research Career Award of the United States Public Health Service. The original research reported here was supported by the United States Air Force Office of Scientific Research (F-49620-81-C-0007), the United States Public Health Service (HL-27317, CA-13525), the American Cancer Society (RD-137), the Swedish National Science Research Council (R-RA-4685-100), and the Swedish Medical Research Council (B-81-04R-6065-504106065).*

ISBN 0-12-387560-9

protein end groups but is likely to introduce new types of forces which are directly related to the macromolecular properties of poly ADP-ribose. We have proposed that even mono ADP-ribosylation could have unique consequences by providing adenine groups to proteins which are known to represent high association forces with other adenine moieties, especially after charge neutralization of the ADP-ribose phosphates by basic protein end groups (Kun and Kirsten, 1982). This situation is expected to be even more complex when the protein modifier is the polymer poly ADP-ribose.

Anticipating the deterministic role of poly ADP-ribose in protein-protein and protein-nucleic acid interactions within the chromatin structure, our work is first concerned with the elucidation of the polymeric molecular properties of poly ADP-ribose. Corollary to this problem is the catalytic regulation of the synthesis and degradation of poly ADP-ribose in a quasi physiological system, the relatively intact nucleus of a cell maintained in culture. The present state of knowledge in both of these areas is preliminary, since despite significant advances regarding the primary structure (Hayaishi and Ueda, 1977) no information is available on the macromolecular properties of the polymer and, as discussed here, the kinetics of the poly ADP-ribose polymerase and glycohydrolase in the isolated form most likely exhibit artefacts that permit only limited extrapolations to *in vivo* systems.

The purpose of the present paper is to summarize experimental work concerned with the above outlined problems.

THE SPECIFIC SUBUNIT OF POLY(ADP-RIBOSE): 2'-(5"-PHOSPHORIBOSYL)-5'-AMP (ABBREVIATED: PR-AMP)

Digestion of the polymer by snake venom phosphodiesterase yields AMP and PR-AMP (Hayaishi and Ueda, 1977) the latter representing the specific nucleotide subunit of the homopolymer. Elucidation of the precise structure of PR-AMP has significant predictive power relevant to the molecular structure of the polymer. Nucleomagnetic resonance studies (Miwa *et al.*, 1977; Inagaki *et al.*, 1978; Ferro and Oppenheimer, 1978) identified the $\alpha(1''\rightarrow 2)$ ribofuranosyl-ribofuranoside bond in PR-AMP, but no significant conformational constraints have been envisaged in aqueous solution (Inagaki *et al.*, 1978; Sugimura *et al.*, 1979). As a continuation of our studies with specific phenyl-boronate affinity resins (Minaga *et al.*, 1979; Romaschin *et al.*, 1981), we find that PR-AMP does not form the phenyl-boronate complex characteristic of *cis*-diol containing molecules under conditions when non-specific electrostatic repulsion is experimentally eliminated by high ionic strength (Minaga *et al.*, 1982). Dephosphorylation to ribosyl-adenosine reestablishes the propensity to form the phenylboronate complex, indicating a steric interference between one of the phosphate groups with the *cis*-diol in PR-AMP. Computer simulated analysis of the eleven conformational parameters that specify the conformation of PR-AMP (Minaga *et al.*, 1982) clearly indicate a unique steric interaction between the AMP phosphate and the *cis*-diol of PR-AMP, as illustrated in Figure 1. It is evident that this conforma-

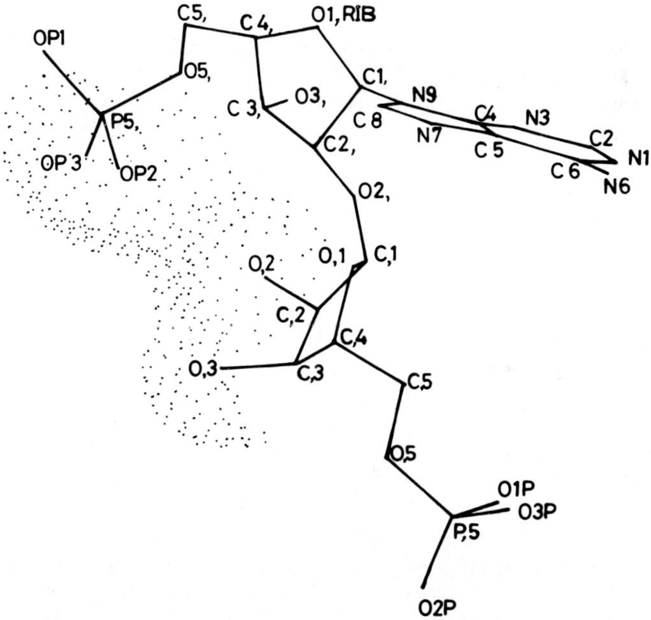

FIGURE 1. Specific computer derived conformation of PR-AMP. In this conformation steric interaction is apparent between the oxygen atoms of the phosphate groups of AMP (OP_2, OP_3) and the oxygen atoms of cis-diol (0.2,0.3) present in the phosphoribosyl portion of the molecule. The dots depict the accessible surfaces of the interacting oxygen atoms. The numbering system reflects identification of atoms by computer programming only. The steric hindrance imposed by the above interaction inhibits the approach of a boronate group that has to be coplanar with the cis-diol.

tional factor was missed by speculative assignment of a molecular structure of PR-AMP in aqueous solution (Inagaki *et al.*, 1978; Sugimura *et al.*, 1979), yet the interaction between the AMP-phosphate and *cis*-diol has predictive value relevant to the structure of the polymer. Both phosphates of PR-AMP are bonded as pyrophosphates in poly ADP-ribose, releasing the inhibitory effect on phenyl-boronate complexation of the *cis*-diols of the polymer, as also demonstrated experimentally (Minaga and Kun, 1983). However, the distinct possibility remains that the residual phosphate-*cis*-diol interactions in the polymer are significant enough to result in an inhibition of deprotonation of certain phosphoric acid groups in the polymer, rendering it less acidic than predictable from a putative random and linear macromolecular interpretation of its structure (Hayaishi and Ueda, 1977). The steric factor imposed by the *cis*-diol phosphate interaction may have significance in determining the macromolecular electric charge of the poly ADP-ribose protein adducts. This charge may be less negative than anticipated from a random structure, not taking into account the *cis*-diol phosphate interactions.

NUCLEIC ACID STRUCTURE OF POLY(ADP-RIBOSE)

Isolation of uncontaminated poly-disperse poly ADP-ribose provided the experimental basis for spectral studies of the secondary structure of the polymer (Minaga and Kun, 1983 and unpublished data). As shown in Figure 2, increasing the concentration of phosphate has a significant effect on the A_{280}/A_{260} value, clearly indicating that this optical test which has often been applied in the past is an incorrect measure of the purity of the polymer and the spectral phenomenon tends to suggest a secondary structure. If % hypochromicity and A_{280}/A_{260} values are correlated to the chain length of the homopolymer, results as shown in Figure 3 are obtained. The critical transition at about n=9 demonstrates that significant conformational transition occurs at this chain length predicting secondary structure of long polymers. The temperature dependence of characteristic CD spectra of long chain polymers, illustrated in Figure 4, exhibits at 5°C positive Θ bands at 205 and 272.5 nm and one negative band at 249.5 nm with a small shoulder at 222 and 225 nm. At 75°C the Θ signals diminish at 272.5 nm and at 249.5 nm, being replaced by a negative band at 267.5 nm and the small shoulder seen at 5°C was enlarged into a negative band at 220 nm at 75°C. A

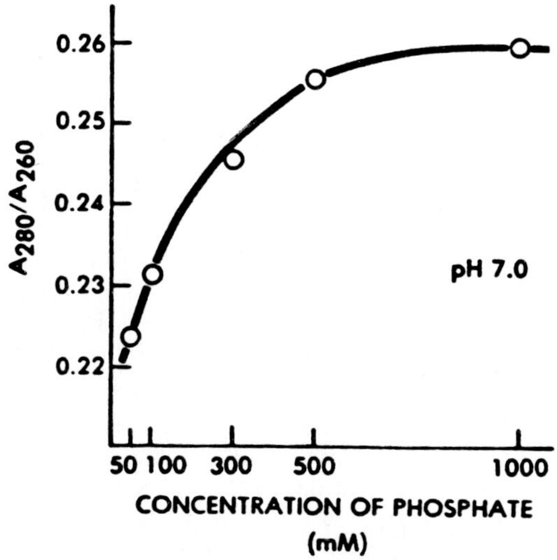

FIGURE 2. The effect of increasing concentration of phosphate on the A_{280}/A_{260} of polydisperse poly ADP-ribose. Polymer concentration was 0.191 A_{260} units (12.4 nmol, 2000 cpm) per ml.

temperature-dependent red shift and a characteristic melting curve (measured at 249.5 nm) were recorded (Figure 4). The sum of more extensive studies (Minaga and Kun, 1983 and unpublished data) which cannot be given in detail in the present report, provided strong evidence that indicate a nucleic acid type conformation of poly ADP-ribose. More specifically, the interaction with inorganic cations and polyamines as well as pH effects, influencing CD spectra, distinctly identify a helical conformation that is reminiscent of poly A, poly (A-U), and DNA rather than RNA (Minaga and Kun, 1983 and unpublished data). It is at present unknown whether or not this helicity represents a single or double helical structure, but the probability of the latter cannot be ruled out. X-Ray diffraction analyses presently in progress are likely to answer this question. It is clear that the molecular basis of the biological function of poly ADP-ribose depends on the specificity of protein acceptors for the polymer. We identify these as primarily non-histone proteins (Minaga *et al.*, 1979; Romaschin *et al.*, 1981; Jackowski and Kun, unpublished data; Jackowski and Kun, 1982).

Equally significant is the specific interaction mechanism provided by the helical conformation of the homopolymer. The

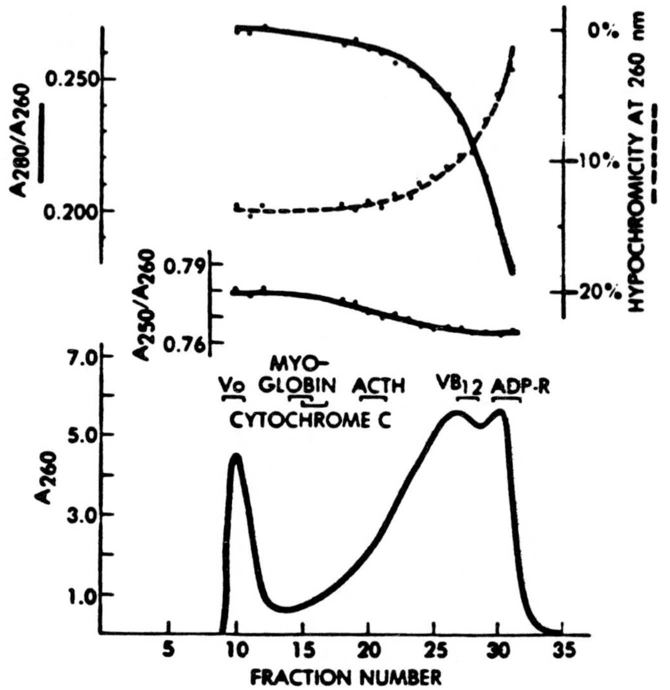

FIGURE 3. Correlation between chain length and spectral properties of poly ADP-ribose , measured at room temperature. The upper section shows spectral properties (left ordinate: A_{280}/A_{260}, right ordinate: hypochromocity); the lower section indicates molecular filtration of polydisperse polymer on Sephadex G-50 and chain lengths determined by direct analyses.

unique possibility exists that specific binding of proteins to DNA-sites is conferred by the macromolecular nature of the homopolymer, which may thus critically regulate protein-DNA interactions.

REGULATION OF POLY(ADP-RIBOSE) SYNTHESIS AND DEGRADATION IN THE CELL NUCLEUS

Isolation of proteins that apparently catalyze the synthesis and degradation of poly ADP-ribose represent a significant advancement (Hayaishi and Ueda, 1977). On the other

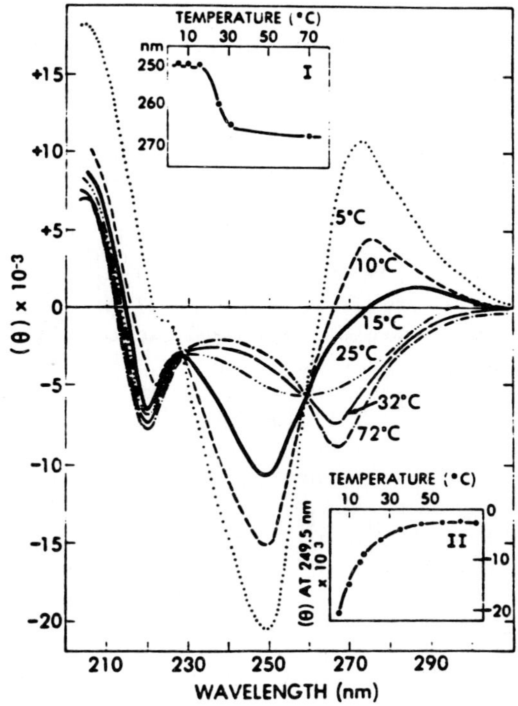

FIGURE 4. The influence of increasing temperature on the CD spectrum of long chain poly ADP-ribose. Inset I: variation of λ in n m of Θ_{max} ordinate) at varying temperature; insert II: melting curve determined at 249.5 nm.

hand, considerable uncertainty prevails regarding quantitative aspects of these enzymatic reactions, especially the regulation of polymerization appears to be poorly understood.

A probable model that describes polymerization can be summed up as follows. Polymer initiation occurs by covalent addition of ADP-ribose at nucleophilic protein sites, followed by successive ADP-ribosylation of the protein-ADP-ribose adduct at the ribose-2-sites, a process equivalent to chain elongation. When all ADP-ribose acceptor sites are saturated (with poly ADP-ribose), the enzyme that performs the catalysis is auto-ADP-ribosylated, resulting in enzyme inhibition, and the entire process is terminated. Reasonable descriptive evidence supports the above mechanism. We have shown in 1976 (Kun *et al.*, 1976) that ADP-ribose readily forms Schiff base-type protein adducts, most probably involving lysine termini. It was subsequently demonstrated in Kyoto (Ueda *et al.*, 1979) that ADP-ribose-protein adducts of basic nuclear proteins can serve as elongation sites for formation by the purified polymerase. That auto-ADP-ribosylation of the polymerase inevitably inhibits the enzyme has been documented in several laboratories (Mandel *et al.*, 1982).

A combination of the three deterministic steps regulating poly ADP-ribose synthesis leads to the inescapable prediction--seemingly not appreciated by the majority of workers in this field (Purnell *et al.*, 1980; Mandel *et al.*, 1982)--that steady state polymerization kinetics can only be studied under conditions that provide a sufficient number of reactive initiation sites on proteins, *other* than the polymerase protein, in order to protect the polymerase from instantaneous self-inhibition. If this requirement is not met, a kinetically artificial condition may prevail that limits the observed velocity by the variable quantity of ADP-ribose acceptor sites and bears no relationship to the molar concentration of the polymerase enzyme protein. It follows that the kinetics of the isolated enzyme, without added initiator sites containing protein is fraught with uncertainties because the auto-inhibition of the enzymatic reaction catalyzed by a single protein obscures the determination of uninhibited reaction rates, unless an excess of protein acceptors, present in form of other proteins, is also added to the system. Added histones are known to increase the rate of the poly-ADP ribose polymerase reaction by the purified enzyme (Hayaishi and Ueda, 1977; Mandel *et al.*, 1982). However, it is uncertain whether or not these basic proteins do not represent an artifactual condition by serving as chemical, not enzymatically competent "second substrate"-type ADP-ribose acceptors in the polymerization reaction. This doubt is supported by the experimental observation that non-histone proteins, rather than histones, appear to be the physiologically relevant polymer acceptors in cellular or nuclear systems (Minaga *et al.*, 1979, 1982; Romaschin *et al.*, 1981; Jackowski and Kun, unpublished data; Jackowski and Kun, 1982).

In view of this unresolved situation, we investigated the kinetics of polymerization in intact cell nuclei in order to circumvent the kinetic artifact introduced by the use of the isolated enzyme, in the absense of, at least initially, excess quantity of ADP-ribose acceptors.

Because of a prevailing proposition, claiming that rates of poly ADP-ribosylation in nuclei are determined by DNA fragmentation (Purnell *et al.*, 1980), we also studied the effects of enzymatically active, inhibited, and denatured

DNA-ase I on the process of poly ADP-ribosylation of nuclear proteins (Kun *et al.*, unpublished data).

Cultured rat brain 9L-gliosarcoma cells (Barker *et al.*, 1973) were treated with lysophosphatidyl choline (Florey, 1953) producing cell ghosts that contain intact nuclei, accessible to externally added NAD and proteins. In this experimental model first a correlation between the quantity, molecular size (expressed in S units) of DNA and v_{init} of poly ADP-ribose polymerase were determined, as shown in Table I.

In controls, DNA exhibited a DNA size of 59 S, and the corresponding v_{init} of poly ADP-ribose polymerase was 0.6. Upon addition of DNA-ase I (containing both activating cations: Ca^{2+} and Mg^{2+}), 98.5% of DNA was degraded to nucleotides and polymerase activity increased about 10 fold. When DNA-ase activity was restricted to a single strand scission, by Ca^{2+} complexation, and addition of Mg^{2+} as the single cationic activator, DNA size was halved (to 28 S) yet v_{init} of the polymerase was still maximal, indicating no quantitative relationship between the degree of DNA degradation and the enzymatic rate of poly ADP-ribose polymerase.

As illustrated in Figure 5, the catalytic effect of DNA-ase I on poly ADP-ribose polymerase is nearly the same when enzymatically active DNA-ase I (curve a.) or enzymatically

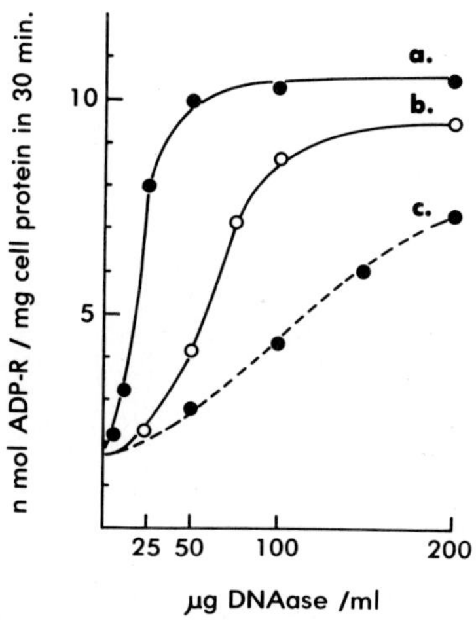

FIGURE 5. Correlation between rates of poly ADP-ribose formation by 9L-gliosarcoma cell ghosts and the concentration of various DNA-ase I proteins: a) fully active DNA-ase I; b) inactivated DNA-ase I by Ca^{2+} depletion; c) base denatured DNA-ase I.

TABLE I. The Effect of Added DNA-ase I on the Molecular Size and Cellular Concentration of DNA and on Polyadenosine Diphosphoribose Polymerase Activity of Permeabilized 9L-Gliosarcoma Cell Ghosts.

No.[a]	Experimental conditions	DNA size in Svedberg units (S)	DNA concentration: (μg DNA/mg cell protein)	Polyadenosine diphosphoribose polymerase (nmol) ADP-R incorporated in 5 min/mg protein
1	Gliosarcoma cells	59.0	25.0	0.6
2	1 + 100 g/ml DNA-ase I + Ca^{2+} + NAD added simultaneously	not measurable after 5 min. incubation	0.3	7.7
3	1 + preincubation with DNA-ase I for 10 min	"	0.3	6.7
4	1 + preincubation with DNA-ase I for 20 min	"	0.3	6.4
5	1 + Ca^{2} depleted DNA-ase I (200 g/ml) + Mg^{2+} added simultaneously	28.0	24.0	5.7
6	1 + preincubation with Ca^{2+} depleted DNA-ase I for 5 min	28.0	24.0	6.2
7	1 + preincubation with Ca^{2+} depleted DNA-ase I for 10 min	28.0	24.0	6.6
8	1 + preincubation with Ca^{2+} depleted DNA-ase I for 20 min	28.0	24.0	7.2

[a]In Exps. 3, 4, 6, 7 and 8 the reaction was started by adding [^{14}C]-NAD.

restricted by Ca^{2+} removal (curve b.) were tested and a significant catalytic effect of base denatured DNA-ase I protein (curve c.) was still demonstrable. These results illustrate a catalytic effect of DNA-ase I protein that is distinct from its hydrolytic effect on DNA.

When the degradation of nuclear DNA was directly determined after *in vivo* labelling of DNA by [H^3]-thymidine, it became apparent that addition of inactive DNA, which increases poly ADP-ribose polymerase 10 fold, actually inhibited the degradation of DNA by endogenous Mg^{2+} dependent nuclease. Therefore, instead of a positive correlation between DNA degradation and poly ADP-ribose polymerase activity, an inverse relationship appeared to emerge. It was therefore concluded that the catalytic effect of DNA-ase I on the poly ADP-ribose polymerase activity of intact nuclei had no connection with the nucleolytic catalytic effect of this enzyme. Further experiments showed that the effect of DNA-ase I was abolished by added actin, a well known specific binding protein of DNA-ase I present in chromatin. As discussed elsewhere (Kun *et al.*, unpublished data), it is probable that the catalytic action of DNA-ase I on nuclear poly ADP-ribose polymerase is related to its binding to nuclear actin, and the consequence of this DNA-ase-actin binding makes otherwise occluded initiator sites of ADP-ribosylation

TABLE II. The Degradation of Cellular DNA by Endogenous Nucleases in Permeabilized 9L-Gliosarcoma Cell Ghosts and the Inhibition of this Process by DNA-ase I

No.	*Experimental conditions*	*μg DNA/mg protein*	*% of DNA remaining*
1	Control (prior to incubation)	25.0	100
2	Incubation with 5 mM EGTA + 10 mM Mg^{2+}	10.7	43
3	Incubation with 5 mM $CaCl_2$ + 10 mM Mg^{2+}	9.7	39
4	Incubation with 200 μg/ml inactivated DNA-ase I + 10 mM Mg^{2+}	23.0	92
5	Incubation with 6.2 μg/ml active DNA-ase I + 5 mM $CaCl_2$ + 10 mM Mg^{2+}	15.7	63

TABLE III. Inhibition of Protein-Polyadenosine Diphosphoribosylation in Permeabilized 9L-gliosarcoma Cell Ghosts by Basic Amino Acids.

No.	Experimental conditions	ADP-R incorporated nmol/mg protein/ 5 min	% Inhibition
1	Control, no DNA-ase I 5 min incubation	0.49	
2	Control, DNA-ase I 200 μg/ ml 5 min incubation	5.25	
3	as 1, + 50 mM Arginine	0.40	18
4	as 2, + 50 mM Arginine	2.71	48
5	as 1, + 50 mM L-Lysine	0.35	29
6	as 2, + 50 mM L-Lysine	3.01	43
7	as 1, + 50 mM Arginine-methyl ester	0.16	68
8	as 2, + 50 mM Arginine-methyl ester	0.53	90

available. According to this mechanism, initiation of ADP-ribosylation should be the rate limiting step in the poly ADP-ribose polymerization process.

Two sets of experimental data support this view. First, if the availability of nucleophilic ADP-ribose acceptor protein sites are to be rate limiting in initiation, there should be competition between protein acceptor sites and externally added basic substances for poly ADP-ribose initiation. This is shown in Table III.

The second experimental evidence that supports the hypothesis proposing initiation as rate limiting step was obtained from direct analysis of chain elongation kinetics (Kun *et al.*, unpublished data).

The time course of polymer formation at 50 and 500 μM [NAD] is illustrated in Figures 6 and 7. After 3 min incubation significant quantities of monomers and rapid accumulation of long chain polymers occurred when [NAD] was 50 μM. At 30 min, long chain polymers were largely degraded and the predominance of monomers was sustained, equivalent to continued chain initiation. Clearly at 50 μM [NAD] maintenance of long chains was limited by poly ADP-ribose glycohydrolase. At 500 μM [NAD] a more regular, synchronous initiation and elongation proceeds, and at 50 min (Figure 7) an almost homogenous population of long chain polymers accumulates. High

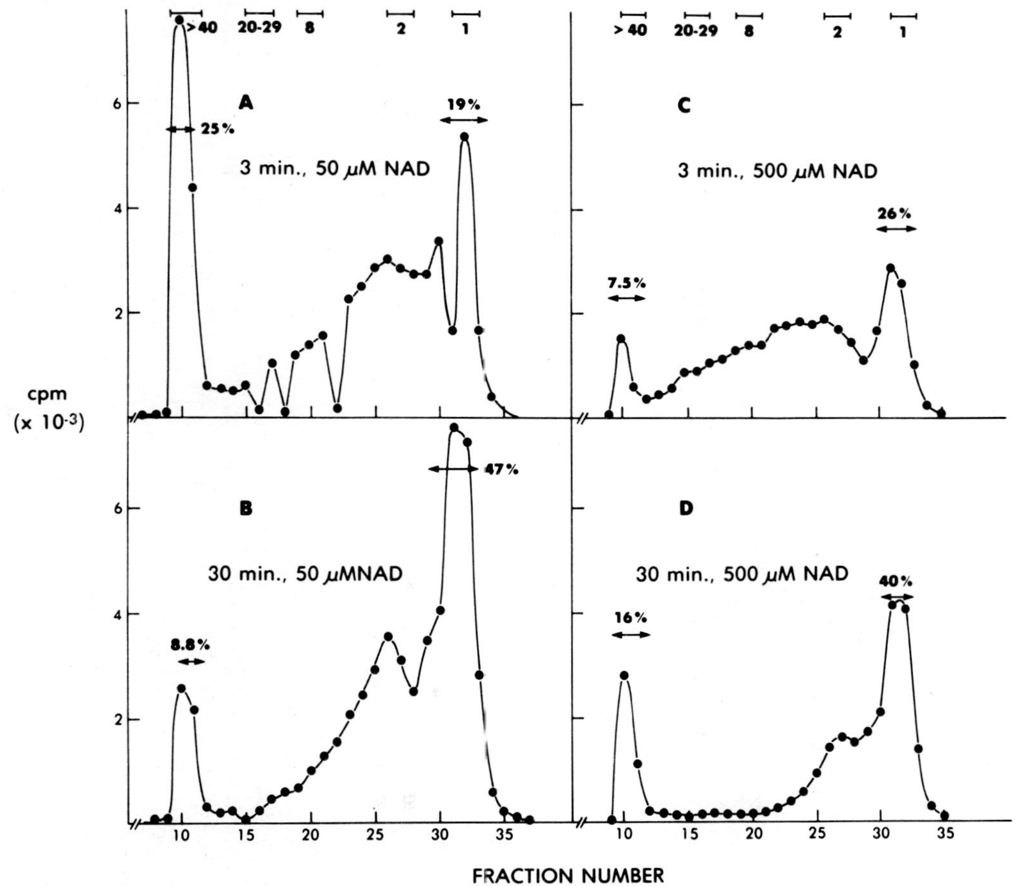

FIGURE 6. Rates of poly ADP-ribose chain formation and decay at 50 and 500 μM NAD concentration. The chain length of isolated polymers was determined by molecular filtration.

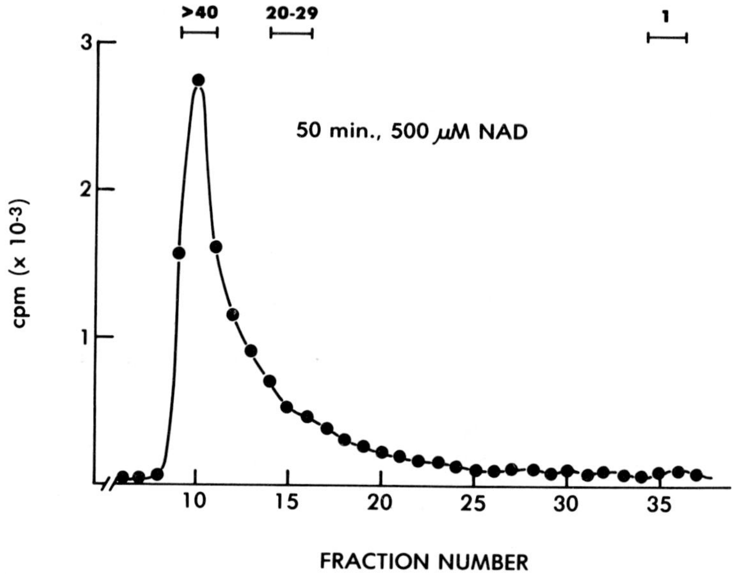

FIGURE 7. Chain length of poly ADP-ribose, determined after 50 minutes incubation.

[NAD] evidently abolishes the interferring action of glycohydrolase. The question was further analyzed as shown in Table IV.

When poly ADP-ribose glycohydrolase was assayed by externally added poly ADP-ribose, a relatively slow decay occurred, uninfluenced by DNA-ase I induced conformational change of the nucleus (Expts. 1,2). On the other hand, when poly ADP-ribose was generated from added NAD at 50 or 500 μM concentrations, yielding 0.6 or 14 μM intracellular polymer concentration, the rate of glycohydrolase activity was inversely proportional to the intranuclear concentration of the *in situ* generated polymer. In any case, added DNA-ase I had no effect on glycohydrolase activity even when the polymer was synthesized intranuclearly.

It can be predicted from these results that at higher NAD concentrations, glycohydrolase activity is negligible, most likely attributable to a significant product inhibition of the glycohydrolase by *in situ* generated ADP-ribose.

TABLE IV. Enzymatic Degradation of a) Externally Added and b) Intranuclearly Synthesized Polyadenosine Diphosphoribose in Permeabilized 9L-gliosarcoma Cell Ghosts

No.	*Externally added substrate*	*Concentration of polymer (as ADP-ribose) synthesized in the nucleus from added NAD*	*$t_{1/2}$ of degradation*
1	*30 μM polyadenosine diphosphoribose*	--	*40 min*
2	*1 + inactivated DNA-ase I (200 μg/ml*	--	*40 min*
3	*50 μM NAD*	*0.6 μM*	*3 min*
4	*500 μM NAD*	*14.0 μM*	*18 min*

The situation illustrated in Figure 6 indicates that after a certain period of polymerization, the controlling factors must be sought in the enzymatic regulation of polymerization itself, not in the catabolism of the polymer, i.e., glycohydrolase.

Consequently, a mathematical analysis of results shown in Figure 6 was undertaken by L. Peller following Florey's formulation of the Poisson-Boltzmann polymerization kinetic scheme (Florey, 1953) with certain simplifications (Margenan and Murphy, 1948) that consisted of replacement of the factorial n (n!) in the Poisson distribution (ν^n ($\exp^{-\nu}/n!$) by Stirling's approximation $n! \approx n^{n+1/2}$ ($\exp -n)\sqrt{2\pi n}$ for arbitrary n and for $n=\nu$. Figure 6 was recalculated by replacing maximal chain length by weight fraction maximum, $W_{n\ max}$ and this was plotted as the ratio $W_n/W_{n\ max}$ against n (chain-length). The results shown in Figure 8 were obtained.

The solid curve was obtained by solving $W_n/W_{n\ max} \approx (\nu/n)^{n+1/2} \exp(n-\nu)$ with n=70 that is an experimentally estimated quantity. Except for shorter chains, calculated and observed data were in close conformity. The most plausible explanation--that follows also from the experimental results indicating minimal interference by glycohydrolase activity--is that the departure from theoretical values at short chain length indicates initiation sites of differing reactivity (towards ADP-ribose). This implies that a main regulatory mechanism of poly ADP-ribosylation is to be sought in the

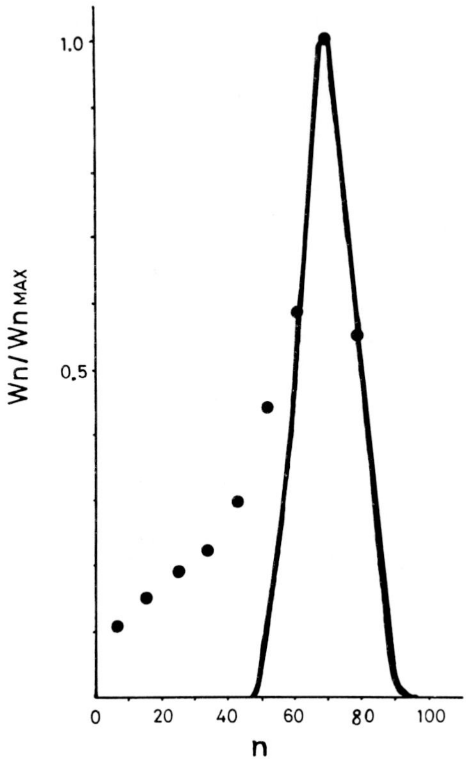

FIGURE 8. Poisson-distribution of polymers as calculated from results shown in Figure 7.

differing reactivity (nucleophilicity?) of initiation sites on acceptor protein molecules. Once a chain passed a critical chain length (4-5 ADP-ribose units), chain elongation is time invariant and no longer reflects the point of origin of the polymer.

In conclusion, our results point to the kinetic significance of initiation sites on acceptor proteins as the most important regulatory sites for the poly ADP-ribose polymerase process.

SUMMARY

1) Chemical evidence is presented that shows steric interaction between *cis*-diol and AMP-phosphate in PR-AMP. 2) Spectral analysis of heterodisperse and homodisperse poly ADP-ribose polymers indicate a helical conformation, similar to poly (A) DNA, but not RNA. 3) The catalytic activation of poly ADP-ribose polymerase of intact nuclei by DNA-ase protein was shown to be unrelated to its action as a hydrolytic enzyme on DNA. The effect is best explained by DNA-ase binding to nuclear actin, causing conformational alteration in the nucleus which exposes hitherto unavailable initiator sites for polymer synthesis. 4) Kinetic evidence was obtained that points to the availability and reactivity of protein initiator sites as the main kinetic control sites of polymerization.

ACKNOWLEDGEMENTS

The typing of the manuscript by Ms. Carrie Paulsen is acknowledged with thanks.

REFERENCES

Barker, M., Hoshino, T., Gurcay, O., Wilson, C.B., Nielson, S.L., Downie, R., and Eliason, J. (1973). *Cancer Res.* *33*, 976.

Ferro, A.M., and Oppenheimer, N.J. (1978). *Proc. Natl. Acad. Sci. USA 75*, 809.

Flory, P.J. (1953). *In* "Principles of Polymer Chemistry," p. 317. Cornell Univ. Press, Ithaca, N.Y.

Hayaishi, O., and Ueda, K. (1977). *Annu. Rev. Biochem. 46*, 95.

Inagaki, F., Miyazawa, T., Miwa, M., Saito, H., and Sugimura, T. (1978). *Biochem. Biophys. Res. Commun. 85*, 415.

Jackowski, G., and Kun, E. (1982). *J. Mol. Cell. Cardiol. 14, Suppl. 3*, 65.

Kun, E., Chang, A.C.Y., Sharma, M.L., Ferro, A.M., and Nitecki, D. (1976). *Proc. Natl. Acad. Sci. USA 73, 3131.*

Kun, E., and Kirsten, E. (1982). In "ADP-ribosylation Reactions" (O. Hayaishi, and K. Ueda, eds.), p. 193. Academic Press, New York.

Mandel, P., Okazaki, H., and Niedergang, C. (1982). *Prog. Nucleic Acid Res. Mol. Biol. 27*, 1.

Margenau, H., and Murphy, G.M. (1948). *In* "The Mathematics of Physics and Chemistry", Chapter 12, p. 423. D. Van Nostrand, Inc., New York.
Miller, M.R., Castellot, J.J., Jr., and Pardee, A.B. (1978). *Biochemistry 17*, 1073.
Minaga, T., and Kun, E. (1983). *J. Biol. Chem., 258*, 725.
Minaga, T., McLick, J., Pattabiraman, N., and Kun, E. (1982). *J. Biol. Chem. 257*, 11942.
Minaga, T., Romaschin, A.D., Kirsten, E., and Kun, E. (1979). *J. Biol. Chem. 254*, 9663.
Miwa, M., Saito, H., Sakura, H., Saikawa, N., Watanabe, F., Matshushima , T., and Sugimura, T. (1977). *Nucleic Acids Res. 4*, 3997.
Purnell, M.R., Stone, P.S., and Wish, W.J.D. (1980). *Biochem. Soc. Trans. 8*, 215.
Romaschin, A.D., Kirsten, E., Jackowski, G., and Kun, E. (1981). *J. Biol. Chem. 256*, 7800.
Sugimura, T., Miwa, M., Saito, H., Knai, Y., Ikeyana, M., Terada, M., Yamada, M., and Utakoyi, T. (1980). *Enzyme Regul. 18*, 195.
Ueda, K., Kawaichi, M., Okayama, H., and Hayaishi, O. (1979). *J. Biol. Chem. 254*, 679.

DISCUSSION

Berger: You stated that DNAase is strongly ADP-ribosylated. Did you actually isolate the DNAase and have you been able to confirm that there is a covalant linkage? Also, were you able to determine how long the polymer chains were and whether or not the ADP-ribosylation of DNAase takes place in the presence or absence of divalent cations?

Kun: Yes, we have actually isolated and shown the covalent linkage of poly-ADP-ribose. If you wait for 30 minutes, the chain is mostly monomers. If you wait 3 minutes, it is mostly polymers. The poly-ADP-ribosylation of acid proteins depends on the source of the cell nucleus. For example, in hamster nuclei we find that actin itself is also poly-ADP-ribosylated. In other nuclei, DNAase is not very well poly ADP-ribosylated.

Berger: Does the ADP-ribosylation of DNAase take place in the absence of cations?

Kun: You need cations. In our experience, the best combination of cations is calcium plus spermidine. This is much better than magnesium.

Jacobson: First, I was relieved to see that there are conditions in your laboratory where not all ribose containing compounds bind to boronyl resins. Secondly, I am not aware of any evidence to the contrary for an additional possible mechanism of poly-(ADP-ribose) synthesis. One possibility is that the first ADP-ribosyl residue is added and then the polymer adds on to the end, but equally likely is that the polymerase may function by making the polymer by always adding ADP ribose residues internally, sort of pushing the polymer out.

Kun: I'm not quite sure what you mean by pushing the polymer out. There is no physical force, this is a chemical addition. By "pushing the polymer out", do you mean "move it along"?

Jacobson: Right, a mechanism in which residues are being transferred back and forth from the active site and that the incoming ADP-ribosyl is always covalently attached to the polymerase. The first residue added ultimately ends up on the end of the chain. Isn't this as equally likely as the one you postulated? I do not know of any evidence that distinguishes between the two.

Kun: Hayaishi, in 1979, published a paper in J.B.C. It shows that the chemical adduct of ADP-ribose serves as a template for elongation. In the absence of contrary evidence, this is at least a reasonable experiment which is in agreement with the initiation and elongation process.

Jacobson: I am unclear as to the evidence that this forms a double helix? Also, if the formation of triribosyladenosine or diribosyladenosine is an artifact of venom phosphodiesterase, this could be a truly novel mechanism. Do you have a postulated mechanism whereby venom phosphodiesterase catalyzes the formation of a glycosidic linkage?

Kun: Concerning the double helix, it is mostly the effect of cations on the optical signals and temperature signals, which are to be compared to results known with other nucleic acids. About the mechanism, phosphodiesterase may phosphorylate some ribose residue at the end of the polymer which is liberated by alkaline hydrolysis.

Jacobson: It is clear from our studies that diribosyladenosine is derived from that polymer and for that to be an artifact of venom phosphodiesterase, venom phosphodiesterase would have to catalyze the formation of a glycosidic linkage, because we are not dealing with phosphorylated compounds.

Kun: Except you isolate your fragment as a result of phosphodiesterase digestion. You cannot get around that.

Jacobson: Right, but you are saying that it is not there before the phosphodiesterase is added. If that is the case, then the phosphodiesterase must have to create it, and to do that it has to create a glycosidic linkage.

Kun: That seems to be a confusion of the facts. The residue can exists as a free residue and it is merely phosphorylated, similar to trace phosphorylations catalyzed by other pyrophosphatases.

Moss: When you discussed initiation, you mentioned that poly (ADP-ribose) was attached to a basic residue. The only evidence that I am aware of is by Koide's group and Hayaishi's group. They suggest that ADP-ribose is attached to a glutamic acid (in histones, at least). Do you have any evidence in non-histone proteins?

Kun: The initiation at the glutamic residue has only been shown on histone H1 (and H2, maybe). We don't see any histone ADP-ribosylation in significant quantities *in vivo*. On the other hand, we can produce a number of proteins, ADP-ribosylate them chemically, and get an initiator site.

Index

D

V

W

Y